ISBN 978-1-330-78209-5
PIBN 10104694

THE PRINCIPLES OF QUALITATIVE ANALYSIS

FROM THE STANDPOINT OF

THE THEORY OF ELECTROLYTIC DISSOCIATION AND THE LAW OF MASS ACTION

BY

WILHELM BÖTTGER

PRIVATDOCENT IN THE UNIVERSITY OF LEIPZIG

TRANSLATED WITH THE AUTHOR'S SANCTION AND REVISED WITH HIS CO-OPERATION

BY

WILLIAM GABB SMEATON

INSTRUCTOR IN GENERAL CHEMISTRY IN THE UNIVERSITY OF MICHIGAN

WITH TEN ILLUSTRATIONS

PHILADELPHIA

P. BLAKISTON'S SON & CO.

1012 WALNUT STREET

1906

PRESS OF
THE NEW ERA PRINTING COMPANY
LANCASTER, PA

DEDICATED

TO

GEHEIMRATH PROFESSOR DR. W. OSTWALD

AUTHOR'S PREFACE.

By compiling the present text I have sought to supply a want to which W. OSTWALD, amongst others, refers in the preface to the "Scientific Foundations of Analytical Chemistry." In the practical teaching of elementary chemistry due regard has not been paid heretofore to the dissemination of new experiences and general views which are important for the comprehension of chemical phenomena. Since the appearance of the "Principles of Inorganic Chemistry" the need for a text-book which introduces modern conceptions into the *teaching* of *qualitative analysis* has become more apparent. This has led me to carry out a long cherished intention.

I have kept in view the problem of presenting the relations between the laws governing chemical reactions in an easily comprehensible manner and, as far as possible, with all exactness. By consistently applying these relations as they have been developed I have aimed to thoroughly establish their significance and also the advantages to be gained from knowing them.

Special attention has been paid to the *theory* of *electrolytic dissociation* and the *law* of *mass action*, the fundamental law governing chemical equilibria. Whenever opportunity has offered, the great importance of this law, even for qualitative experiments, has always been emphasized.

I have also aimed to define the most common conceptions that are used in general chemistry, by means of suitable illustrations, in such a way that even the beginner can soon learn to apply them correctly. As occasion demanded I have also pointed out differences in the use of certain conceptions, for instance, basicity (§ 129) and valence (§ 167).

No attempts have been made to set up constitutional formulae and to explain the nature and mechanism of chemical processes. All such questions have been left out of consideration, for to understand them *without the necessary preliminary knowledge of facts* makes greater demands upon the imagination than does the understanding of simple relations between the different phenomena, to which the more detailed study of the facts directly leads. The phenomena met

with in qualitative exercises will serve not only to distinguish the *meanings* of *facts* but also as a fundament of *acquired facts.*

In order to fill up gaps in the presentation of the subject-matter, which in part is not yet worked out in this sense, I have made numerous references to the corresponding text in the two above-mentioned works of OSTWALD, which are stored with a vast wealth of illustrations. In the matter of other references to literature I have cited original papers when their contents are specially important or not widely known. In case I should have unwittingly made any mistakes in this connection I beg beforehand that they be excused and that I be informed of them.

The table of contents presents a subdivision of the subject-matter which has been chosen in accordance with successful laboratory practice. The laboratory experiments are begun with the analysis of *solutions* which contain only *one* metallic component. When the nature of the component (ion) has been determined, the solution of a salt containing this constituent is prepared and the reactions described in a special paragraph are carried out with this solution. By this means the student has acquired a sufficient command of the methods of investigating the various group precipitates and of the distinguishing characteristics of the individual components when the complete analysis of a given unknown for the metallic and non-metallic constituents is undertaken. The use of the book is, moreover, by no means limited to this procedure and the subdivision into paragraphs readily permits the introduction of changes in the method.

The *tables* present a general view of the analytical procedure and the references to corresponding paragraphs in the text make a more detailed description of the various operations unnecessary.

Certain paragraphs of *theoretical* or *general* nature have been printed in small type, rather out of regard for the immediate aims of the book than with the idea of representing them as less important. For the choice and arrangement of these I am partly indebted to the interchange of ideas with students during practical instruction. As far as possible I have connected the *theoretical* and *descriptive* paragraphs by numerous references.

In the paragraphs which deal with the properties of the ions of different elements, no description of the physical properties of the elements themselves and no enumeration of their natural compounds

has been made. By this it is not aimed to represent the knowledge of these relations as *generally* unnecessary, but to avoid impairing the views which have been placed in the foreground.

The rarer elements are not included in the first course of analysis described. I hope, however, to have facilitated the detection of compounds of the rare elements by classifying them according to the group reagents, which also precipitate the ions of these elements. The methods of separating and detecting compounds of the more closely *related rare elements* may be obtained through the references to recent literature in those cases where the description of the methods would lead us beyond our prescribed bounds.

A number of experimental innovations have been introduced which have been proven valuable through manifold applications.

In this first exposition of the subject I have had to struggle with the overcoming of difficulties arising from the fact that qualitative analysis has been only partly exploited in the sense undertaken. Hence I beg that any lack of clearness or in simplicity of presentation be excused on these grounds.

To facilitate practical work for the student and furthermore to increase his interest and hold it beyond the ordinary limits of qualitative exercises has been my chief aim. The text is presented with the request that my attention be called to all errors and to possible improvements.

The figures have been partly taken from the " Principles of Inorganic Chemistry " with the permission of Prof. OSTWALD, and in part they have been prepared from drawings by Dr. BRAUER.

I would express my thanks to Prof. OSTWALD for his kindness in placing the figures at my disposal and for the interest he has shown throughout the preparation of the book.

Dr. BRAUER and Messrs. Walter NEUMANN of KATTOWITZ and Alexander TITOFF of Moscow have rendered invaluable assistance in reading the proof-sheets. For this service and for numerous suggestions which I have been able to use, I would here express my indebtedness to them.

WILH. BÖTTGER.

TRANSLATOR'S PREFACE.

Dr. Böttger has specially revised the German manuscript for the English translation and has made numerous emendations and additions, particularly in Parts I and II. With his sanction I have further emended some sections and have added tests and references here and there, so far as was compatible with the aims of the book. Two new sections, on bromates and iodates, have been incorporated in the translation. At the author's request the additions are enclosed in square brackets.

The longer ionic nomenclature has been adopted because it shows more clearly the relations between ions and the substances from which they are derived.

One special feature of the book, to which attention should be called, is the definite system used for separating anions into groups, from which the individual members can be identified with a minimum of labor.

I would express to Dr. Böttger my hearty thanks for his suggestions and for the interest he has taken in the revision of the manuscript. My thanks are also due to Dr. S. C. Lind for aid in revising the earlier sections, and in particular to Mr. H. H. Willard for permission to use some of his tests and also for suggestions and assistance in reading the proofs.

WILLIAM GABB SMEATON.

ANN ARBOR, June 15th, 1906.

CONTENTS.

PART III. THE COMPLETE ANALYSIS OF A GIVEN UNKNOWN.

PRINCIPLES OF

QUALITATIVE ANALYSIS

INTRODUCTION.

1. **Object of chemical analysis.** Chemical analysis consists in the systematic treatment of a given substance or mixture with selected substances, in order to determine its simpler *components* by the phenomena accompanying this treatment. When the analysis is limited to the designation of the components it is termed *qualitative,* and has to do solely with the determination of their nature. On the other hand, *quantitative* analysis, as the name implies, deals with the *amounts* of the individual constituents.

Not *chemical elements alone* are ranked as simple constituents, but under this head are also included *radicals,* i. e. groups of elements of definite structural composition and characteristic properties. The different elements and radicals are further classified into *metallic* and *non-metallic.* In their general character some radicals resemble metallic components; for example, the radical ammonium, (NH_4), is like the metallic component of potassium salts: others, for instance, the sulphate, (SO_4), nitrate, (NO_3), and cyanide, (CN), radicals resemble the non-metallic components, (Cl, Br, I), of the halide salts.

It is perhaps surprising that the simplest substances known to the chemist, the *chemical elements,* do not always serve to give the composition of a mixture, and that radicals may express it more definitely. This will be understood, however, by the consideration of a particular example. Imagine that the elements Na and Cl, and the radicals NH_4 and SO_4 had been found in a given mixture. It would not be more reasonable to indicate the composition by the elements Na, Cl, N, H, S and O. For under the conditions of analysis the radicals are *not decomposed into their respective elements,* but they have characteristic properties by which they are

recognized during the analytical operations. Furthermore, in all those cases where several radicals, made up of the same elements, as (SO_4), (SO_3), are known, merely to name the elements present would give *less* information concerning the composition than is given by the radicals. Evidently, then, a more definite expression of the composition of a substance or mixture can be made by means of radicals than with elements alone.

2. **Limitations.** Since a very large number of elements and radicals are known, qualitative analysis is customarily limited to those elements which are widespread in nature, or which have important compounds, for the sake of simplifying the study of analytical methods. Hence we shall have to do largely with the *analysis* of *natural* or *artificial* combinations of the *better known elements.* The most important properties of the components of mixtures which contain rare elements, or radicals composed of rare elements, are given in §§ 195–228. Mention is also made of a few organic radicals (derivatives of hydro-carbons), which have special importance, and in § 194 are given the methods of determining the elements present in organic compounds.

Since the investigation of gases demands practical experience which is usually acquired after a course of analysis of solids and liquids, gas mixtures are not taken up. We shall begin therefore with the analysis of *homogeneous liquid* mixtures, and from this will proceed to the more difficult problem of analyzing drv mixtures which have to be brought into solution before the analysis can be carried out.

Our study of homogeneous liquid mixtures, or solutions (Principles, p. *8*),[1] has to do chiefly with *aqueous solutions,* because water dissolves most of the substances with which we are dealing more readily than any other solvent, as for example, ethyl alcohol.

3. **Reactions.** The analysis of a solution is effected through an appeal to our senses by means of easily recognizable changes that are brought about through addition of substances to the solution, or in other ways, as by heating. These purposely effected changes are called *reactions,* and they are marked by a *change of color,* or by the *separation* of a solid from a solution (precipitation), or by the *dis-*

[1] "Principles" is used in abbreviation of Ostwald's Principles of Inorganic Chemistry, translated by Alexander Findlay, to which reference is frequently made.

solving of a solid. The occurrence of a definite *reaction* on addition of a definite substance, *the reagent*, to a solution of unknown composition, is the criterion by which we judge whether the solution contains the component for which the reaction is characteristic, or not. Hence substances can serve as reagents for a given component only when we *know* in advance their reaction with the component in question. Of the different reagents which show characteristic reactions with a certain component, that one is most important which gives the most easily recognized change in its reaction with the smallest possible quantity of the component. Moreover the reagent used must not give the same or a similar reaction with any *other* component.

4. General method of testing a solution for metals. On testing for the metals in a solution a reagent is added which is able to transform not merely one, but *several* components simultaneously into solid *difficultly soluble compounds.* In case a precipitate forms, it indicates that the solution contains one or more of the constituents which react in the same way with the added substance. After separating the precipitate from the solution, the former is tested for all possible compounds that may be present, while to the latter is added a second reagent, which again can form a precipitate that may contain several other components. Proceeding in this way a system of reactions can be built up, which show unequivocally whether or not any members of the different groups of constituents are contained in the given solution in the smallest quantities detectible by the reactions employed.

Manifold schemes of analysis have been worked out, but, in accordance with the character of this text-book, these various methods are not exhaustively treated. On the contrary, only one scheme which has been found practical in the laboratory, is given in detail. When the student comprehends this one scheme, and has mastered the properties of individual metals, which are fully described in the text, he will be in a position to introduce any changes that may be desirable, or to comprehend and apply any other method leading to the same end.

The scheme for separating the metals, as described in the text, is based upon the systematic application of the following reagents:

(1) Hydrochloric acid,
(2) Hydrogen sulphide,

(3) Ammonia,
(4) Ammonium sulphide
and (5) Ammonium carbonate.

By successively adding these reagents the metallic constituents of the salts of all the better known metals are removed from solution, except magnesium, potassium and sodium, and there remain in solution only salts of these three metals, if the separations have been carried out correctly.

By these reagents five groups of metals are classified, to which is appended a sixth group containing ammonium in addition to magnesium, potassium and sodium.

PART I.

TESTING A SOLUTION FOR METALS.

GROUP I.

5. Precipitation with hydrochloric acid. Dilute hydrochloric acid is added to a small portion (2–3 cc.) of the solution in a test-tube. In case a white precipitate forms, about 10 cc. of the solution are poured into the test-tube and hydrochloric acid is added drop by drop until precipitation ceases. This is readily seen if the supernatant liquid remains clear, or if no cloud forms in the clear liquid that has run through a filter, when a few more drops of hydrochloric acid are added. By this means it is demonstrated that the metals precipitated by hydrochloric acid are removed as far as possible from the solution. Too great an excess of the precipitant should be avoided, otherwise the subsequent testing of the solution would be disturbed (see § 19).

The analyst dare not form the habit of pouring at haphazard any quantity of the precipitant or reagent into a solution, but he must follow the fixed rule of adding the precipitant in *small* portions as long as a visible reaction occurs.

The solution that is being tested for the first group metals should be kept at *room temperature,* since *lead salts* are either not at all or only very incompletely precipitated by hydrochloric acid at higher temperatures. In itself this would not occasion an analytical error, as in this case lead salts would be precipitated with the second group, but it occasionally gives rise to the impression that the result obtained by the addition of hydrochloric acid is uncertain, depending upon unknown circumstances, and is thus accidental.

Separation of *precipitate* from the *liquid* is effected by *filtration.* The filter is placed in a glass funnel so that one fold lies against the glass on one half, on the other, three folds. It is moistened with water and smoothed tight against the glass. The upper edge of the filter should be 1 cm. below the edge of the funnel. The process of filtration separates the precipitate only partially from the liquid, because a considerable quantity of the latter remains behind, wetting

precipitate and filter, and this might give rise to errors during the later investigation of the precipitate if it is not removed. This can easily be done by *washing* the precipitate, namely by pouring distilled water over it on the filter several times and letting it run off completely each time before renewing. The *wash water* that runs through the filter is added to the solution filtered off from the precipitate, *the filtrate from the first group,* and this is laid aside for further investigation.

The above described operation of washing applies to nearly all precipitates that have to be further *analyzed.* It is advisable to use as *little* water as possible in washing the first group precipitate (see § 7). The student can easily construct for himself a *wash bottle,* generally serviceable in the laboratory, from a flask, glass tube and cork (*Fig. 1*). The amount of water issuing from the flask is regulated to any desired degree through the quantity of air blown in, and by means of the movable nozzle the stream of wash water may be directed upon any particular object. The *washing* is done with *distilled water.* A little practice will enable the student to loosen the precipitate from the side of the filter and cause it to collect at the apex, which is of importance, since less water is used to wet it under these conditions than would be required in case the filter had to be completely filled up with water.

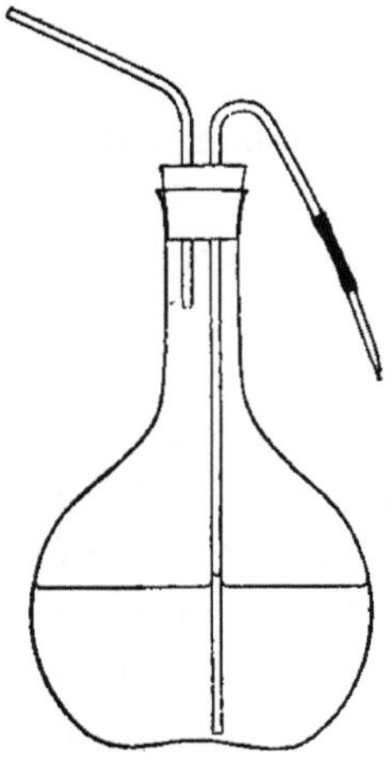

FIG. 1.

The first group precipitate may contain chlorides of *lead, silver* and *monovalent mercury,* the composition of which is expressed by the formulae, $PbCl_2$, AgCl, HgCl. In § 18 are considered certain other compounds which might be precipitated by hydrochloric acid, and therefore could be mistaken for the above chlorides (continued in § 7).

6. Solubility. As has already been pointed out in § 5, the property of a metallic component to be precipitated or not precipitated by hydrochloric acid is not arbitrary or accidental, but depends directly upon the *solubility* of the metallic chlorides in question, a numerically defined property of all chlorides. By *solubility* is meant the number of grams or gram formula-weights of a substance that are contained in unit volume, 1 liter, of the solution saturated with the substance

at a given temperature. In expressing solubility it is essential to assign a definite temperature (Principles, pp. 213–216); in other words, it is a function of temperature.

Chlorides of different metals have quite different solubilities. In particular, the three above-mentioned chlorides are very slightly soluble. On shaking up any one of them with water, only a very small quantity goes into solution. Moreover, when hydrochloric acid reacts with a solution of silver nitrate and forms silver chloride according to the equation:

$$AgNO_3 + HCl = AgCl + HNO_3,$$

then separation of the solid, AgCl, takes place if the amount of *silver chloride* so formed is in *excess* of the quantity that corresponds with its *solubility* in the presence of hydrochloric and nitric acids at the temperature of the experiment. This separation continues until the same condition is reached that would result if solid silver chloride were brought in contact with the *solution containing the same other constituents.* The condition is known as *equilibrium.* This term expresses the fact that no further changes are apparent between solution and precipitate.

Solutions which are in *equilibrium* with the solid substance are said to be *saturated.* When more of the substance is present in the solution than corresponds to equilibrium the solution is *supersaturated,* while it is *unsaturated* when less is present. *In contact with the solid substance* supersaturated and unsaturated solutions are *not in a stable condition,* but go over into saturated solutions by separating or dissolving solid salt respectively.

Chlorides of the other metallic components are more soluble in water. Hence no precipitation occurs on adding hydrochloric acid to a solution, for example, of nitrates, which contains no salts of lead, silver or monovalent mercury. It does not follow, however, that no reaction takes place in the sense of the equation:

$$MNO_3 + HCl = MCl + HNO_3,$$

where MNO_3 is the formula for the nitrate of any monovalent metal. On the contrary, it must be assumed, in view of the phenomena observed with the more difficultly soluble chlorides, that chlorides are also formed in these cases. Ordinarily the quantities of the metallic components in the solution, and the amount of acid added,

are too small to produce supersaturation with respect to these chlorides. In § 90, however, are considered the conditions for precipitating other chlorides with hydrochloric acid. The non-appearance of a precipitate upon addition of hydrochloric acid, therefore, does not mean that the metallic components have not reacted with hydrochloric acid, but merely shows the absence of metals, the chlorides of which are difficultly soluble, and that the chlorides of the other metals which may be present are more readily soluble (compare § 143, footnote 2).

7. **Analysis of the hydrochloric acid precipitate** (consult also the appendix, analytical table, group 1). In separating and identifying the three chlorides (§ 5), the first step is to *wash* the *precipitate* with *hot water, collecting* the water that *runs through the filter.* In this process *lead chloride* goes into solution, while *silver* and *mercurous chlorides* practically all remain behind on the filter, being almost entirely insoluble in hot water. The separation of long fine needles on cooling the filtrate indicates lead chloride in the first group precipitate. These are filtered, or are redissolved by warming again, and a confirmatory test is made by adding dilute sulphuric acid or a solution of a sulphate drop by drop. *Lead chloride* in the solution gives a *white fine-grained* precipitate of *lead sulphate,* according to the equation:

$$PbCl_2 + H_2SO_4 = \underline{PbSO_4} + 2HCl.$$

(What inference concerning the relative solubilities of lead sulphate and lead chloride can be drawn from this fact?) The formula of lead sulphate has been *underlined* in the equation to point out that a *precipitate is formed* in this reaction. Henceforth the *dissolving* or *forming* of a precipitate (solid phase) will always be indicated in this way.

It is noteworthy that the solution of any soluble sulphate, for example, sodium- or ammonium sulphate, may be used in place of sulphuric acid in the above experiment. This fact, the truth of which the student is recommended to test for himself experimentally, is very important, and we shall see in § 8 that valuable deductions can be drawn from it.

If the first group precipitate contains only *lead chloride,* it *dissolves completely* in hot water. Hence a residue left after washing with hot water several times is to be tested for silver and mercurous

chlorides. This is done by washing the residue with dilute *ammonia,* whereby *silver chloride* goes into solution and *mercurous chloride* becomes black. To concentrate the silver chloride in the ammonia solution as much as possible, the filtrate is repeatedly poured over the precipitate on the filter. It is sometimes found that the ammonia filtrate has a clouded appearance. Since the test for silver chloride *demands* a *perfectly clear* ammonia solution, it may be necessary to filter it through a second filter until clear.

When the residue after removing lead chloride is completely soluble in ammonia, it consists of silver chloride alone. A residue insoluble in ammonia indicates that lead chloride had not been completely removed with hot water, or that *mercurous chloride* is present, or that compounds which are referred to in § 18 are present. Since *mercurous chloride* turns *black* with *ammonia,* there is no uncertainty with regard to it.

The solubility of silver chloride in ammonia depends upon the formation of easily soluble *silver ammonia chloride,* $Ag(NH_3)_2Cl$. From this solution silver chloride is reprecipitated on removing *ammonia.* Since ammonia is very volatile, i. e. is easily vaporized when heated, it can be driven off by *warming* the solution. Addition of an *acid,* as *hydrochloric, nitric* or *acetic acid,* affords a still more *convenient* means of removing ammonia, which is thereby chemically changed. Hence precipitation of silver chloride from the ammonia solution shows that silver is one of the metallic components of the original solution. The reaction is accompanied by a considerable evolution of heat, and is represented by the equation:

$$Ag(NH_3)_2Cl + 2HA = \underline{AgCl} + 2NH_4A,$$

where HA is used for any monobasic acid to indicate that the nature of the component A (Cl, NO_3, Br, etc.) is a matter of indifference.

8. The behavior of acids, bases and salts. The agreement between the reactions of different acids with silver ammonia chloride is by no means the only property with respect to which acids show at least qualitative agreement. Their aqueous solutions have a characteristic acid taste, whence the name is derived. They redden blue litmus paper in common, dissolve magnesium, zinc and other metals with an evolution of gas, etc.

On the other hand certain substances, which in solution have an

alkaline taste, turn red litmus blue and form salts with acids, are classified under the head of bases.

This agreement in properties is, however, limited to the *aqueous solutions* of acids or bases, as is shown by experiments, for example, with perfectly dry hydrochloric acid, anhydrous sulphuric acid, etc. This fact, taken in conjunction with numerous others, for instance, the agreement in the behavior of dilute sulphuric acid and sulphates towards a solution of lead chloride, has given rise to an hypothesis which explains many of the different relationships involved.

According to this conception the agreement in properties is not due to the acids or bases themselves, but to products resulting from changes which the above-mentioned (solid, liquid or gaseous) substances undergo when they are dissolved in water, whereby each class of substances *with similar properties* gives rise to a *common product.* The following example will illustrate this point. Solutions of different lead salts, as lead acetate or -nitrate, show no *qualitatively* recognizable differences in their reactions with hydrochloric or sulphuric acid. The changes which these salts undergo on dissolving, whereby their agreement is explained, may be formulated as follows:

$$Pb(NO_3)_2 = Pb + 2NO_3 \qquad (1),$$

or

$$Pb(C_2H_3O_2)_2 = Pb + 2C_2H_3O_2 \qquad (2).$$

The common product in (1) and (2) is Pb. Similarly the changes to which hydrochloric and nitric acids are subjected on being dissolved in water are represented by:

$$HCl = H + Cl; \quad HNO_3 = H + NO_3 \qquad (3),$$

where H is the common product, while in the cases of potassium- or sodium hydroxide:

$$KOH = K + OH; \quad NaOH = Na + OH \qquad (4),$$

OH is the common product.

At first sight it would seem that this method of expressing these relations is opposed to our everyday experience, for a solution of lead nitrate or -acetate does not contain metallic lead, which is practically insoluble; and an acid, for instance, concentrated sulphuric acid, does not evolve hydrogen, which would have to escape as gas when the acid is diluted with water. The inconsistency is, however,

only apparent and can be readily gainsaid. The product common to all lead salts is assumed to be not ***metallic lead,*** but it is mutually transformable with lead. The difference lies in their different energy content, just as ice and water possess different amounts of energy, yet can be mutually interchanged, the transformations being accompanied by a change of energy. Similarly the product common to acids is not identical with gaseous hydrogen, but it may be formed from hydrogen and is transformable into hydrogen. Analogous transformations are met with everywhere in chemistry and come under the head of isomerism (Principles, p. *190*).

A special nomenclature is made use of to indicate the distinctions and also the mutual transformability of these classes of isomers. That part of an acid in solution which is ***isomeric*** with hydrogen is called ***hydrogen ion,*** while the product ***isomeric*** with lead, that is common to ***lead salts,*** is termed ***lead ion.*** Ions isomeric with ***metals*** or ***metallic radicals*** are represented by the corresponding symbols, to the upper right hand of which are appended one or more dots, according to the valence (Principles, p. *371*), while for ions isomeric with non-metallic elements or radicals dashes are similarly appended to the ordinary symbols, thus: Cl', SO_4'', NO_3', $K^{\cdot}$, $Ca^{\cdot\cdot}$, $Bi^{\cdot\cdot\cdot}$, etc.

Accordingly the above equations would be expressed in the following form:

$$Pb(NO_3)_2 = Pb^{\cdot\cdot} + 2NO_3' \qquad (1b),$$

$$Pb(C_2H_3O_2)_2 = Pb^{\cdot\cdot} + 2C_2H_3O_2' \qquad (2b),$$

$$HCl = H^{\cdot} + Cl' \ ; \ HNO_3 = H^{\cdot} + NO_3' \qquad (3b),$$ [1]

$$KOH = K^{\cdot} + OH'; \ NaOH = Na^{\cdot} + OH' \qquad (4b).$$

On the basis of this conception the term *acid* can now be defined with greater exactness. Since the common properties of acids are ascribed to hydrogen ion, it is necessary that a substance which acts as an acid must form hydrogen ion on dissolving. Hence ***hydrogen compounds,*** the solutions of which do ***not*** contain ***hydrogen ion,*** are ***not acids.***

Similarly the ***basic*** properties of substances which are called bases are characteristic of the ***ion,*** OH′, ***hydroxyl ion.*** Accordingly a base is a ***substance*** the solutions of which contain ***hydroxyl ion.***

[1] Hydrogen has a metallic character in the chemical sense, but not in a physical.

Close examination of the relations brought out in *3b* above shows that the qualitative *differences* in the properties of *different* acids depend upon *differences in the properties of the ions* Cl', NO_3', etc., just as the *agreement* in their properties is to be ascribed to *hydrogen ion.* Similar considerations hold for solutions of lead salts and of bases, as is seen in equations *1b, 2b* and *4b* respectively.

9. The relations of chemical phenomena with physical. Theory of Arrhenius. Components. The hypothesis discussed in § 8 has not been reached from the side of chemical phenomena, but through entirely different considerations which we can mention only briefly in view of the aims of this text-book. Aqueous solutions of different substances do not behave alike when they are closed in an electric circuit, but they exhibit essential differences, which are utilized to classify them. On the one hand *some* solutions *conduct* the *electric current,* while others are *non-conductors,* or more correctly, they are incomparably worse conductors of electricity. The former have the above-mentioned characteristic that the properties of their solutions can be ascribed, at least partly, to two (or more) components, the ions. Non-conductors do not have this essential feature.[1]

In accordance with their behavior towards the electric current two classes of substances are distinguished, *non-electrolytes* and *electrolytes* or *salts.* The latter are conductors. In this category are included salts in the narrow sense of the term, i. e. substances formed by the action of acids on bases, and also acids and bases. These

[1] It would not be justifiable to base the question of the existence of free ions in the solution of a salt upon an *exclusive* appeal to the observed chemical phenomena. The agreement in the behavior of, for instance, lead acetate, -chloride and -nitrate towards potassium- or sodium sulphate or sulphuric acid, is also comprehensible on the assumption that the reaction, which is recognized in these cases by the formation of solid lead sulphate, proceeds with *great velocity.* In fact isolated reactions are known, where all the characteristics which have just been described, are met with, and yet there is no other indication of a decomposition into ions (Kahlenberg, Journal of Physical Chemistry, 6, *1–14*, 1902).

Our conception of the phenomena in question does not appear any *less certain* in view of these remarks, which expressly point out that the chemical phenomena *suggest,* but do not thoroughly *establish,* the hypothesis that *salts* in *aqueous solution* are (partially) decomposed. To establish the hypothesis still other independent supports based on fact must be taken into account; for instance, the phenomena observed on closing a salt solution into the circuit of an electric battery.

It is necessary to sound a note of warning against too far-reaching generalizations of the hypothesis upon which the subject matter of this text-book is based.

latter are regarded as salts, the solutions of which contain hydrogen and hydroxyl ions respectively.

The phenomena respecting the dissolving of salts with which we have been dealing come under the head of *dissociation,* to indicate that several products are formed from a single substance. The above-mentioned hypothesis is included in the *theory of electrolytic dissociation* (Arrhenius, 1887; Principles, p. *196*), which is based upon experiments concerning the behavior of substances towards the electric current. In the light of this theory ions are accompanied by electric charges, one class with *positive,* the other with *negative.* The former are called *cathions* and move in the direction of the *positive current* through the solution. The *negatively charged ions, anions,* move in the *opposite direction* with the current. Cathions are isomeric with metals[1] and metallic radicals, anions with non-metals and non-metallic radicals.

It must be borne in mind that the electrolytic dissociation theory makes the *existence* of ions *independent* of the *passage* of an *electric current,* but also states that their previous *existence* is related to or responsible for their behavior towards the *electric current.* The electric charges produced on dissolving a salt in water cannot be detected, as will be shown later. The assumption of their existence, however, is not contrary to the law of conservation of energy, for this electrical energy may, on the basis of other experiments, be considered as derived from the energy difference between *ions* and the *substances* from which they are formed.

Since equal numbers of positive and negative charges simultaneously appear or vanish in the reactions under consideration, salt solutions will give no indication of positive or negative electric charges. Recourse is frequently had to this relation for the formulation of chemical reactions in which ions take part. It is of further use in formulating equations for the dissociation of different substances, when we know *how many charges one* or *other* of the ions has, as the following examples show:

$$K_2SO_4 = 2K^{\cdot} + SO_4'',$$

$$FeCl_3 = Fe^{\cdot\cdot\cdot} + 3Cl'.$$

[1] For the sake of simplicity the hydration of ions is not taken into consideration. It suffices to point out that certain phenomena support the assumption that ions, like many salts which contain water of crystallization, are combined with water.

Our choice of *metallic* and *non-metallic components* (see § 1), to express the composition of a substance or mixture, is also based upon the above considerations, since analytical results are obtained from *chemical reactions* which are characteristic of *ions*. Hence it is *practical* and *advisable* to select the components in accordance with the composition of the ions, although these cannot be transformed into the solid state as such without change, but may form solid salts only in pairs.

We shall use the dissociation theory as set forth in the above discussion and shall apply it to all cases where it helps us *understand* facts and brings them into closer *connection* with one another.

10. Reactions of the ions of group 1: $Ag^{\cdot}$, $Pb^{\cdot\cdot}$, $Hg^{\cdot}$. The operations described in § 7 suffice to *identify* these ions. Certain of their other properties and the properties of the solid salts of the corresponding metals will now be dealt with according to their importance. To gain insight into the relative distinctness of reactions, solutions of known concentration should be employed. If necessary[1] test solutions are made by weighing out a definite quantity of the given salt on an ordinary side-table balance and dissolving it in a definite volume of water, which is measured in a graduated cylinder for this purpose.

11. Silver ion. A test solution of suitable strength for the following reactions is prepared by dissolving $\frac{1}{10}$ formula weight[2] of silver nitrate, ($AgNO_3$)—see table of combining weights—in 1000 cc. of water, or $\frac{1}{100}$ formula weight in 100 cc. For each test 2–3 cc. of this solution are used.

Alkali hydroxide[3] precipitates *silver hydroxide*, white, turning brown to black almost instantly.[4] The change of color is due to the formation of silver oxide, Ag_2O, from silver hydroxide:

$$Ag^{\cdot} + NO_3' + K^{\cdot} + OH' = \underline{Ag(OH)} + K^{\cdot} + NO_3'.$$

$$2Ag(OH) = Ag_2O + H_2O.$$

[1] [In most American laboratories stock solutions of the more common salts are provided. As a rule these have a concentration suited for the special tests.]

[2] The formula weight of a compound is the sum of the products of the combining weights of the individual elements with their indices; for example, the formula weight of silver nitrate is $1 \times 108 + 1 \times 14 + 3 \times 16 = 170$ g. The combining weights are rounded off, since the concentrations need be known only approximately for qualitative tests.

[3] The term '*alkali hydroxide*' is used to indicate sodium- or potassium hydroxide. Similarly 'alkali sulphate' would stand for sodium- or potassium sulphate.

[4] The white precipitate is observed only when the caustic potash or soda used is very dilute.

Ammonia,[1] added drop by drop, yields the same precipitate, but in this case the precipitate dissolves in an excess of the reagent. Similarly the precipitate obtained with potassium hydroxide is soluble in ammonia.

The solubility of silver oxide in ammonia, like that of silver chloride, is due to the formation of a cathion containing two combining weights of NH_3 to one of Ag, $Ag(NH_3)_2^{\cdot}$, the compounds of which are more readily soluble. Ammoniacal solution of silver oxide contains the compounds $Ag(NH_3)_2OH$ and $Ag_2(NH_3)_4O$.

With *potassium-* or *sodium carbonate* solution, i. e. alkali carbonate, silver ion gives a yellowish white precipitate of *silver carbonate:*

$$2Ag^{\cdot} + 2NO_3{}' + 2K^{\cdot} + CO_3{}'' = \underline{Ag_2(CO_3'')} + 2K^{\cdot} + 2NO_3{}'.$$

On warming, silver carbonate decomposes into carbon dioxide, CO_2, which escapes as a gas, and silver oxide, Ag_2O, which is black.

Potassium chloride, like *hydrochloric acid,* precipitates *silver chloride,* white, flocculent, which turns bluish-violet in the light.[2]

Potassium bromide and *-iodide* form *silver bromide* and *-iodide* respectively, the latter having a characteristic yellow color. Silver bromide is considerably less soluble in ammonia than the chloride, and the iodide is practically insoluble in ammonia.

This is readily shown by precipitating the three halide salts separately, washing and shaking up with equal volumes of ammonia (not enough to completely dissolve the silver halides). The residue is filtered off in each case and the *clear filtrate* acidified. It is observed that different quantities of precipitates form, in accordance with the varying solubility of the chloride, bromide and iodide respectively, in ammonia.

The behavior of silver ion towards *cyanide ion* demands special

[1] The formula of ammonia, (NH_3), does not indicate that an ammonia solution may contain hydroxyl ion, which, however, must be the case according to § 8, since the solution shows the typical reactions of a base. The formula, NH_3, expresses the composition of gaseous ammonia. On dissolving in water this goes over partly into NH_4OH,—($NH_3 + H_2O = NH_4OH$)—which in turn can form the ions $NH_4^{\cdot}$ and OH', according to the equation:

$$NH_4OH = NH_4^{\cdot} + OH'.$$

[2] The change of color produced by light is due to the formation of compounds containing less chlorine, probably Ag_2Cl or Ag_4Cl_2. These are changed back to ordinary chloride through the action of chlorine (Principles, p. 676).

consideration. On adding *potassium cyanide* solution drop by drop to a solution of silver nitrate, silver cyanide precipitates at first. The amount of precipitate increases until a certain definite quantity of potassium cyanide has been added, and then decreases and finally vanishes when two formula weights of potassium cyanide are present for each formula weight of silver nitrate.

12. **Solubility of silver cyanide in potassium cyanide.** This property of silver cyanide depends upon the formation of a *complex* ion, (§ 13), between silver cyanide and cyanide ion. On introducing an excess[1] of potassium cyanide into a saturated solution of silver cyanide, for instance, the solution of silver nitrate in which silver cyanide has been precipitated, the dissolved silver cyanide reacts with the ions of potassium cyanide:

$$\underline{AgCN} + K^{\cdot} + CN' = Ag(CN)_2' + K^{\cdot} \qquad (1).$$

If we take into account that only very little silver cyanide is dissolved and that this decomposes into the ions $Ag^{\cdot} + CN'$, the above reaction may also be formulated as follows:

$$Ag^{\cdot} + CN' + K^{\cdot} + CN' = Ag(CN)_2' + K^{\cdot} \qquad (2).$$

Through this *transformation* the solution becomes *unsaturated* with respect to *silver cyanide.* Hence the precipitated cyanide must go into solution until the saturation point is again reached, according to § 6. When a sufficient amount of potassium cyanide has been added, the precipitate of silver cyanide entirely disappears. From the above equation it is evident that *one formula weight of potassium cyanide* will bring into solution *one formula weight of silver cyanide.*

The cyanide is by no means the only silver salt that is soluble in potassium cyanide. Nearly all difficultly soluble silver salts will dissolve. Thus silver chloride goes into solution as follows:

$$\underline{AgCl} + 2K^{\cdot} + 2CN' = K^{\cdot} + Ag(CN)_2' + K^{\cdot} + Cl' \qquad (3).$$

Silver sulphide, however, is exceptional in its behavior and is precipitated by passing hydrogen sulphide, § 19, into potassium silver cyanide solution. Since hydrogen sulphide is an especially sensitive reagent for silver ion, it follows that the above reaction, (*2*), is not complete or that potassium silver cyanide decomposes to a slight, but quite definite, extent according to the equation:

$$K^{\cdot} + Ag(CN)_2' = AgCN + CN' + K^{\cdot}$$
$$AgCN = Ag^{\cdot} + CN' \qquad (4).$$

13. **Complex ions.** Silver cyanide ion belongs to the class of *complex ions.* These are distinguished by characteristic properties, differing from those of the ions or salts from which they are formed; for instance, $Ag(CN)_2'$ from AgCN and CN'. Complex ions, however, always exhibit the properties of their simpler component ions to a slight extent (compare § 12).

The increase in the solubility of one substance by the addition of a second substance is frequently due to the formation of complex ions; for instance,

[1] A little more than enough to effect precipitation.

the solubility of silver cyanide in potassium cyanide. Similarly the solubility of the oxide, chloride and bromide of silver in ammonia also depends upon the formation of silver ammonia ion, $Ag(NH_3)_2^{\cdot}$.

We have seen in § 7 that silver ammonia ion decomposes and a precipitate of silver chloride forms, when the solution of silver chloride in ammonia is acidified. Similarly by adding hydrochloric acid to other silver ammonia salt solutions, precipitation of silver chloride also takes place. On the other hand ammonia that has stood in contact with silver iodide gives no precipitate when acidified, even with hydrochloric acid. From this we may conclude that silver iodide is but very little soluble in water.

The equations:

$$\underline{AgCl} = Ag^{\cdot} + Cl' \qquad (1a),$$

and

$$Ag^{\cdot} + 2NH_3 = Ag(NH_3)_2^{\cdot} \qquad (1b),$$

represent the formation of the complex silver ammonia ion in the case of silver chloride (compare § 11 for the solubility of the three silver halides in ammonia). The extent to which the complex ion forms, depends upon the amount of silver ion present in unit volume of the saturated solution of silver salt. If a very small quantity of silver ammonia ion is produced, as is indicated by the absence of a precipitate on acidifying or on vaporizing the ammonia, it follows that the saturated aqueous solution of the silver salt in question contains a *negligible amount* of silver ion, e. g. AgI.

From the fact that silver iodide is not noticeably soluble in ammonia, it is evident that potassium iodide will precipitate silver iodide from an ammoniacal solution of silver chloride.

$$Ag(NH_3)_2^{\cdot} + I' = \underline{AgI} + 2NH_3 \qquad (2).$$

This equation is the inverse of (*1b*) above.

With silver cyanide ion potassium iodide does not yield a precipitate of silver iodide. Hence we are justified in concluding that silver cyanide ion is more complex than silver ammonia ion. This conclusion as to the relative complexity of these ions, however, is only approximate, since it is based upon *qualitative*, i. e. approximate experiments.

14. Lead ion. For the following tests a solution is used containing $\frac{1}{20}$ formula weight of lead nitrate, $Pb(NO_3)_2$, or lead acetate, $Pb(C_2H_3O_2)_2 + 3H_2O$,[1] per liter.

From this solution *alkali hydroxide* precipitates white *lead hydroxide:*

$$\underbrace{Pb^{\cdot\cdot} + 2NO_3'} + \underbrace{2K^{\cdot} + 2OH'} = \underline{Pb(OH)_2} + 2K^{\cdot} + 2NO_3' \quad (1).$$

By omitting the ions which stand on both sides of equation (*1*) it becomes:

$$Pb^{\cdot\cdot} + 2OH' = \underline{Pb(OH)_2} \qquad (2).$$

[1] Account must be taken of the three formula weights of water in calculating the amount of lead acetate to be dissolved. In case the lead salt does not yield a clear solution with water, a few drops of nitric acid are added.

This shortening of the method of formulating equations will be used frequently, since it indicates the essential changes only, and hence gives greater clearness.

Moreover this shortened method of formulation presents an important *generalisation* of the reaction between lead nitrate and potassium hydroxide, for it also expresses the reaction between lead acetate and potassium hydroxide:

$$Pb^{\cdot\cdot} + 2C_2H_3O_2' + 2K^{\cdot} + 2OH' = \underline{Pb(OH)_2} + 2K^{\cdot} + 2C_2H_3O_2' \qquad (3),$$

as well as the reactions of lead acetate and -nitrate with sodium hydroxide, barium hydroxide, etc. In other words the precipitation of lead hydroxide is not *limited* to lead nitrate and caustic potash, as might be inferred from equation (*1*), but any soluble base will give the same reaction with any *lead compound,* provided lead ion is present. However, it must be emphasized particularly that such *generalisations* are permissible only when they are based upon experimental evidence. If no precipitate is formed, it indicates that a secondary reaction has taken place, which removes lead ions from the solution.

The precipitated *lead hydroxide* dissolves in an *excess* of *alkali hydroxide:*

$$\underline{Pb(OH)_2} + 2OH' = PbO_2'' + 2H_2O \text{ (compare § 42)} \qquad (4).$$

Ammonia also yields a precipitate of *lead hydroxide* which, however, dissolves to a much less extent in an excess of the reagent.

Alkali carbonates give a white precipitate of *lead carbonate* and *-hydroxide.*

Iodides form a yellow precipitate of *lead iodide.*

Lead iodide and lead chloride are each much more readily soluble in solutions of the corresponding alkali halides than in water, owing to the formation of complex ions, $PbCl_3'$, PbI_3', analogous to the complex mercury halogen ions (§ 33).

Potassium dichromate and *potassium chromate* give a yellow precipitate of *lead chromate,* soluble in nitric acid and in caustic potash or soda.

Lead sulphate is precipitated by *potassium-, ammonium-* or *sodium sulphate* or *dilute sulphuric acid.* The precipitate is appreciably soluble in *hydrochloric* and *nitric* acids, also in solutions of *ammonium salts,* especially *ammonium acetate* and *-tartrate.* The solubility in the latter serves to distinguish lead sulphate from barium sulphate, which is scarcely more soluble in ammonium tartrate than in water. The solution of ammonium tartrate is prepared by adding *ammonia* to *tartaric acid* solution until *alkaline.*

The solubility in ammonium tartrate depends upon the formation of a complex ion of the formula $Pb_2C_8H_4O_{12}''$, stable only in alkaline solution. Ammonia must be used to make the tartrate solution alkaline, and not sodium- or potassium hydroxide, because lead sulphate is also soluble in the latter, even without the presence of tartaric acid.

The probable reasons for the greater solubility of lead sulphate in ammonium acetate and in hydrochloric and nitric acids cannot be stated until later (§ 112).

The above-mentioned difficultly soluble salts, except *lead chromate* and also *lead sulphide,* all dissolve in *sodium thiosulphate,* forming a complex lead thiosulphate ion, quite stable at ordinary temperature. Its composition has not yet been definitely settled. On warming the solution the complex ion slowly decomposes, separating lead sulphide.

By passing *hydrogen sulphide* gas into the solution of a lead salt black *lead sulphide* is precipitated. If, however, the solution has been strongly acidified with hydrochloric acid an *orange red* precipitate, the composition of which is not definitely known ($PbCl_2.PbS$ or $PbCl_2.2PbS$), first forms, but on further addition of hydrogen sulphide this goes over into black lead sulphide.

15. **Preparation of lead nitrate.** Lead nitrate is easily prepared from commercial *minium,* a lead oxide of the formula, Pb_3O_4. From 10–20 g. of this red powder are heated with ***dilute nitric acid*** in a porcelain evaporating-dish until the red color has been transformed into brown, care being taken to stir the mixture frequently with a glass rod. The brown color is characteristic of *lead peroxide,* which is formed according to the equation:

$$\underline{Pb_3O_4} + 4HNO_3 = 2Pb(NO_3)_2 + 2H_2O + \underline{PbO_2} \qquad (1).$$

Lead peroxide remains practically undissolved. The solution containing lead nitrate and any free nitric acid that is left is filtered and evaporated to dryness on a water-bath.[1] The residue of lead nitrate is weighed and the amount obtained is compared with the quantity calculated from the above equation and the weight of red lead taken. Two formula weights of lead nitrate are obtained from one formula weight of red lead; $Pb_3O_4 : 2 \times Pb(NO_3)_2 = 684.7$ g.: 622 g. From two experiments, starting with 20 and 24 g. respec-

[1] A ***water-bath*** is used in cases where a temperature higher than that of boiling water is to be avoided. For this purpose a large beaker may be used, on top of which is placed the porcelain evaporating-dish.

tively, there were obtained 21 and 25.3 g. $Pb(NO_3)_2$, while the quantities calculated from equation (*1*) are 19.3 and 23.2 g. respectively. What is the explanation of this *variation* from the *theoretical* quantity? Lead peroxide is not noticeably soluble in dilute nitric acid.

When heated with concentrated hydrochloric acid, lead peroxide dissolves with evolution of chlorine, which gives a brown stain of iodine on filter paper moistened with *potassium iodide* solution. The equations are:

$$\underline{PbO_2} + 4HCl = PbCl_2 + 2H_2O + Cl_2 \qquad (2)$$

and

$$Cl_2 + 2KI = 2KCl + I_2 \qquad (3).$$

The latter may also be written:

$$Cl_2 + 2I' = 2Cl' + I_2 \qquad (4),$$

taking into account that only dissolved, i. e. ionized potassium iodide is decomposed by chlorine.

16. The ions of mercury. There are two well-known series of salts of this metal, in which one and two combining weights of a monovalent non-metal or radical are respectively combined with one combining weight of mercury. The former are called *mercurous* salts, the latter *mercuric*. These two classes of salts have entirely different properties. Accordingly we distinguish the corresponding ions, mercurous and mercuric. The two *ions* and *metallic mercury* are isomeric, § 8. All three are mutually transformable.

17. Mercurous ion. The solution to be used for the following tests contain $\frac{1}{10}$ formula weight of $HgNO_3$ per liter. To prepare it the weighed-out salt is first dissolved in a little water, previously acidulated with a few drops of *dilute nitric acid,* and then the requisite amount of water is added to this solution. Water decomposes the salt and turns it yellow when the acid is not first added. For the explanation of this decomposition see § 22.

Alkali hydroxide precipitates black *mercurous oxide:*

$$2Hg^{\cdot} + 2OH' = \underline{Hg_2O} + H_2O \qquad (1).$$

Mercurous oxide decomposes into *mercuric oxide* and metallic *mercury* when left standing in the light, or when heated:

$$Hg_2O = HgO + Hg \qquad (2).$$

The reaction with ammonia, already mentioned in § 7, not only occurs with solid mercurous chloride,[1] but also takes place when ammonia is added to the solution of the nitrate:

$$2Hg^{\cdot} + 2NO_3' + 2NH_3 = NH_4^{\cdot} + NO_3' + \underline{NH_2Hg_2NO_3} \quad (3).$$

When not perfectly pure and dry, mercurous chloride decomposes in the light, and, like the oxide, yields mercuric chloride and mercury:

$$2HgCl = HgCl_2 + Hg \quad (4).$$

On account of its slight solubility mercurous chloride is much less dangerous to the living organism than the very poisonous mercuric chloride. Hence this point is of especial importance in medicine.

Iodide ion precipitates green *mercurous iodide*, HgI, which decomposes into HgI_2 + Hg. Mercurous iodide dissolves in an excess of alkali iodide, separating mercury in consequence of this reaction, § 33.

Soluble *cyanides* precipitate mercury and form mercuric cyanide, according to the equation:

$$2Hg^{\cdot} + 2CN' = Hg(CN)_2 + \underline{Hg} \quad (5).$$

Mercuric cyanide solution is an exceptionally poor conductor of electricity. Hence in equation (5) no indication is made of a decomposition into the ions $Hg^{\cdot\cdot} + 2CN'$ (compare § 34).

The reactions of mercuric ion and its relations with mercurous ion will be dealt with later.

18. Some special cases belonging to the first group. On testing a given solution with hydrochloric acid for the first group components, a precipitate is occasionally formed in which no chlorides of the metals of this group can be detected. To avoid mistakes in consequence of this more of the first group precipitant is added in order to try whether the precipitate dissolves in an *excess* of hydrochloric acid or not. Components which have this property are re-

[1] The presence of water has considerable influence upon the reaction with solid mercurous chloride, as is shown by taking two beakers with a little concentrated ammonia, and adding dry mercurous chloride to the one, to the other a moistened sample. In the latter case darkening occurs at once and is more intense than in the former. Although it is known that most reactions proceed more quickly when the substances taking part are in solution, still it may be shown that it is not the case in the present instance that mercurous chloride is dissolved in the water, and, like the nitrate, gives with dissolved ammonia a more difficultly soluble compound, NH_2Hg_2Cl. This point will be considered later.

ferred to in §§ 37 and 46 and in § 184a (methods of bringing cyanides into solution). The last point comes into consideration only for the *complete analysis* of a mixture according to § 168. The phenomena described in §§ 168–174 indicate that an error is possible.

In case the *first group* precipitate has been treated with a great excess of acid to remove precipitated substances which do not rightfully belong in group 1, as SbOCl, BiOCl, it must be remembered (§§ 22, 46) that the precipitation of AgCl, $PbCl_2$, HgCl is also *less complete* under these conditions than when only a small excess of hydrochloric acid is present. The chlorides of the first group metals are more readily soluble in hydrochloric acid than in water, depending upon the formation of complex ions (already mentioned in § 14).

It must therefore be taken into account that $Pb^{\cdot\cdot}$, $Ag^{\cdot}$, $Hg^{\cdot}$ can be carried over into the second group. For details consult § 22.

GROUP II.

19. **Precipitation with hydrogen sulphide.** The filtrate from the first group or the original solution, in case a test with hydrochloric acid has not given a precipitate, thereby proving the absence of the cathions $Hg^{\cdot}$, $Ag^{\cdot}$ and $Pb^{\cdot\cdot}$ up to the limits of the solubility of their corresponding chlorides, is next treated with the second group reagent, *hydrogen sulphide.* This is a gas that is usually generated from ferrous sulphide with hydrochloric or sulphuric acid. Its *aqueous solution* gives characteristically colored precipitates with different cathions, from the colors of which certain deductions can be drawn regarding the metallic components present. On account of the unpleasant odor and poisonous character of the gas care must be taken to avoid generating it in the open laboratory as far as possible. It is best to have special hydrogen sulphide rooms off the qualitative laboratory.

For most analytical purposes a stock solution of hydrogen sulphide[1] is not used, but the gas produced in special generators (Principles, p. *270*) is passed into the solution. Generally the laboratory is provided with a large generator and a system of delivery tubes which makes it possible to use the gas simultaneously in several places. A glass tube is attached to one of the delivery tubes of the system and passes through a stopper fitting loosely the neck of the

[1] As a rule the solution is used only for preliminary tests to see if second group cathions are present, and also in washing the second group precipitate.

vessel in which the solution is contained (*Fig. 2*). By this means the unused gas is kept from escaping into the room.

Before introducing the gas the solution is *weakly acidified* with *dilute hydrochloric acid* (the concentration of acid should be approximately that of a fifth normal solution; consult § 31) and heated. Complete precipitation is effected by *saturating* the liquid with hydrogen sulphide, i. e. by passing in the gas until no more will dissolve and the gas begins to escape unchanged. On closing the neck of the vessel with the thumb or palm of the hand and agitating the liquid, a suction indicating a decrease of pressure will be observed in case any more gas goes into solution. When no longer any suction is produced in this way no more gas enters the solution, i. e. the saturation point has been reached.

FIG. 2.

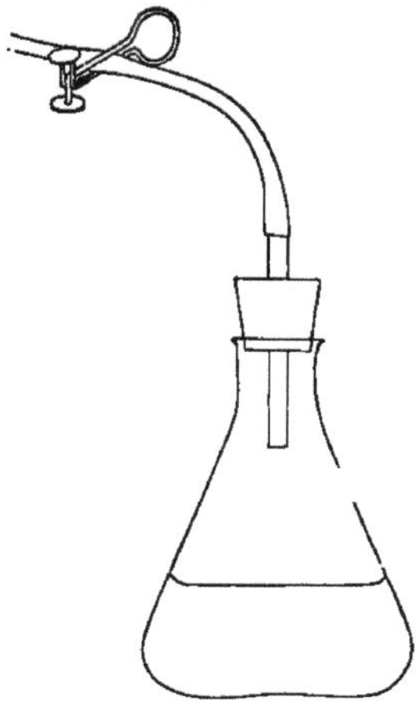

Precipitate and liquid are then separated by filtration. The filtrate, after dilution with water or on addition of some ammonia,[1] is further treated with hydrogen sulphide to see if any more precipitate forms, in *which case* the gas is *added* until precipitation ceases. The filtrate is tested according to § 51 for the remaining metallic components.

The *sulphides* are collected on the filter, washed with a *solution* of *hydrogen sulphide* and tested according to §§ 21 and 24. The precipitation takes place according to the equation:

$$M^{\cdot\cdot} + H_2S = \underline{MS} + 2H' \qquad (1).$$

The divalent metal ion, $M^{\cdot\cdot}$, reacts with *hydrogen sulphide,* forming the *difficultly soluble sulphide,* MS, which precipitates. In the process *hydrogen ion* is also formed. The latter fact has very much to do with the theory of this reaction, to be described later (§ 31).

20. Sulphides precipitated by hydrogen sulphide. The metals, of which the ions react with hydrogen sulphide according to equation (*1*), § 19, are: Ag, Hg, Pb, Bi, Cu, Cd, As, Sb, Sn. Their sul-

[1] Care must be taken here that the solution remains *acid,* otherwise components which are normally precipitated in the third and fourth groups, might be carried down in the second group precipitate.

phides have the formulae: Ag_2S, HgS, PbS, Bi_2S_3, CuS, CdS, As_2S_3, (As_2S_5), Sb_2S_3, Sb_2S_5, SnS, SnS_2.

A number of elements form two or more difficultly soluble sulphides. Of the above list As, Sb and Sn in their different valencies form difficultly soluble compounds with hydrogen sulphide. The sulphides of these three metals differ essentially from the remaining sulphides of group II in their behavior towards ammonium sulphide, a solution made by passing hydrogen sulphide into ammonia. As_2S_3, As_2S_5, Sb_2S_3, Sb_2S_5, SnS, SnS_2 dissolve, while Ag_2S, HgS, PbS, Bi_2S_3, CuS[1] and CdS are insoluble in ammonium sulphide.

21. Separation of the sulphides insoluble in ammonium sulphide. The separation and identification of the sulphides *insoluble* in ammonium sulphide will be dealt with first, as this is the simpler case. The means of showing that only these insoluble sulphides are present, and also the method of *separating* the *insoluble* sulphides from those *soluble* in ammonium sulphide, are considered in §§ 24 and 25. In the present case the sulphides are scraped off the filter directly with a spatula, or the filter may be taken out of the funnel and spread out flat and the sulphides washed off by means of the wash bottle into a porcelain dish in which they are gently heated with dilute nitric acid. Except for mercuric sulphide all dissolve with evolution of hydrogen sulphide, which, however, is in part oxidized to sulphur or sulphuric acid. The following reactions may take place:

$$\underline{MS} + 2HNO_3 = M(NO_3)_2 + H_2S \qquad (1),$$

$$3\underline{MS} + 8HNO_3 = 3M(NO_3)_2 + 4H_2O + 3\underline{S} + 2NO \qquad (2),$$

$$3\underline{MS} + 14HNO_3 = 3M(NO_3)_2 + 4H_2O + 3H_2SO_4 + 8NO \qquad (3).$$

Any undissolved HgS is separated from the solution of nitrates by filtration (see § 33). It is then washed and brought into solution[2] by warming with *aqua regia* (a mixture of 3 volumes concentrated HCl and 1 volume concentrated HNO_3). Mercuric chloride is formed according to the equation:

$$3\underline{HgS} + 6HCl + 2HNO_3 = 3HgCl_2 + 2NO + 4H_2O + 3\underline{S} \qquad (4).$$

[1] CuS is slightly soluble in ammonium sulphide; see § 32.

[2] A residue of sulphur left from the reactions in (*2*) and (*4*) has different colors when it contains small amounts of various undissolved sulphides. All danger of missing one or other of the sulphides is, however, avoided by longer heating.

To show the presence of mercuric chloride stannous chloride is added to the filtered solution. *Mercuric ion* gives a *white precipitate* which soon becomes *gray*, especially in an excess of stannous chloride. *Stannous ion* and *mercuric ion* cannot exist together in the concentrations with which we are dealing. The *former* goes over into tetravalent *stannic ion*, while the *latter* is simultaneously transformed into *mercurous ion*, and finally *metallic mercury*. The reactions are:

$$Sn^{\cdot\cdot} + 2Hg^{\cdot\cdot} = Sn^{\cdot\cdot\cdot\cdot} + 2Hg^{\cdot} \quad (5),$$

$$Sn^{\cdot\cdot} + 2Hg^{\cdot} = Sn^{\cdot\cdot\cdot\cdot} + 2Hg. \quad (6).$$

The white precipitate is due to reaction (5) and consists of *mercurous chloride*. A sufficient excess of stannous chloride effects the transformation of mercurous chloride into metallic mercury (6), which separates as a finely divided gray powder. Occasionally the gray precipitation follows immediately after adding stannous chloride. In this case the second reaction (6) takes place very rapidly, so that the formation of mercurous chloride escapes observation.

22. **Continuation. Separation of lead- and bismuth ions.** The nitric acid solution, which may contain $Ag^{\cdot}$, $Pb^{\cdot\cdot}$, $Bi^{\cdot\cdot\cdot}$, $Cu^{\cdot\cdot}$ and $Cd^{\cdot\cdot}$, is first tested with dilute hydrochloric acid for $Ag^{\cdot}$, and then with dilute sulphuric acid for $Pb^{\cdot\cdot}$.[1] Since the precipitation of lead sulphate is hindered either *completely* or in any case *partially* in the presence of an excess of nitric acid, the solution must be evaporated to dryness on a water-bath in order to remove the excess of nitric acid before dilute sulphuric acid is added. The residue from the evaporation is then treated with water. In case it does not all dis-

[1] Only when the solution which has been in contact with the first group precipitate contains a substance that *dissolves silver chloride* or *hinders* the *precipitation* of silver ion by hydrochloric acid, does *silver ion* go over into the *filtrate* from the first group and thus is found as *silver sulphide* in the second group *precipitate*. This is, for example, the case if the solution contains *mercuric ion*, or if the behavior of the hydrochloric acid precipitate towards an *excess* of *hydrochloric acid* (i. e. concentrated HCl) is tested according to § 18. In the presence of $Hg^{\cdot\cdot}$ the following reaction takes place:

$$2\underline{AgCl} + Hg^{\cdot\cdot} = HgCl_2 + 2Ag^{\cdot}.$$

Precipitation of AgCl is *hindered* because *part* of the *chloride ions added to the solution* form *undissociated mercuric chloride* with the *mercuric ions*. Since only free *chloride ions* react with silver ions to form *silver chloride*, it follows that *more hydrochloric acid* is required to precipitate a given amount of *silver ion* when the solution also contains *mercuric ion*. Thus the precipitation is less complete unless an excess of hydrochloric acid is used.

The precipitation of *lead ion* in group II is due to the solubility of lead chloride which is more soluble to a very remarkable degree at temperatures higher than room temperature (see § 5).

solve, and a white insoluble portion remains behind, this is filtered, washed, dissolved in hydrochloric acid and tested for $Bi^{\cdots}$ in the usual way (see below). To the filtrate or the aqueous solution dilute *sulphuric acid* is added *drop by drop.* Lead ion gives a white precipitate of lead sulphate, which is filtered. In the filtrate or, when no precipitation took place, in the aqueous solution *acidulated* with *sulphuric acid,* $Bi^{\cdots}$ is indicated by a white precipitate on addition of ammonia. The precipitate is *bismuth hydroxide* or the basic nitrate or sulphate:

$$Bi^{\cdots} + 3OH' = \underline{Bi(OH)_3} \quad (1).$$

Copper and *cadmium ions* may also be precipitated under these conditions, especially when only a small quantity of ammonia has been added. By adding more ammonia these precipitates vanish, however. In the presence of copper ion the *solution assumes* a *deep blue color,* due to the formation of a complex *copper ammonia ion* (§ 40).

The *precipitate* of $Bi(OH)_3$ or basic bismuth nitrate or -sulphate, obtained by adding *ammonia* to the aqueous solution, is filtered. This, together with any *insoluble residue* from *evaporation* of the *nitric acid solution,* after washing, is dissolved in a *very small amount* of warm dilute *nitric acid.* Since an excess of acid lessens the delicacy of the test, *dilute ammonia* must be added to this solution *drop by drop* until a *slight permanent opalescence* results. Then just enough dilute nitric acid is dropped into the solution from a drawn-out glass tube to barely remove the opalescence. On adding *sodium-* or *potassium chloride* drop by drop to the clear solution a white precipitate of *basic bismuth chloride,* or *bismuth oxychloride,* BiOCl, forms if bismuth ion is *present.* When warmed with the solution the precipitate does not dissolve, which distinguishes it from *lead chloride*[1] that might otherwise be mistaken for the bismuth precipitate.

The following equations make clear the detrimental influence of an excess of acid upon the precipitation of bismuth:

$$Bi^{\cdots} + H_2O = BiO^{\cdot} + 2H^{\cdot} \quad (2a),$$

and

$$BiO^{\cdot} + Cl' = \underline{BiOCl} \quad (2b).$$

[1] This means of identifying bismuth is much more delicate than the customary method of dissolving the precipitate in hydrochloric acid, neutralizing the excess of acid with ammonia and finally diluting this solution with a large amount of water. In the presence of bismuth separation of the oxychloride takes place

Hydrogen ion limits the formation of bismuthyl ion, BiO˙ (see § 28), and since chloride ion reacts with bismuthyl ion the delicacy of the test is thereby lessened.

The formation of a white precipitate on adding ammonia to the nitrate solution (*1*) is quite often *wrongly* regarded as a sufficient indication of the presence of bismuth ion in the solution, and thus it is considered just as certain a test as the formation of *bismuth oxychloride.* However this is not the case. When lead has been only incompletely precipitated by dilute sulphuric acid, the filtrate contains *lead ion* which gives a white precipitate with ammonia, but which does *not produce* a precipitate with neutral chlorides. The reason is that the quantities of lead with which we are dealing in these cases are at most only very small. Hence the amount of lead chloride formed does not exceed the quantity present in the saturated solution of this salt. In any event it should be shown by warming the precipitate with the solution that it is *not lead chloride.*

It is essential that the analyst follow the fixed rule of identifying the components of a precipitate by an *especially characteristic reaction,* thus bismuth in bismuth hydroxide in the present instance.

From the above equations (*2a*) and (*2b*) we are able to follow the influence which an increase of any of the substances taking part in the reaction will have upon the condition of the system. If a solution of bismuth chloride is diluted with water, equilibrium (consult §§ 6, 27 and 28 for the explanation of this term) establishes itself in the system after a definite time, i. e. the system undergoes no change except by the addition of other substances. It is thus easy to show on the one hand what effect an increase of water will have on the system, on the other hand the effect of an increase of acid. In case the initial content of water is not too great, i. e. if all bismuth ion has not been precipitated as BiOCl, it is readily seen that addition of more water increases the quantity of precipitated BiOCl, while addition of acid decreases it, until finally all the precipitate vanishes. These phenomena are very important for the above considerations.

23. **Continuation. Separation of copper- and cadmium ions.** A *colorless ammoniacal* solution or filtrate shows the *absence* of *copper.* In this case the solution is slightly *acidified,* and is then saturated with hydrogen sulphide. *Cadmium ion* gives a yellow precipitate of *cadmium sulphide,* CdS. To ensure the success of this test a large excess of acid must be avoided.

When copper ion is present the ammoniacal solution has a deep blue color. *Solution* of *potassium cyanide* is added until the *blue color disappears* and then hydrogen sulphide is passed into the colorless liquid. Under these conditions a precipitate forms only when *cadmium ion* is present. The fact that *copper ion* does not react with hydrogen sulphide in this case goes hand in hand with the striking change of color. Since the blue color of all copper salt solutions and

under these conditions. Beginners frequently experience difficulties in carrying out the usual test, always due to the excess of acid, which is not sufficiently removed with ammonia. By increasing the amount of chloride ion through addition of a *neutral chloride* this disadvantage is for the most part overcome (compare § 37).

the agreement in their chemical properties are *characteristic* of *copper ion*, it is evident that the *solution decolorized* by *potassium cyanide* contains *practically no copper ion*. Copper remains in the solution, however. Hence a *complex copper ion* must be formed, which is considerably *more complex* (§ 13) than *copper ammonia ion*. The transformation of the latter into the complex cuprocyanide ion[1] is expressed by the following equations:

$$2Cu(NH_3)_4^{\cdot\cdot} + 6CN' = 2Cu(CN)_2' + C_2N_2 + 8NH_3 \quad (1).$$

Cupric cyanide or *cupricyanide ion*, which may be formed as an intermediate product under certain conditions, break down into *cyanogen* and *cuprous cyanide* or *cuprocyanide ion* respectively, according to the equations:

$$2Cu(CN)_2 = 2CuCN + C_2N_2 \quad (2),$$

$$2Cu(CN)_3' = 2Cu(CN)_2' + C_2N_2 \quad (3).$$

The identification of $Cd^{\cdot\cdot}$ in the presence of $Cu^{\cdot\cdot}$[2] can be carried out as follows: 2 cc. of a neutral solution which may contain both $Cd^{\cdot\cdot}$ and $Cu^{\cdot\cdot}$ are acidified with dilute hydrochloric acid. Then 20 cc. of a *saturated* solution of *sodium chloride* are added. The solution is next warmed and hydrogen sulphide is passed into it, precipitating cupric sulphide, black. This is filtered through a dry filter into a dry test-tube. On adding a few drops of water or dilute ammonia to this filtrate a yellow precipitate of cadmium sulphide forms if the original solution contained $Cd^{\cdot\cdot}$.

In case the filtrate is not clear, owing to the formation of a colloidal solution of copper sulphide, a small quantity of a copper salt solution is added. Colloidal solution of copper sulphide will be formed when very little $Cu^{\cdot\cdot}$ is present (see § 77).

To test for cadmium in a precipitate of sulphides the precipitate is heated with dilute hydrochloric acid and an excess of saturated sodium chloride solution. The resulting solution is filtered from any residue and the filtrate is diluted with water or dilute ammonia. On passing hydrogen sulphide into this solution a yellow precipitate indicates that cadmium sulphide was contained in the original precipitate.

[1] It is noteworthy that the *complex cuprocyanide ion* is an anion, although it contains a metallic component which can form elementary cathions (§ 9). Similar cases are quite frequently observed.

[2] Cushman, Am. Chem. J. 17, *379*, 1895.

The presence of lead and bismuth ions in the solution, or of lead and bismuth sulphides in the precipitate, may give rise to errors. Hence this test applies only when compounds of these metals are absent or have been already removed.

Sodium chloride causes a complex compound to form, probably of the composition Na_2CdCl_4. This is, however, easily decomposed by water.

24. Testing for sulphides which dissolve in ammonium sulphide. The analysis becomes considerably more complicated if the second group precipitate contains members that are *soluble* in *ammonium sulphide* (§ 20). To show if any of this group of sulphides are present a portion of the hydrogen sulphide precipitate is digested in a test-tube with ammonium sulphide solution. The separation and identification follows according to § 26 in case only *soluble sulphides* are contained in the precipitate. A residue left after treatment with ammonium sulphide is *filtered* from the *supernatant liquid*. The filtrate is acidified with dilute hydrochloric acid which causes an evolution of hydrogen sulphide (equation *2*) together with the formation of a *milky precipitate* of *sulphur*. The simultaneous formation of a *flocculent colored precipitate* is the critical indication of the presence of soluble sulphides.

Along with other ions the solution of ammonium sulphide[1] contains chiefly the ion S″, which reacts, for example, with arsenic trisulphide as follows:

$$\underline{As_2S_3} + 3S'' = 2AsS_3''' \qquad (1).$$

The *reprecipitation* of the *sulphide* from this solution by hydrochloric or nitric acid is expressed by the equation:

$$2AsS_3''' + 6H^{\cdot} = \underline{As_2S_3} + 3H_2S \qquad (2).$$

Acid should never be poured directly from a *reagent bottle* into ammonium sulphide solution, but from a *test-tube*. Any acid remaining in the test-tube must then be thrown away. Under no circumstances should it be poured back into the reagent bottle, for it will be contaminated with hydrogen sulphide which would spoil the stock solution.

Yellow ammonium sulphide, which is most generally used, contains dissolved sulphur which gives rise to divalent ions of the gen-

[1] Compare Ostwald, Foundations of Analytical Chemistry, translation by McGowan, p. *176*.

eral formula, S_n'',[1] where "n" is a whole number varying from 2 to 7. The ion, S_n'', reacts as follows with *hydrogen ion,* producing a milky separation of very finely divided sulphur which consequently appears white (Principles, p. *13*):

$$S_n'' + 2H^{\cdot} = H_2S + \underline{(n-1)S} \qquad (3).$$

25. Separation of the soluble sulphides from those insoluble in ammonium sulphide. Provided a *preliminary test* (§ 24) has shown that the second group precipitate is a mixture of the soluble and insoluble sulphides, the entire precipitate is digested with ammonium sulphide, as above described. The undissolved residue is filtered, washed and further tested for members of the copper sub-group according to §§ 21–23.

26. Separation of the sulphides soluble in ammonium sulphide. Tin Sub-group. The ammonium sulphide solution is precipitated with dilute hydrochloric acid and the precipitate well *washed* with *hot water.* It is then boiled for 2 or 3 minutes with concentrated hydrochloric acid of specific gravity 1.2 in a porcelain evaporating dish or casserole. *Sulphides of antimony and tin are completely soluble in concentrated hydrochloric acid,* while *arsenic sulphide*[2] remains for the most part undissolved. The filtrate, or, in case no residue of arsenic sulphide was left, the clear hydrochloric acid solution is tested for *antimony ion* by placing a few drops on a clean platinum foil and dipping one end of a U-shaped strip of *tin foil* in the liquid, leaving the other end in contact with the platinum beyond the drop. *Tin* displaces *antimony* from the solution (§ 34), and a dark brown stain appears on the platinum.

[The test for antimony may also be made on a bright silver coin. All hydrogen sulphide must be previously expelled from the solution

[1] Solutions of ammonium sulphide which contain ions S_n'', have a *characteristic* yellow color. The transformation of S''-ions into S_n''-ions is effected by a reaction with atmospheric oxygen, which may be summarized as follows:

$$nS'' + (n-1)O = S_n'' + (n-1)O''.$$

Consult Principles, pp. *505–6*.

[2] Contrary to the statements in most text-books, arsenic sulphide is not inconsiderably soluble in hydrochloric acid, as well as in water (compare § 42). That the above method nevertheless gives a separation sufficient for practical purposes is probably due, in part at least, to the great slowness with which the reaction between *arsenic sulphide* and *hydrochloric acid* proceeds.

by boiling, otherwise the solution alone would form a stain on silver. In the presence of antimony ion the tin causes a brown to black stain to form after a few minutes.] G. Denigès, Chem. News, 88, 262, 1903.)

To identify tin ion a few pieces of granulated zinc are placed in the hydrochloric acid solution.[1] *Antimony* and *tin* both *precipitate,* while *zinc* goes into *solution* and *hydrogen escapes.*

$$\underline{Zn} + 2H^{\cdot} = Zn^{\cdot\cdot} + H_2 \quad (1),$$

$$\underline{Zn} + Sn^{\cdot\cdot} = Zn^{\cdot\cdot} + \underline{Sn} \quad (2).$$

The first of these reactions takes place when the solution contains a large excess of acid. Hence it is advisable to neutralize the *greater part* of the acid with *alkali hydroxide* before making the test, in order to obtain the second reaction almost exclusively. The zinc is left in the solution until its entire surface is coated with a spongy deposit which prevents any further dissolving of the zinc, or until an increase in the amount of metallic deposit is no longer observed.

The zinc together with its spongy coating is washed with water and then heated with dilute hydrochloric acid. The presence of stannous ion, $Sn^{\cdot\cdot}$, is indicated by the formation of a white precipitate of mercurous chloride upon addition of *mercuric chloride* solution (compare § 21, (5)).

[A simpler means of detecting tin consists in warming the hydrochloric acid solution with clean iron wire in a test-tube until it becomes colorless. Antimony will deposit on the iron in black spongy form, while tin remains in solution as stannous ion, which gives the above reaction with mercuric chloride.]

The *residue*[2] *insoluble* in concentrated *hydrochloric acid* is filtered and washed. It is then heated with concentrated nitric acid, whereby arsenic sulphide oxidizes to arsenic acid and goes into solution. This solution is made *strongly alkaline* with *ammonia.* Presence of arsenic is then shown by the formation of a white precipitate of *magnesium ammonium arsenate,* ($MgNH_4AsO_4$), on addition of *magnesium sulphate* or *chloride.* Magnesium ammonium arsenate

[1] The few drops of solution used in the test for antimony with tin foil are of course thrown away.

[2] The precipitate obtained by adding dilute acids to an ammonium sulphide solution is made up partly of sulphur (§ 24, (3)). Hence the partial solubility of this precipitate in concentrated hydrochloric acid is not a satisfactory criterion of the presence of arsenic. Direct proof is therefore always necessary.

will not dissolve when a few cc. of an ammonium salt solution are added (compare § 102).

[Another method of detecting arsenic consists in dissolving the residue of sulphide left after treatment of the precipitate with hydrochloric acid in *warm ammonia*. The alkaline solution is then oxidized by means of ***hydrogen peroxide*** to ammonium arsenate. Precipitation of $MgNH_4AsO_4$ follows in the above manner. By this method precipitation of sulphur is avoided.]

27. Reversible reactions. The equations for the precipitation of the ions of the second group (§ 19) show that hydrogen ion is a product of the reaction. This agrees entirely with the actual conditions, as already stated in § 19, (*1*). On the other hand the precipitated sulphides are soluble in acids (§ 21, (*1*)), particularly in nitric acid. Since HNO_3 and $M(NO_3)_2$ each yield the same ion, NO_3', which may be omitted from the equation (§ 14), it is evident that the latter reaction is the inverse of the former:

$$M^{\cdot\cdot} + H_2S \rightleftarrows \underline{MS} + 2H^{\cdot}.$$

By ***varying*** the ***conditions*** it is possible to make the reaction proceed practically to an end from left to right, →, or conversely ←; that is, until one or other of the substances taking part are used up. For the first case, from left to right, it is necessary that hydrogen sulphide be passed into the solution uninterruptedly; for the other, that the resulting hydrogen sulphide constantly escapes in gaseous form.

Reactions of this type are termed "*reversible.*" Inverted arrows ⇄ are used instead of the ordinary symbol of equality to connect the two sides of the equation, in order to indicate that the reaction may proceed in either direction.

Those reactions which proceed only in ***one direction*** until one or other of the original substances entering into the reaction practically disappear completely—i. e. to the limits of their analytical detection—thereby giving rise to new substances, are usually, but not always, only ***special limiting cases.*** The conditions which determine the course of a reversible reaction will be taken up in detail at this point, since it will help us to understand the above reaction of the precipitation of sulphides and also other cases.

28. Continuation. To settle the question as to the conditions which determine the direction of a reversible reaction, this particular case of the second group sulphides is specially adapted. Experience has taught us that by continuously passing hydrogen sulphide into the system, i. e. by adding a substance which is ***used up*** in the reaction →, ***the transformation involving a consumption of this substance is increased.*** Conversely the *dissolving* of a sulphide by heating it with an acid, takes place only when the hydrogen sulphide that is *formed* escapes continuously (through heating). It is evident, therefore, that the amount of dissolved hydrogen sulphide has an essential influence upon the direction in which the reaction will proceed. This depends, however, not upon the ***absolute*** but upon the ***relative quantities*** of hydrogen sulphide in the solution; that is, upon the amount present in unit volume of the liquid. This quantity is called the ***concentration.***

The following experiments will show clearly the influence of the concentrations of substances that take part in a given reaction. From a graduated cylinder equal volumes, say 10 cc., of a zinc nitrate or -sulphate solution containing about 1.5 g. salt in 100 cc. solution, are poured into several test-tubes. To the one 2 cc. dilute acetic acid are added, to another 2 cc. dilute nitric acid, to a third 2 cc. water in order to have the concentration of the zinc salt, and thus the concentration of zinc ion, the same in all three cases.[1] On saturating these three solutions with hydrogen sulphide it is observed that a considerable amount of *zinc sulphide* precipitates in the non-acidulated solution and in the solution containing acetic acid, while in the solution acidulated with nitric acid a very much smaller amount of precipitate forms, and only after standing for some time. That the precipitation is much *less complete* in this last case than in the others is still more clearly shown by filtering, adding *sodium acetate* to the filtrate and again passing hydrogen sulphide into the solution. The precipitate is now formed as abundantly as in the other two cases. Practically all of the *sodium acetate* reacts with *nitric acid,* forming *sodium nitrate* and *acetic acid* (§ 38). Hence the conditions become the same as when the zinc solution was originally acidulated with acetic acid.

The difference between the action of acetic and nitric acids (§ 49) is due to the fact that the specific acid properties of an acetic acid solution are the same as those of a nitric acid solution which contains (in formula weights) about eighty times less acid than the first solution. Hence in place of 2 cc. acetic acid one eightieth of 2 cc. nitric acid solution and seventy-nine eightieths of 2 cc. water might have been used without appreciably altering the result.

If on the other hand hydrogen sulphide is passed into 10 cc. of a *saturated zinc nitrate solution,* acidulated with 2 cc. nitric acid, a much greater amount of zinc sulphide separates, and in much shorter time than in the former case, where a more dilute solution of zinc nitrate is used. With respect to this reaction it is justifiable to conclude that an *increase* in the *concentration* of one of the substances used up in the reaction is followed by an *increase* of the *transformation,* and conversely, although the conclusion is based solely upon qualitative, i. e. approximate, experiments. Similarly an *increase* in the *concentration* of substances that are *formed* by the reaction, e. g. $H^{\cdot}$, is accompanied by a *decrease* of the transformation.

These considerations hold generally for the precipitation of all second group cathions. Zinc ion has been selected to illustrate these relations only because the influence of different acids is more clearly shown in this case than in others.

29. **The Law of Mass Action of Guldberg and Waage** expresses the results of the *quantitative* study of numerous reversible reactions. In a chemical reaction of the general form

$$A + B \rightleftarrows C + D \qquad (1),$$

where A, B, C and D represent the formulae of substances taking part in the reaction, the concentrations C_A, C_B, C_C and C_D, i. e. the quantities contained in unit volume, of the four components are related by the following mathematical expression when the system has reached a state of equilibrium:

[1] This is, however, only an approximation. The solutions of acetic and nitric acids should contain the same number of formula weights of acid per unit volume.

[2] Reference to the literature will be found in § 183.

$$\frac{C_A \cdot C_B}{C_C \cdot C_D} = k \qquad (2)$$

When changes of temperature and pressure are excluded this expression holds for all cases. K is a constant, the value of which is characteristic for every reaction. In words equation (*2*) reads as follows: When equilibrium is reached in a system of four substances, the product of the concentrations of the initial substances, divided by the product of the concentrations of the components that are formed by the reaction, must have a constant value. If this is not the case the system is not in a state of equilibrium, and a reaction will proceed between A, B, C and D according to equation (*1*) until the law is satisfied.

Thus each component is a factor in determining the equilibrium. In other words the concentration of each substance taking part in a reaction is an equilibrium condition. For the sake of a more complete explanation we will suppose that equilibrium has been reached in a system of the above type. Suppose further that a little of A is brought into solution without changing the total volume. Its concentration will now be C_A'. The concentrations of B, C and D remain as before, C_B, C_C and C_D respectively. Since $\frac{C_{A'} \cdot C_B}{C_C \cdot C_D} > k$ (for $C_A' > C_A$), the equilibrium has been disturbed. Consequently a change must take place in the system whereby C_A' and C_B become smaller, C_C and C_D larger. Hence a reaction will proceed in the direction $\rightarrow$ according to equation (*1*), assuming that other possible ways of reaching the equilibrium, as, for example, through the dissolving of C or D, which would increase C_C or C_D, are excluded. It is seen that this result obtained by applying the quantitative relations agrees with the rule that has been deduced from qualitative experiments.

The following is a more general statement of the above relation. *When an equilibrium is disturbed by increasing or decreasing the concentration of one of the components of the system, the disturbance is manifested by a reaction which causes a decrease or increase respectively in the concentration of the substance in question* (compare Principles, p. *130*).

The mass action equation assumes another form for a reaction in which different numbers of formula weights of the component substances take part. The equilibrium condition of the general equation $mA + nB \rightleftarrows pC + qD$ has the form:

$$C_A^m \cdot C_B^n = kC_C^p \cdot C_D^q \qquad (3).$$

The factors m, n, p, q express the number of combining weights of the individual substances taking part in the reaction. They appear as exponents in the expression of the mass action law. Equation (*2*) is a special case of this general type, which derives its simpler form from the fact that $m = n = p = q = 1$. K is the quotient of products which are made up of as many factors as there are formula weights of the substances used up or formed in the reaction.

30. Influence of solids. Special consideration is necessary for the case in which one of the substances is present in solid form. It has been shown in the last paragraph that the equilibrium condition varies according as the concentration of any one of the substances taking part in the equilibrium is altered. The concentration of a substance that is present in solid form may

be varied, however, only within certain definite limits, i. e. between the value zero and the saturation point of the solid. For a system in equilibrium it is an impossible condition that the concentration of any one of the components be zero. This may, however, be true of the initial state of the system for any of the substances in question which are formed by the reaction between the initial components.

If a substance is present in solid form, other concentrations than that of its saturated solution (see § 6) cannot be realized. Hence in this case the concentration of the substance present in solid form is *not variable.* It is therefore possible to simplify formula (*2*) by including this fixed concentration in the constant 'k.' Thus through the presence of a solid substance the number of *variables* upon which the equilibrium depends is lessened by one. Applications of this important case will be made frequently.

It must be borne in mind that the quantities of the substances present in solid form are without influence upon the equilibrium (compare Principles, pp. *100–101*).

31. **Theory of the precipitation with hydrogen sulphide.** We can now make use of the views just set forth in the last paragraphs to assign reasons for the rules mentioned in § 19 to precipitate the ions of the second group as completely as possible with hydrogen sulphide.

The first group filtrate must be only moderately acidulated before passing hydrogen sulphide into it, owing to the influence exerted by hydrogen ion upon the transformation (§ 28). Solutions which had originally an acid reaction, should be neutralized with ammonia after the first group ions have been tested for with hydrochloric acid. Since *other ions* which would *normally* be *precipitated* in the *fourth group,* form sulphides that are insoluble in neutral or alkaline solutions, it is expedient to add a slight excess of hydrochloric acid, just enough to keep them in solution and at the same time give a complete precipitation of the second group sulphides, before introducing hydrogen sulphide. No attention need be paid to any permanent precipitate that may form through addition of ammonia. It will either be transformed into a second group sulphide or will go into solution eventually in the excess of acid. By this means the student personally controls the quantity of acid present, which should approximate the concentration of a fifth normal solution, and thus avoids many difficulties, particularly with reference to the precipitation of cadmium sulphide. For still other reasons (§ 17) a *small excess* of *acid* is of benefit, because many sulphides precipitate from neutral solutions in a condition which makes their filtration difficult. The necessity for having a large excess of acid to ensure the complete precipitation of arsenic sulphides will be referred to later.

If the solution was saturated with the precipitant, an equilibrium exists in the hydrogen sulphide filtrate between the substances $M^{\cdot\cdot}$, H_2S,[1] MS and $H^{\cdot}$, on the assumption, however, that no loss of hydrogen sulphide occurs through contact of the solution with the atmosphere during filtration. This assumption is never *strictly* justified. By decreasing the concentration of the prod-

[1] For these considerations it is assumed that hydrogen sulphide is not dissociated, since in this way the formation of hydrogen ion will be more readily understood. Strictly speaking the concentration of the S''-ion, resulting from the dissociation of hydrogen sulphide, ($H_2S \rightleftarrows 2H^{\cdot} + S''$), is of essential influence. This point will be dealt with later (§ 74).

ucts MS or $H^{\cdot}$, an increased transformation in the sense of more complete precipitation results. In practice, however, only the latter means is employed, viz. the *partial* neutralization of the hydrogen ion, with the above-mentioned limitations (§ 28).

It is more convenient to dilute the filtrate with water and again saturate the diluted solution with hydrogen sulphide. The following relation exists between the substances in solution, according to § 29 (*3*):

$$\frac{C_{M^{\cdot\cdot}} \ C_{H_2S}}{C_{MS} \cdot C^2_{H^{\cdot}}} = k.$$

By diluting n times, where n > 1, $C_{M^{\cdot\cdot}}$, C_{MS} and $C_{H^{\cdot}}$ sink to 1/nth of their original values. C_{H_2S}, however, remains unaltered, because the filtrate is again saturated with hydrogen sulphide after dilution. Since the quotient $\dfrac{\dfrac{C_{M^{\cdot\cdot}}}{n} \cdot C_{H_2S}}{\dfrac{C_{MS}}{n} \cdot \dfrac{C^2_{H^{\cdot}}}{n}} >$ k, it follows that the system is no longer in equilibrium. The change that necessarily follows consists in an increase of $\frac{C_{MS}}{n}$ and $\frac{C_{H^{\cdot}}}{n}$, and in a decrease of $\frac{C_{M^{\cdot\cdot}}}{n}$. As hydrogen sulphide is constantly introduced during the precipitation, C_{H_2S} is not altered. Hence more complete precipitation results.

It is to be expressly noted that the *increased precipitation* is not conditioned by the longer period during which hydrogen sulphide is passed into the solution, but that this is a consequence of the dilution and of the renewed saturation of the solution with hydrogen sulphide.

32. Use of alkali sulphide in place of ammonium sulphide to separate the second group. In certain cases it is advisable to use *potassium-* or *sodium sulphide in place of ammonium sulphide* for separating the copper and tin sub-groups (§ 26). These solutions are made by passing hydrogen sulphide into sodium- or potassium hydroxide. They serve the same purpose as ammonium sulphide and *exclude errors* which might arise from the *slight solubility of copper sulphide in ammonium sulphide.*

In case the second group precipitate contains mercuric sulphide, a solution of ammonium sulphide is always used.

The following equation shows how S″-ion is formed from alkali hydroxide and hydrogen sulphide:

$$2OH' + H_2S = 2H_2O + S''.$$

It appears paradoxical to use potassium- or sodium sulphide for the separation when the precipitate contains copper sulphide, and on the other hand to avoid the use of alkali sulphide in case mercuric sulphide is present, for this presupposes a partial knowledge of the components making up the precipitate

before the method of separation is selected. The change of color undergone by the *solution* on passing hydrogen sulphide into it, indicates with considerable certainty, however, whether the precipitate will contain copper sulphide or not. Copper salt solutions have a distinct blue color. When the *blue salts* are removed from the solution by transformation into difficultly soluble sulphide, the filtrate from the *second group* will have a color essentially different from that of the first group filtrate.

Furthermore certain phenomena are observed on heating *solid* mercury compounds (§ 170, **B**), which give indications of the probable presence of mercuric sulphide in the second group precipitate. The above remarks relate chiefly to the question of the *complete* analysis of a dry mixture, which will be taken up in detail in part **III**. Moreover the disadvantages arising from neglect of these precautions are not very great. Possible errors may be even entirely avoided when it is known how they manifest themselves. Suppose, for example, that the second group precipitate contains copper sulphide, and it is digested in the usual way with ammonium sulphide. Some copper sulphide will go into solution. On acidifying the solution it will be reprecipitated along with the other sulphides (§ 24) and goes into solution with antimony and tin in concentrated hydrochloric acid. It may happen that a red separation of metallic copper is observed when this solution is tested for antimony (§ 26). The copper deposit is sometimes wrongly taken for antimony, although the latter has a quite different appearance when deposited.

If it is desirable copper may be removed from the hydrochloric acid solution that is to be tested for antimony and tin, by warming it and adding potassium hydroxide drop by drop (see below). Copper oxide forms and is removed by filtration. The filtrate is then weakly acidified with hydrochloric acid and is again saturated with hydrogen sulphide. The precipitated sulphides, now practically free from copper, are tested for antimony and tin according to § 24.

To completely precipitate copper ion (§ 40) enough potassium hydroxide must be added to give the solution a decided alkaline reaction. Otherwise stannous ion might be inadvertently precipitated. A *very slight excess* of *potassium hydroxide,* however, suffices to bring any precipitated *stannous hydroxide* back into solution (see § 38 (*2*)).

Reactions of the Second Group Ions.

33. Mercuric ion. The test solution is prepared by dissolving $\frac{1}{20}$ formula weight of *mercuric nitrate,* $Hg(NO_3)_2$, in a liter of water, slightly acidulated with nitric acid. The nitrate may be made by dissolving mercuric oxide in nitric acid and evaporating the solution on a water-bath (compare § 36). In case *mercuric chloride* solution, 'corrosive sublimate,' is used instead of the nitrate, certain deviations, which will be referred to later, are observed.

Alkali hydroxide precipitates yellow mercuric oxide:

$$Hg^{\cdot\cdot} + 2OH' = \underline{HgO} + H_2O.$$

When an excess of OH' is present along with $Hg^{\cdot\cdot}$, yellow basic salts (see § 22) are formed.

Ammonia yields with mercuric ion white insoluble compounds, known in pharmacy as "*white precipitate.*" The composition varies according as mercuric chloride or nitrate is used. It corresponds with the formulae $HgNH_2Cl$ and $Hg_2O(NH_2)NO_3$ respectively.

Alkali carbonate forms red brown *basic carbonate,* which goes over into *oxide* on boiling the solution in contact with the precipitate.

Potassium iodide precipitates red *mercuric iodide,* HgI_2, which dissolves in an excess of potassium iodide to a colorless solution that contains the complex ion HgI_4'':

$$\underline{HgI_2} + 2I' = HgI_4'' \qquad (1).$$

This solution gives none of the above-mentioned reactions of $Hg^{\cdot\cdot}$, in accordance with the complex nature of the ion HgI_4''.

Black mercuric sulphide (§ 21) is *precipitated* by passing hydrogen sulphide into the solution of any mercury salt. From mercuric nitrate a white compound usually separates. It changes into the sulphide on longer treatment with hydrogen sulphide. This compound is comparable with a basic salt, having sulphur in place of oxygen. It has the formula $Hg_3S_2(NO_3)_2$ and is also formed when mercuric sulphide is heated with concentrated nitric acid. Mercuric sulphide dissolves in concentrated nitric acid, however, under some conditions (§ 21). Consequently mercury is sometimes not detected in the second group, either because the sulphide goes into solution on digesting the precipitate with nitric acid, or because the above-mentioned white compound forms and is taken for sulphur and therefore is not tested for mercuric sulphide.

Mercuric sulphide is soluble in potassium sulphide, especially in the presence of alkali hydroxide, owing to the formation of the complex ion HgS_2'':

$$\underline{HgS} + S'' = HgS_2'' \qquad (2).$$

Metallic mercury will deposit on a strip of copper, iron or zinc from any solution containing mercuric ion, and ions of copper, iron or zinc respectively will be found in the solution whenever an appreciable amount of mercury has been deposited. The following equation represents this reaction:

$$\underline{Cu} + Hg^{\cdot\cdot} = \underline{Hg} + Cu^{\cdot\cdot} \qquad (3).$$

The reciprocal displacement of metals from their salt solutions is by no means a property due to chance. It is possible to arrange the metallic elements in one series, each member of which can displace ions of all succeeding members from solutions of their salts, transforming them to the metallic state. In cases of *abnormal dissociation,* where the concentration of the ion being displaced from solution is very small, *exceptions* to this general *behavior* are noted. From the properties of galvanic combinations of metals and their salt solutions, this series is known as the *electric potential series.* The following table contains the most important metals arranged according to this point of view: K, Mg, Al, Zn, Fe, Co, Pb, H, Sb, Cu, Hg, Ag, Pt, Au.

The metals standing at the left side of the series are termed *base metals,* those at the right hand, *noble metals.* The following case illustrates the above-mentioned variation in the mutual positions of members of the series. Zinc normally precipitates copper ion. By adding potassium cyanide to a copper salt solution, however, the concentration of copper ion is so much decreased that a film of copper will no longer separate on zinc. On the contrary *copper* goes into *solution* and *zinc* deposits on immersing a *copper* strip in a *zinc salt solution* that contains *potassium cyanide.* It is therefore evident from this experiment that the position of the elements in the electric potential series is not determined solely by the *nature* of the *metal,* but is also dependent upon the salts dissolved.

34. Difference in the behavior of solutions of mercuric nitrate and -chloride. With the above tests it makes no difference whether mercuric nitrate or mercuric chloride solution is used. *Potassium sulphocyanate,* however, precipitates *mercuric sulphocyanate* from a mercuric nitrate solution, but does not yield a precipitate with mercuric chloride. The following is the reason for this difference in the behavior of mercuric chloride and nitrate. The latter is much more dissociated than the former, as may be shown independently by the difference between the electrical conductivities of the two solutions. Of two solutions, one of the chloride, the other of the nitrate, containing each the same number of formula weights in the same volume, for instance 1/10 formula weight per liter (or 27.12 g. $HgCl_2$ and 32.44 g. $Hg(NO_3)_2$ respectively), the solution of the *nitrate* will be much *more concentrated* with respect to *mercuric ions* than the solution of the chloride. As the formation of mercuric sulphocyanate according to the equation:

$$Hg^{\cdot\cdot} + 2CNS' = \underline{Hg(CNS)_2} \qquad (1),$$

depends upon the concentration of mercuric *ion* and is not directly dependent upon the concentration of the mercuric salt, it is at once evident that the nitrate yields a precipitate and the chloride gives none, if we remember what has been stated in § 28. To conclude that the chloride does not react with sulphocyanate ions at all would, however, be unwarranted. In this case so small an amount of mercuric sulphocyanate forms that no precipitation takes place.

If potassium- or ammonium sulphocyanate is added to a precipitate of mercuric sulphocyanate, the precipitate dissolves readily, forming a complex anion containing mercury according to the equation:

$$\underline{Hg(CNS)_2} + 2CNS' = Hg(CNS)_4'' \qquad (2).$$

The corresponding potassium- or ammonium salts are easily soluble. Other

complex anions containing mercury and other elements, as chlorine, in addition to the sulphocyanate radical, also exist. Such an anion forms to a certain extent when a solution of potassium or ammonium sulphocyanate is added to a solution of mercuric chloride:

$$HgCl_2 + 2CNS' = [Hg(Cl)_2(CNS)_2]'' \qquad (3).$$

Taking this secondary reaction between mercuric chloride and sulphocyanate ion into account, the difference in the behavior of mercuric chloride and mercuric nitrate towards sulphocyanate solutions becomes even more obvious. For, by this reaction (*3*) the concentration of mercuric chloride, and correspondingly the concentration of mercuric ion, as well as that of sulphocyanate ion, are diminished. The two latter ions are used up by reaction (*1*), as was stated above.

The first-mentioned influence seems, however, to predominate.

Mercuric bromide, -iodide and -cyanide, as well as the chloride, exhibit anomalous dissociation. The cyanide is practically non-dissociated, and the iodide is considerably less dissociated than the chloride.

With regard to the secondary effect of the formation of a complex anion, it should be remembered that a similar reaction takes place when mercuric iodide is dissolved in a solution of potassium iodide (see § 33).

A similar reaction between mercuric chloride and soluble chlorides, e. g. $HgCl_2 + 2Cl' = HgCl_4''$, affords an explanation of the fact that mercuric chloride solutions containing potassium- or sodium chloride are less poisonous than solutions of mercuric chloride alone which contain the same relative amount of mercury. It has been shown by Kroenig and Paul that the poisonous properties depend also upon the concentration of $Hg^{\cdot\cdot}$ When a solution of mercuric chloride has reacted with Cl' to a certain definite extent, it behaves like a correspondingly more dilute solution of corrosive sublimate, i. e. it becomes less poisonous.

35. Experiments with the bulb-tube. All mercury compounds volatilize when heated in a bulb-tube.[1] The chloride, $HgCl_2$, yields a white deposit, which may be driven along the tube with the flame, always depositing again on the colder portions of the walls. This *transport* of *solids* through *transformation* into *vapor* is termed *sublimation* and the deposit is called a *sublimate*. Other mercuric salts, and also a mixture of mercuric chloride with sodium carbonate, form a deposit of oxide:

$$Na_2CO_3 + HgCl_2 = 2NaCl + \underline{HgO} + CO_2.$$

At higher temperatures a further decomposition of mercuric oxide into mercury and oxygen takes place, giving rise to a gray deposit which is seen to consist of globules of metallic mercury under the microscope.

[1] The bulb-tube is constructed from ordinary clean glass tubing by sealing off one end. The closed end may be blown out to a small bulb. The tube should be about 10 cm. long and 8–10 mm. in diameter. For bulb-tube experiments small quantities of solids are used, approximating the size of a grain of wheat.

36. Preparation of oxygen, mercurous nitrate and mercuric nitrate. In the above-mentioned decomposition of mercuric oxide oxygen is evolved and mercury remains behind. The oxygen may be collected if the mercuric oxide is heated in a tube of the form shown in Fig. 3. The tube has a knee-shaped bend in which the

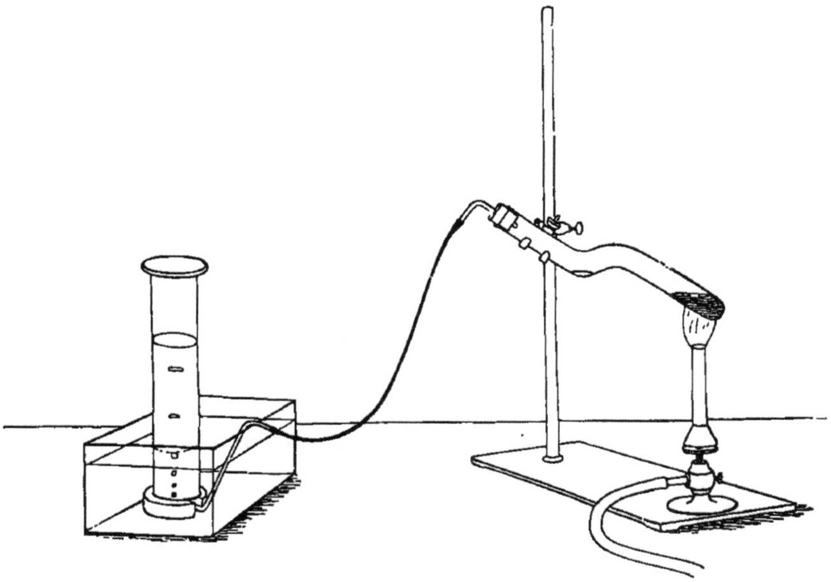

FIG. 3.

condensed mercury collects. A delivery tube is fitted to the apparatus and consists of a short piece of glass tube passing through the stopper and connected by rubber tubing to a second glass tube. As is evident, the evolved gas can be passed by this means into an inverted cylinder filled with water. It may be shown that the gas collected in the cylinder is oxygen by testing it with a glowing splinter. It supports the combustion of the splinter much more vigorously than does air. From 2–3 g. mercuric oxide are taken for the experiment. The volume of oxygen may be approximately estimated in advance from the following data: *216.3* g. mercuric oxide yield *16* g. oxygen, and at room temperature *32* g. oxygen occupy approximately *24* liters (compare Principles, p. *88*).

The mercury condensed at the bend of the generator tube may be transformed into *mercurous* or *mercuric* nitrate by heating it with *dilute* or *concentrated* nitric acid. A high temperature and concentrated acid favor the formation of the mercuric salt. An excess of

mercury is, on the other hand, essential to the formation of mercurous salt. One portion of the mercury is converted into mercurous nitrate and another portion into mercuric nitrate. The solutions obtained are evaporated to dryness and the residue left in each case is tested to see if it contains any of the other salt. Thus the mercuric nitrate is tested for mercurous nitrate by means of hydrochloric acid (§§ 5 and 21). To detect small amounts of mercuric nitrate in the residue of mercurous nitrate the salt is brought into solution and the mercurous ion is precipitated with a soluble chloride. The filtrate is then tested with hydrogen sulphide.

37. Bismuth ion. The test solution is made from *basic bismuth nitrate,* $Bi(OH)_2NO_3$,[1] by dissolving $\frac{1}{80}$ formula weight in a liter of water, acidulated with nitric acid. It is best to first *warm* the *salt* with *just enough acid to dissolve it* and then *dilute with water* to a liter. In case a cloudy precipitate forms on diluting, a little more acid is added until the solution becomes clear. The precipitation of bismuth salts from their solutions by diluting them with water is used as a test for bismuth ion. It is due to a reaction between bismuth ion and water according to the equation:

$$Bi^{\cdot\cdot\cdot} + H_2O \rightleftarrows BiO^{\cdot} + 2H^{\cdot}.$$

The ion, $BiO^{\cdot}$, is called *bismuthyl ion.*

The formation of an insoluble compound by diluting a bismuth salt solution is more pronounced with the chloride than with the nitrate. This is shown by adding *very dilute* hydrochloric acid drop by drop to a nitrate solution which contains just enough nitric acid to prevent precipitation of the basic nitrate. A white precipitate of bismuth oxychloride, BiOCl, forms. The addition of very dilute nitric acid, however, causes no precipitation. Consequently the basic chloride is less soluble than the basic nitrate.

The following consideration will make this more obvious. The precipitation begins whenever the solution becomes supersaturated with respect to bismuth oxychloride or bismuth oxynitrate, $BiONO_3$, which are formed from $BiO^{\cdot}$ and Cl' or NO_3' respectively. Since the above solution contains fewer Cl'-ions than NO_3'-ions, it follows that a smaller amount of bismuth oxychloride forms than of bismuth oxynitrate. Nevertheless bismuth oxychloride precipitates. Therefore it is more difficultly soluble than bismuth oxynitrate.

According to what has been stated in § 28 it is clear that the addition of

[1] Basic bismuth nitrate is really a mixture, the composition of which corresponds with the above formula.

chloride ion to a solution of bismuth nitrate favors the precipitation of the oxychloride, while the addition of hydrogen ion acts in the contrary sense, if we take into account that the reaction proceeds in the following steps:

$$Bi^{\cdot\cdot\cdot} + H_2O \rightleftarrows BiO^{\cdot} + 2H^{\cdot},$$

$$BiO^{\cdot} + Cl' \rightleftarrows \underline{BiOCl}.$$

Accordingly no precipitate forms unless the hydrochloric acid is sufficiently dilute, because otherwise the influence of hydrogen ion preponderates. If, however, potassium- or sodium chloride is added in place of hydrochloric acid, *only* the action of chloride ion, which favors the precipitation, comes into play. Hence this affords an especially delicate test. The method of testing for bismuth mentioned in § 22, is based upon the above considerations.

38. Reactions of bismuth ion. The test solution (§ 37) forms with *alkali hydroxide* or *ammonia* a precipitate of difficultly soluble bismuth hydroxide. It oxidizes to a light brown higher oxide when treated with *bromine* or *hydrogen peroxide.*

Potassium dichromate precipitates yellow *bismuthyl dichromate,* $(BiO)_2Cr_2O_7$, easily soluble in dilute hydrochloric or nitric acid, practically insoluble in acetic acid, *insoluble* in alkali hydroxides (distinction from lead chromate).

In case a bismuth salt solution contains too much free hydrochloric or nitric acid, it is not precipitated with potassium dichromate. The conditions for precipitating $(BiO)_2Cr_2O_7$ are established by adding sodium acetate to the solution. Acetate ion and hydrogen ion form undissociated acetic acid:

$$H^{\cdot} + C_2H_3O_2' \rightleftarrows C_2H_3O_2H \qquad (1).$$

Similar differences in the behavior of acids will be considered in §§ 49, 89 and 151.

A solution of *alkali stannite* precipitates *black metallic bismuth.* This reagent is prepared just before use by adding to 1 cc. *stannous chloride* solution enough potassium or sodium hydroxide to barely redissolve the precipitate of stannous hydroxide which is at first formed:

$$\underline{Sn(OH)_2} + 2OH' = 2H_2O + SnO_2'' \qquad (2).$$

Bismuth hydroxide reacts with the stannite ion (§ 47), yielding stannate ion, SnO_3'', and metallic bismuth.

$$\underline{2Bi(OH)_3} + 3SnO_2'' = \underline{2Bi} + 3SnO_3'' + 3H_2O \qquad (3).$$

In case too great an excess of alkali hydroxide is present, or if the solution is heated, the test may be spoiled.

Potassium iodide forms a *reddish black* precipitate of bismuth iodide, BiI_3, soluble in an excess of potassium iodide. The solution is yellowish red and contains the complex ion, BiI_4' (?). Addition of stannous chloride or other substances that destroy free iodine does not take away the color (distinction from copper, etc., § 40).

Dry mixtures of bismuth compounds with sodium carbonate, when heated on charcoal before the blowpipe (see § 39), yield a characteristic *brittle bead* (very easily broken up). A yellow deposit of bismuth oxide forms on the charcoal.

39. **Blowpipe tests.** Many metallic compounds when mixed dry with sodium carbonate and heated in the blowpipe flame show characteristic reactions. It is frequently possible to determine directly the components of a given mixture by means of blowpipe tests (compare §§ 168 and 171).

Conditions for decomposing many metallic compounds into the corresponding metals or metallic oxides are established with the blowpipe. In order that reduction to the metal takes place, some substance must be present to combine with the oxygen of the metallic oxide.[1] By cutting off the air supply to a *bunsen burner* a *luminous flame* is produced which contains very finely divided carbon particles in consequence of an incomplete combustion of the gas fed into the burner. *This carbon* acts as a reducing agent in blowpipe experiments. A dry mixture of approximately equal quantities of oxide (or salt) and sodium carbonate is placed in a hollow on a piece of charcoal. By holding the blowpipe in the outer edge of the bunsen flame and gently blowing, a luminous pointed flame, the *reducing flame,* is directed against the mixture. The reduced *metallic globules* are directly visible in some cases, while in others there is no visible reduction of the *fused mass.* The metallic grains are seen by breaking up the fusion in a mortar after it solidifies and washing away the lighter particles, mostly carbon.

[1] In general metallic oxides are formed by a reaction between sodium carbonate and a salt of the metal in question. They are reduced to metal by carbon. In addition sodium carbonate acts as a flux. With lead chloride the following reactions take place:

$$PbCl_2 + Na_2CO_3 = 2NaCl + PbO + CO_2$$

and

$$2PbO + C = 2Pb + CO_2, \text{ or } PbO + C = Pb + CO.$$

On the other hand, if it is desired to form metallic oxides, the bunsen flame is made *non-luminous* by *increasing* the *air supply* and a *non-luminous blowpipe flame, oxidizing flame,* is directed against the mixture. Certain metals may be distinguished by the color of the oxides deposited on the colder portions of the charcoal, others from the characteristics of the non-oxidized metallic globules.

A brief description of the blowpipe reactions of the most important metals and their compounds will be found in § 171.

For blowpipe practice the bunsen burner is generally provided with a special tube flattened to a slit at the top and cut off obliquely to permit the operator to direct the flame readily against any object (see **Fig.** 7, § 106, Flame Reactions).

40. **Cupric- and cuprous ion.** Copper forms two series of salts (compare Mercury) with the corresponding ions, *cuprous* and *cupric.* Under ordinary conditions the latter is more stable, i. e. cuprous ion *tends* to pass into cupric ion, separating metallic copper simultaneously. Hence the following reactions relate to cupric ion exclusively.

The test solution may be made of the chloride, nitrate or sulphate, and contains $\frac{1}{20}$ formula weight of the salt per liter. In making up the solutions account must be taken of the water of crystallisation of the solid salts. They have the formulae: $CuCl_2.2H_2O$, $Cu(NO_3)_2.3H_2O$ and $CuSO_4.5H_2O$ respectively.

Alkali hydroxide precipitates blue *cupric hydroxide,* $Cu(OH)_2$, which passes into the black *oxide,* CuO, when heated, or on standing in an open vessel.

For the behavior of copper salt solutions towards *ammonia* consult § 23. The oxide and hydroxide both dissolve in *ammonia* and in *ammonium salt solutions.* These solutions have a characteristic deep blue color. The corresponding solid salts, copper ammonia complexes, are precipitated from the solutions by means of alcohol.

Potassium ferrocyanide precipitates brown cupric ferrocyanide, the composition of which varies somewhat. The reaction is an especially delicate test for copper.

With *potassium sulphocyanate* cupric ion forms a dark precipitate of cupric sulphocyanate, $Cu(CNS)_2$. [Dilute copper salt solutions, however, give no precipitate, but the color changes from blue to green.] By adding sulphurous acid or a sulphite solution a white precipitate of *cuprous sulphocyanate* forms:

$$\underline{2Cu(CNS)_2} + SO_3'' + H_2O = \underline{2Cu(CNS)} + 2CNS' + SO_4'' + 2H^{\cdot} \quad (1).$$

The instability of cupric sulphocyanate is contrary to the ordinary behavior of copper salts. Other still more pronounced cases are found with the halide salts and the cyanide. Apparently a brown precipitate forms on adding *potassium iodide* to a cupric salt solution. By filtering and washing, however, it is found to be white, the filtrate being brown through the presence of free iodine:

$$CuI_2 = CuI + I.$$

If substances which transform free iodine into iodide ion are added, for instance, ferrous sulphate (§ 159) or sodium sulphite (§ 121), the white precipitate of cuprous iodide is seen without filtering (compare § 38)

Cuprous ion is colorless. With alkali hydroxide it forms yellow *cuprous hydroxide* which passes into red *cuprous oxide* on boiling the solution. Both the hydroxide and oxide dissolve in sulphuric acid, yielding metallic copper and cupric sulphate:

$$\underline{Cu_2O} + \underbrace{2H^{\cdot} + SO_4''} = Cu^{\cdot\cdot} + SO_4'' + \underline{Cu} + H_2O \quad (2).$$

From this reaction it follows that cupric sulphate is more stable than cuprous sulphate. Reverse relations hold for the halide salts.

A solution containing cuprous ion may be made by heating an acidulated cupric salt solution with strips of copper foil, for instance, a solution of cupric chloride acidulated with hydrochloric acid, until the color disappears. Air must be excluded.

Copper salts yield small metallic grains of characteristic red color when heated with soda on charcoal in the reducing flame.

41. Cadmium. A solution containing $\frac{1}{20}$ formula weight per liter of the chloride, $CdCl_2.2H_2O$, the nitrate, $Cd(NO_3)_2.4H_2O$, or the sulphate, $CdSO_4.\frac{8}{3}H_2O$, is used for the tests.

Alkali hydroxide and *ammonia* yield a white precipitate of cadmium hydroxide, $Cd(OH)_2$. The precipitate is soluble in an excess of ammonia owing to the formation of a complex *cadmium ammonia ion*, $Cd(NH_3)_4^{\cdot\cdot}$.

Alkali carbonate precipitates white basic carbonate.

With *potassium cyanide* cadmium ion reacts like many other di-

valent cathions. A white precipitate of cadmium cyanide forms at first. By adding more potassium cyanide this precipitate goes into solution, forming the complex anion $Cd(CN)_4''$ (2).

$$Cd^{\cdot\cdot} + 2CN' \rightleftarrows \underline{Cd(CN)_2} \qquad (1),$$

and

$$\underline{Cd(CN)_2} + 2CN' \rightleftarrows Cd(CN)_4'' \qquad (2).$$

From this solution *hydrogen sulphide* precipitates yellow *cadmium sulphide*. Hence the complex formation is not entirely complete under ordinary conditions, and the solution contains some cadmium ion. It is evident, on the other hand, that cadmium sulphide is difficultly soluble. Nevertheless it belongs to the more readily soluble sulphides of the second group, and is sometimes not precipitated, especially when the filtrate from the first group was too strongly acidified.

42. Arsenic. It is uncertain whether arsenic and antimony belong to the metals or non-metals. The uncertainty is partly founded on the behavior of their compounds. The characteristic property of a metal is its ability to form cathions. With arsenic and antimony this property is only weakly developed, although the evidence that they do exist as cathions in solutions is well founded.

Arsenic has a much greater tendency to form anions containing oxygen, as AsO_3''' and AsO_4'''. The separation and identification of these anions will be dealt with in §§ 132 and 133 in connection with the other anions.

The following considerations will serve to make the double role of arsenic clear. In water solution arsenic trioxide is partly hydrated into arsenic hydroxide, which dissociates into ions in the following two ways:

$$As(OH)_3 \rightleftarrows AsO_3''' + 3H^{\cdot} \qquad (1),$$

and

$$As(OH)_3 \rightleftarrows As^{\cdot\cdot\cdot} + 3OH' \qquad (2).$$[1]

Solutions of pure arsenous oxide contain these ions to only *very slight* extent. By adding certain substances, however, conditions may be established under which only one kind of dissociation takes place. The nature of the additions required to produce either dis-

[1] In such cases the hydroxide is said to have *amphoteric* character (Bredig). Compare also $Sb(OH)_3$, $Sn(OH)_2$, $Sn(OH)_4$, $Al(OH)_3$, etc., §§ 46, 47, 48, 63.

sociation may be deduced from the rule given in § 28. A decrease in the concentration of hydrogen ion will favor the dissociation according to (*1*), while a decrease in the concentration of hydroxyl ion favors (*2*). In alkaline solutions the concentration of hydrogen ion is very small (compare § 50), and in acid solutions only a very little hydroxyl ion is present. Hence alkaline solutions of arsenic trioxide contain the ion AsO_3''', while an acid solution, for instance, a hydrochloric acid solution, contains a definite but small amount of the cathion $As^{\cdot\cdot\cdot}$.

This conception, deduced from actual facts, proves very valuable when applied to the various phenomena observed with arsenic compounds. Thus it explains why an acid solution of arsenic trioxide yields a precipitate of arsenic trisulphide with hydrogen sulphide, and why an alkaline solution gives no precipitate. In alkaline solutions hydrogen sulphide forms the same product that is obtained by dissolving arsenic trisulphide in alkali- or ammonium sulphide (§ 24, (*1*)):

$$AsO_3''' + 3H_2S \rightleftarrows AsS_3''' + 3H_2O \qquad (3).$$

The precipitation from acid solution follows the equation:

$$2As^{\cdot\cdot\cdot} + 3H_2S \rightleftarrows \underline{As_2S_3} + 6H^{\cdot} \qquad (4).$$

As has been already mentioned in § 26 the reaction (*4*) may proceed in the reverse direction $\leftarrow$, provided the concentration of hydrogen sulphide in the solution is very slight. While the fact mentioned in § 26, footnote, places a limit upon the insolubility of arsenic trisulphide in hot concentrated hydrochloric acid, the extent to which arsenic trisulphide dissolves is also limited on the other hand by the hydrogen sulphide liberated in the reaction between antimony and tin sulphides with hydrochloric acid. The latter effect becomes less pronounced according as the hydrogen sulphide escapes from the system on warming it. Although the dissolving of arsenic trisulphide can never be entirely avoided, still the amount that goes into solution within a *short* time is only very slight.

Arsenic trisulphide also goes into solution when heated with water or when left standing in contact with water for a long time. This fact indicates that the system $As_2S_3 + 3H_2O$ is not always more stable than the system $As_2O_3 + 3H_2S$ of the same elementary composition. The transformation in the sense

$$As_2S_3 + 3H_2O \rightarrow As_2O_3 + 3H_2S \qquad (5)$$

takes place to a noticeable extent when the concentration of the resulting hydrogen sulphide is decreased (for example, by volatilization). By leaving water standing in contact with arsenic trisulphide in a *closed vessel* for some time and then filtering and passing hydrogen sulphide into the acidulated filtrate, it is readily shown that the *hydrolysis* of arsenic trisulphide is only

very slight under these conditions. On the other hand water that has been left in contact with arsenic trisulphide in an *open vessel*, gives a much greater amount of precipitate after it has been filtered, acidulated and treated with hydrogen sulphide as above.

A solution of arsenic trioxide in concentrated hydrochloric acid or a solution of an arsenite yields a black precipitate of arsenic when heated with stannous chloride (Bettendorf's reaction). The equation is:

$$2As^{\cdots} + 3Sn^{\cdot\cdot} = \underline{2As} + 3Sn^{\cdot\cdot\cdot\cdot} \qquad (6).$$

43. Ions of pentavalent arsenic. The properties and dissociation relations of arsenic pentoxide are very similar to those of the trioxide. The ion $As^{\cdot\cdot\cdot\cdot\cdot}$ is a still weaker cathion than $As^{\cdots}$, while AsO_4''' has more pronounced anion characteristics than AsO_3'''.

The solution of arsenic acid required for the following tests is most conveniently made by heating a small quantity of arsenic trioxide with nitric acid and evaporating the solution on a water-bath (under the hood) until acid vapors can no longer be detected with moistened blue litmus paper.

Acid solutions of arsenates are precipitated *slowly* by hydrogen sulphide, more quickly when warmed. They give rise to arsenic trisulphide and sulphur. Hydrogen sulphide reduces the ion $As^{\cdot\cdot\cdot\cdot\cdot}$ (§ 52) as follows:[1]

$$As^{\cdot\cdot\cdot\cdot\cdot} + H_2S = As^{\cdots} + 2H^{\cdot} + \underline{S} \qquad (1).$$

The reduction takes place more rapidly when the solution is warmed; hence the precipitation follows more quickly in this case. Furthermore by adding other substances, particularly by warming with sulphurous acid (displacing any excess of sulphur dioxide afterwards with carbon dioxide), the reduction is more conveniently effected:

$$As^{\cdot\cdot\cdot\cdot\cdot} + SO_3'' + H_2O = As^{\cdots} + SO_4'' + 2H^{\cdot} \qquad (2).$$

[1] The reaction of hydrogen sulphide with this solution may also be formulated as follows:

$$AsO_4''' + H_2S = AsO_3''' + H_2O + \underline{S} \qquad (1b).$$

According to this latter equation one formula weight of hydrogen sulphide is used up in reducing one formula weight of arsenic. Experiments relating to the influence of hydrochloric acid or alkali hydroxide upon this reaction readily show which of the two equations applies to a given case. By adding acid the concentration of AsO_4''' is decreased, that of the cathion $As^{\cdot\cdot\cdot\cdot\cdot}$ increased. Alkali hydroxide has the opposite effect.

To prepare *arsenic pentasulphide* advantage is taken of the fact that the trisulphide dissolves in yellow ammonium sulphide, which contains S_n''-ions according to § 24, yielding AsS_4'''.

$$\underline{As_2S_3} + 3S'' + 2S_n'' = 2AsS_4''' + 2S_{n-1}'' \qquad (3).$$

The pentasulphide is precipitated on acidifying this solution:

$$2AsS_4''' + 6H^{\cdot} = 3H_2S + \underline{As_2S_5} \qquad (4).$$

Arsenic pentasulphide is also precipitated by passing a strong current of hydrogen sulphide gas into an arsenic acid solution containing a large excess of concentrated hydrochloric acid.

Stannous chloride separates arsenic from a strongly acidified solution of arsenic acid, but the reaction proceeds much more slowly than is the case with arsenic trioxide.

44. Behavior of arsenic compounds when heated on charcoal. Dry mixtures of arsenic compounds with sodium carbonate, when heated in the *reducing flame,* yield vapors having a *garlic* odor. *Great care must be taken to avoid inhaling these vapors,* since they give rise to arsenical poisoning. The garlic odor is characteristic of a particular modification of arsenic (Principles, p. *706*) that is very unstable. In contact with air it yields arsenic trioxide, which forms white fumes depositing on the colder portions of the charcoal.

45. Detection of very small quantities of arsenic by Marsh's test. By reason of its delicacy the following test is more important than any of those mentioned in §§ 42 and 43. It is always applied in criminal cases of arsenical poisoning. The test is based upon the production of arsine gas, AsH_3, which decomposes into hydrogen and arsenic when heated in a tube. The former escapes, while the latter forms a velvety black deposit on the colder portions of the tube walls. To prepare arsine an arsenical solution or a solid substance that is being tested for an arsenic content is placed in a flask in which hydrogen is being generated from the action of dilute sulphuric acid on zinc. The ordinary commercial acids and also the commercial metals very frequently contain arsenic. Zinc and sulphuric acid are used because they may be obtained practically free from arsenic. It is necessary, however, to test both the sulphuric acid and the zinc used in each experiment for an arsenic content by running a blank experiment (see below).

Fig. 4 shows how the Marsh apparatus is set up. The tube "t" is filled with dehydrated calcium chloride in order to dry the evolved gases.

In carrying out the experiment the bottom of a 200 cc. Erlenmeyer flask is covered with granulated pure zinc,[1] over which is poured dilute sulphuric

[1] If the zinc is coppered before beginning the experiment, by dipping it for a short time into copper sulphate solution, a more rapid evolution of gas is ob-

acid (1 vol. concentrated sulphuric acid slowly added to 8 vols. water). After the gas has been generating for about 15 minutes, a dry test-tube is held inverted over the end of the delivery tube "e." It is kept there for several minutes until all air has been displaced by gas from the generator, and is then closed with the thumb and brought to a gas flame. As soon as the gas collected in this way will ignite without explosion, it is certain that all air has been driven out of the apparatus. The hydrogen issuing from the end of the tube "r" (of hard glass) is then ignited and the tube is heated with a bunsen

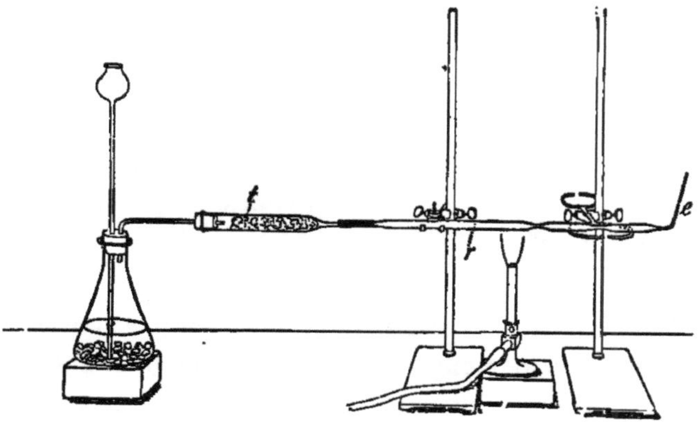

FIG. 4.

burner just in front of one of the constricted portions (see diagram). This blank test is continued for about 15 minutes. If at the end of this time no dark deposit of arsenic is observed in the narrow part of the tube just behind the burner, the sulphuric acid and zinc used are free from arsenic and will serve for testing the solution or substance under observation.

The arsenical solution[1] is then poured into the flask through the safety tube. Care must be taken to avoid letting air into the erlenmeyer flask while pouring in the solution. After a few minutes the jet of burning hydrogen assumes a pale lilac color and a brown black deposit of arsenic settles beyond the heated part of the tube "r."

A further test for arsenic consists in holding a cold object, for example, the cover of a porcelain crucible, in the flame of burning arsine. A brilliant

tained, which best serves the requirements of the test. This is due partly to the fact that the surface of copper is rougher and that hydrogen gas escapes more easily from rough than from smooth surfaces. Hence the reaction

$$Zn + 2H^{\cdot} = Zn^{\cdot\cdot} + H_2$$

proceeds faster.

Furthermore the deposit of copper on part of the zinc surface produces the effect of a short-circuited galvanic element with zinc and copper electrodes. Under these conditions *zinc dissolves* much more rapidly in acids.

[1] The solution must contain no nitric acid or nitrates. In case NO_3'-ion is present, a few drops of concentrated sulphuric acid are added and the solution is heated under a hood until white fumes of sulphur trioxide escape freely. Then it is certain that the volatile nitric acid has been displaced.

black mirror of arsenic deposits on the cold surface. This is due to the fact that the cold body produces a lowering of the temperature and only hydrogen burns under these conditions, while arsenic deposits in metallic form. The characteristic color of the arsine flame and the formation of arsenic spots on a cold porcelain surface will be observed, however, only if not all of the arsine has been already decomposed in the heated tube.

Under similar conditions antimony compounds also form a black deposit of antimony. Hence it is necessary to have tests to distinguish arsenic and antimony spots.

Arsenic spots are soluble in a solution of *sodium hypochlorite* or *bleaching powder*, while *antimony spots* do not dissolve. The solubility of the arsenic mirror is due to oxidation. The antimony spots are less easily oxidized than arsenic spots.

Both the arsenic and antimony mirrors dissolve in ammonium sulphide. On evaporating this solution *arsenic* leaves a *yellow* residue, *antimony* an *orange*. [Furthermore the arsenic mirror deposited in the tube "r" is more volatile than an antimony mirror.]

Lately a very careful investigation of the Marsh test has been made by G. Lockemann, Zeitschrift für angewandte Chemie, **18**, *416–29*, **1905**, to which the student is referred in cases where the greatest care is desirable.

46. The ions of antimony. Antimony resembles arsenic in its behavior in so far as it exists in *trivalent* and *pentavalent* forms, to which likewise two series of cathions and anions correspond. *Trivalent* antimony forms the ions $Sb^{\cdot\cdot\cdot}$ and SbO_3'''. The latter exists in solutions of the salts of antimonous acid, *antimonites*, the former in hydrochloric acid solutions of antimony trioxide.

A solution of antimony trioxide in hydrochloric acid has the following characteristics. When diluted with water a white precipitate of antimony oxychloride[1] forms. It is distinguished from bismuth oxychloride (§ 37) by its solubility in tartaric acid. The ion $SbO^{\cdot}$ is called *antimonyl ion*, the chloride, SbOCl, being correspondingly antimonyl chloride. The formation of antimonyl ion takes place according to the equation:

$$Sb^{\cdot\cdot\cdot} + H_2O \rightleftarrows SbO^{\cdot} + 2H^{\cdot} \qquad (1).$$

Antimonous chloride solution yields with *alkali hydroxide* or *ammonia*, and also with *alkali carbonates*, a white precipitate of *antimonous hydroxide:*

$$Sb^{\cdot\cdot\cdot} + 3OH' = \underline{Sb(OH)_3} \qquad (2).$$

Antimonous hydroxide is soluble in acids and in alkalies, owing to a double dissociation analogous to that of arsenous acid (§ 42).

[1] The composition of the precipitate varies with the amount of water present. As more water is added it becomes relatively poorer in chlorine and richer in antimony.

$$Sb(OH)_3 \rightleftarrows Sb^{\cdot\cdot\cdot} + 3OH' \qquad (3a),$$

$$Sb(OH)_3 \rightleftarrows SbO_3''' + 3H^{\cdot} \qquad (3b).$$

From acid solutions *hydrogen sulphide* precipitates *antimony trisulphide,* Sb_2S_3, orange, soluble in *acids* and in *sodium-, potassium-* and *ammonium sulphide.*

For identifying antimonous ion it is preferable to deposit antimony with *tin* (according to § 26) from a solution which may contain *stannous ion,* rather than to use zinc, because the latter also precipitates tin which is frequently mistaken for antimony, although the tin deposit is much brighter.

The *ions* of *pentavalent antimony* are $Sb^{\cdot\cdot\cdot\cdot\cdot}$[1] and SbO_4'''. Neither has any special definitive properties.

A solution of antimony pentoxide in hydrochloric acid may be used for the following tests. When diluted with water the solution precipitates difficultly soluble *oxychloride,* SbO_2Cl, which becomes *antimonic acid* when heated in contact with the solution.

Hydrogen sulphide precipitates *antimony pentasulphide,* which is not essentially different from the trisulphide in its solubility relations.

An *ammoniacal solution* of *silver* is used to *distinguish* the *trivalent* ions of antimony from the *pentavalent* ions.[2] It transforms *alkali antimonite* (see above) into *antimonate,* separating metallic *silver.* In rendering the *antimony* solution *alkaline* for this experiment it is essential to use *ammonia* (hence also ammoniacal silver solution) and *not alkali hydroxide,* because in this latter case *silver oxide* would form simultaneously with *metallic silver.* Hence the test is applicable only when ammonia is present to redissolve the silver oxide (§ 11). Provided sufficient ammonia is present, a black deposit can only be due to the reaction with antimonite:

$$SbO_3''' + 2Ag^{\cdot} + H_2O = SbO_4''' + 2H^{\cdot} + \underline{2Ag} \qquad (4).$$

To distinguish the ions of pentavalent antimony the solution is acidified and *potassium iodide* solution is added. In case $Sb^{\cdot\cdot\cdot\cdot\cdot}$ or SbO_4''' were present, a *brown red color,* due to free iodine, results.

[1] Compare Principles, p. 702.

[[2] Another method is to add an excess of alkali hydroxide in order to form antimonite ion, and then a drop of mercuric chloride, which yields mercuric oxide under these conditions. Mercuric oxide is changed by antimonite ion to gray or black mercury.]

In each case ions of trivalent antimony are formed, as shown by the equations:

$$Sb^{\cdot\cdot\cdot\cdot\cdot} + 2I' = Sb^{\cdot\cdot\cdot} + 2I \quad (5),$$

and $$SbO_4''' + 2H^{\cdot} + 2I' = SbO_3''' + H_2O + 2I \quad (6).$$

Equation (*6*) shows unequivocally that *hydrogen ion must be present,* while in (*5*) the effect of the presence of hydrogen ion may be understood from what has already been mentioned concerning the existence of antimony cathions (compare also § **43**, footnote).

The solubility of nearly all difficultly soluble antimony salts in tartaric acid or tartrate solutions depends upon the formation of the complex ion $C_4H_4O_6.SbO'$ from the anion $C_4H_4O_6''$ and antimonyl ion, $SbO^{\cdot}$. This complex ion is less stable in acid solutions than in neutral or alkaline solutions, since hydrogen ion reacts with it as follows:

$$C_4H_4O_6.SbO' + H^{\cdot} \rightleftarrows C_4H_4O_6'' + SbO^{\cdot} \quad (7).$$

Now antimonyl salts, as SbOCl, $SbONO_3$, are decomposed by water, i. e. *hydrolysed* (§ 71), separating SbO.OH according to the equation:

$$SbO^{\cdot} + H_2O \rightleftarrows \underline{SbO.OH} + H^{\cdot} \quad (8).$$

Hence solutions which contain the complex ion $C_4H_4O_6.SbO'$ *separate* SbO.OH when *weakly acidified.*

On adding *more acid* the *precipitate vanishes* because the ion $SbO^{\cdot}$ resulting from SbO.OH forms the ion $Sb^{\cdot\cdot\cdot}$ under these conditions according to (*1*) ←.

Dry mixtures of *antimony compounds* with *sodium carbonate* when heated in the *reducing flame* on charcoal yield a *brittle metallic bead,* and a *white sublimate* of *antimony trioxide* deposits on the charcoal (consult § **171**, Charcoal Tests).

47. Ions of tin. Tin likewise forms two series of salts in which it is *tetravalent* and *divalent* respectively. The corresponding ions are *stannic ion,* $Sn^{\cdot\cdot\cdot\cdot}$ and *stannous ion,* $Sn^{\cdot\cdot}$.

Stannous ion. For the test solution *tin foil* is dissolved in *hydrochloric acid. Alkali hydroxide* precipitates *white stannous hydroxide,* soluble in an excess of alkali hydroxide and also in acids. This indicates that stannous hydroxide dissociates in two ways (according to § **42**):

$$Sn(OH)_2 \rightleftarrows Sn^{\cdot\cdot} + 2OH' \quad (1a),$$

and $$Sn(OH)_2 \rightleftarrows SnO_2'' + 2H^{\cdot} \quad (1b).$$

The alkaline solutions of "*stannites*" are rather unstable. When heated they form "*stannates*," *separating tin* simultaneously:

$$2SnO_2'' + H_2O = SnO_3'' + 2OH' + \underline{Sn} \qquad (2).$$

Stannous hydroxide is considerably less soluble in *ammonia* than in alkali hydroxide, in accordance with the weaker basic properties of ammonia (§ 49).

Hydrogen sulphide precipitates *brown stannous sulphide*, which dissolves in acids more readily than the other sulphides of this group. *Yellow ammonium sulphide* dissolves it much more abundantly than ordinary *colorless* ammonium sulphide, because the ion SnS_3'' is formed more easily than the ion SnS_2''. The general equation for this reaction is expressed as follows:

$$\underline{SnS} + S'' + S_n'' = SnS_3'' + S_{n-1}'' \qquad (3).$$

From this solution acids precipitate the *yellow disulphide:*

$$SnS_3'' + 2H^{\cdot} = \underline{SnS_2} + H_2S \qquad (4).$$

48. Stannic compounds. In accordance with a general rule tetravalent tin has but a slight tendency to form cathions. This is shown by the fact that the different tetravalent tin salts do not exhibit the same analogy in their properties as do the salts of other metals.

The fact that a solution of *stannic chloride* forms a *yellow* precipitate of *stannic sulphide* with *hydrogen sulphide* indicates that the ion $Sn^{\cdot\cdot\cdot\cdot}$ exists in this solution. Stannic chloride is easily prepared by adding bromine water drop by drop to stannous chloride solution until the yellow color of the bromine solution remains permanently:

$$Sn^{\cdot\cdot} + 2Br = Sn^{\cdot\cdot\cdot\cdot} + 2Br'.$$

Solutions of stannic halides are decomposed by water and separate stannic hydroxide, especially when warmed or on addition of certain salts, as sodium sulphate or ammonium nitrate (consult § **77**; also Principles, p. *722–23*).

The acid character of stannic hydroxide is developed to a greater extent than the basic. Hence it is frequently called *stannic acid*. This compound is of special interest, because it exists in two (and probably more) states which are mutually transformable. The one modification is soluble in hydrochloric, nitric and sulphuric acid and

in alkali hydroxide. It is prepared by precipitating stannic chloride solution with hydroxyl ion. The precipitation is hindered by tartaric acid.

The other modification, *metastannic acid,* has exactly the same composition. It is formed by the action of concentrated nitric acid upon tin. This form is soluble in alkalies, insoluble in dilute acids, insoluble in tartaric acid. Concentrated hydrochloric acid changes it into the tetrachloride which dissolves in water, but is not soluble in hydrochloric acid.

Stannic and stannous ions may be distinguished by the different colors of stannic and stannous sulphides, also by means of reactions (*5*) and (*6*) of § 21 and by reaction (*6*) of § 42.

In *dry* mixture with sodium carbonate *tin compounds* yield in the reducing flame on charcoal white *malleable* metallic grains and a white sublimation. Tin compounds are not easily reduced in this way. Hence in case only this dry test is used for the detection of tin, some potassium cyanide should be added to the carbonate mixture and the operation is then carried out under the hood. Potassium cyanide is thereby oxidized to potassium cyanate, KCNO.

The reduction is also easily effected with the aid of sodium or magnesium (consult § 172 on the Hempel method of reduction).

49. The relation of certain exceptions to the difference in the dissociation of the substances in question. In a number of cases where substances with a common ion have been compared, for example, mercuric chloride and -nitrate (§ 34), acetic, hydrochloric and nitric acids (§§ 28 and 38), ammonia and alkali hydroxide (§ 47), differences in behavior have been noted which are apparently contrary to the earlier statements (§§ 8 and 9) relating to the *agreement* in the *properties* of electrolytes with a common ion. Although the variance has already been explained in individual cases, nevertheless it is expedient to consider somewhat more closely the reasons for these relations. Exact investigation shows that these exceptions result only from an *incomplete* (*qualitative*) *formulation* of the dissociation relations of the salts in question.

Dissociation is classed amongst the reversible reactions, for which criteria are cited in § 27. Thus, for example, the reaction

$$NaCl \rightleftarrows Na^{\cdot} + Cl',$$

which takes place on dissolving sodium chloride, proceeds practically to an end in the one sense → only under certain conditions, which will be dealt with later (§ 88). Provided these definite conditions do not exist, the system reaches a stable condition, i. e. the reaction does not proceed any farther if *undissociated* salt is still present *along with the dissociation products. Equilibrium* then exists and *no change* will take place in one direction or the other unless outside influences are brought to bear upon the system. The

entire amount of a substance present in unit volume of the solution, its *total concentration*, thus consists of two parts, the *concentration* of the *undissociated portion* and that of the *ions*. In the case of substances with a common ion, for instance, different salts of a metal, or different acids, although their solutions may have the same total concentration, nevertheless that of the dissociated parts[1] may vary a good deal from one salt or acid to the next. These differences will not affect very delicate reactions that are characteristic of the ion in question, but they will influence less delicate reactions. For example, hydrogen sulphide precipitates solutions of mercuric chloride and mercuric nitrate. On the other hand mercuric sulphocyanate is precipitated only from solutions of the nitrate, and not from those of the chloride. The concentration of $Hg^{\cdot\cdot}$-ions in a solution of mercuric chloride of the same total concentration as that of a mercuric nitrate solution, is too small to yield a precipitate with CNS' (see § 34), while it is sufficient to form so much mercuric sulphide with hydrogen sulphide that a precipitate results. This is due to the fact that only a very small quantity of $Hg^{\cdot\cdot}$-ion is needed to give a perceptible reaction with S''-ion. Hence the difference in the dissociation of mercuric chloride and mercuric nitrate does not interfere with this latter test.

Still more striking differences are observed in the behavior of mercuric cyanide solution. The solution is precipitated by hydrogen sulphide and reacts with stannous ion (§ 21), but does not give any of the other tests mentioned in § 33.

The *neutral salts* have very *similar* dissociation relations. Hence it is easy to bear in mind the comparatively small number of exceptional cases. Moreover other regularities hold with respect to them. These cannot, however, be taken up here. Soluble cyanides of the heavy metals are often dissociated to an extremely small degree. The chloride, bromide and iodide of mercury, cadmium and zinc are also less dissociated than other salts.

According to their relative dissociation acids[2] and bases are distinguished as strong, moderately strong and weak. Thus hydrochloric, nitric and sulphuric acids are strong, acetic acid is moderately strong, potassium and sodium hydroxides are strong bases, ammonia is moderately strong, etc. In dealing with the corresponding cathions of the bases and the anions of the acids, these relations will be taken up in more detail.

50. Dissociation and ions of water. Although it is one of the least dissociated substances, water is dissociated into ions, however, to a certain extent: $H_2O \rightleftarrows H^{\cdot} + OH'$. How hydrogen ion has been recognized as the ion common to acids, while hydroxyl ion plays the same role with the bases (§ 8). Since pure water does not alter the color of red or blue litmus paper, the assumption that equal amounts of hydrogen- and hydroxyl ions are present in water is obvious. The exact quantitative basis for this assumption has been attained by physical methods. Application of the law of mass action to the dissociation of water leads to the result that for *all aqueous solutions* the

[1] Physical methods are used for the exact determination of the concentrations of the ions. Chemical phenomena may be utilized only for *qualitatively* judging these relations.

[2] Compare Ostwald, Scientific Foundations of Analytical Chemistry, translation by McGowan, p. 55.

product of the concentration of hydrogen- and hydroxyl ions is constant within rather wide limits. The value of the product is extremely small.[1] This agrees fully with experimental results.

Thus acid solutions always contain hydroxyl ion, and alkaline solutions contain hydrogen ion, although in very small amounts. If $C_{H\cdot}$ and $C_{OH'}$ represent the concentrations of hydrogen- and hydroxyl ions respectively, then for all aqueous solutions the relation

$$C_{H\cdot} \times C_{OH'} = \text{constant}$$

holds. In the case of acid solutions $C_{H\cdot} > C_{OH'}$; for bases $C_{OH'} > C_{H\cdot}$, while for water alone and for neutral solutions $C_{H\cdot} = C_{OH'}$.

The fact that *acids* and *bases* lose their *typical properties* when they are mixed in a solution is explained on the basis of the above considerations. Their respective hydrogen and hydroxyl ions react to form water, $H^{\cdot} + OH' \rightleftarrows H_2O$, and disappear from the solution until the equilibrium condition is satisfied, i. e. until the product of the $H^{\cdot}$- and OH'-ions remaining in the solution has become a very small value (see § 71).

It is sometimes an advantage to express in the equation for a reaction the fact that a certain substance is but little dissociated in comparison with the other substances present. For this we will make use of the recommendation of Walker by writing the formulae of these compounds in the ordinary way, indicating in the formula the dissociation products as usual. For example, water is formulated $H^{\cdot}OH'$; acetic acid is represented by $C_2H_3O_2'H^{\cdot}$, mercuric cyanide by $Hg^{\cdot\cdot}(CN)_2''$, etc.

GROUP III.

51. *Ammonia* is the third group precipitant. All hydrogen sulphide contained in the filtrate from the second group must be expelled, however, by boiling the solution, before ammonia is added. To show that the removal of hydrogen sulphide is complete, lead acetate paper, made by dipping filter paper in lead acetate solution, is held in the escaping vapors. With hydrogen sulphide this gives a brown stain due to the formation of lead sulphide. A few drops of the solution, freed from hydrogen sulphide, are tested for ferrous ion with freshly prepared potassium ferricyanide solution, $K_3Fe(CN)_6$. This gives a dark blue precipitate with ferrous ion (§ 148). In case the test indicates ferrous ion, the entire solution is then mixed with a few cc. of concentrated nitric acid and the mixture is heated for a few minutes, in order to transform ferrous- into ferric ion (see § 54).

Even if hydrogen sulphide had not been added, i. e. when the preliminary test (§ 19) indicated the absence of second group metals, the above test for ferrous ion is, nevertheless, carried out.

[1] Compare Ostwald, Grundriss der allgemeinen Chemie, 3d edition, p. 460.

The removal of hydrogen sulphide and transformation of ferrous into ferric ion are essential to the third group separation. From 5–10 cc. *ammonium chloride* solution and *ammonia* just sufficient to give the solution an alkaline reaction are next added. *Warming* the solution *before* adding ammonia favors the separation of a precipitate. The components of the well washed precipitate are identified according to § 55, while the filtrate is further tested according to § 72 for the remaining groups.

The use of an excess of ammonia to precipitate the third group components is generally recommended. Some hydroxides (consult § 63) are, however, only incompletely precipitated with an excess of ammonia. Hence it is further recommended in many cases to boil the solution in contact with the precipitate until no longer any ammonia vapors escape. *Ammonium salt solutions* decompose when boiled; for example, $NH_4Cl = NH_3 + HCl$ (see § 71), and thus give rise to *ammonia* which escapes. The free *acid* that is simultaneously formed will dissolve a portion of the precipitate. If a large amount of ammonium salt is present, and in case the boiling is continued for some time, the *precipitate* may finally redissolve completely. Hence to boil off the excess of ammonia is just as disadvantageous as is the use of too large an excess without boiling. It is therefore best to add no more than the requisite amount, or at least a very small excess of ammonia, at first.

The precipitate may contain:

A. Hydroxides of *iron, aluminum, chromium* and *manganese* (also cobalt):

B. *All salts* which are *dissolved* by acids, the *cathions of which* have not been precipitated by the first and second group reagents. The following are most frequently met with: *aluminum-, ferric-* and *magnesium ammonium phosphate;* also the *fluorides, oxalates* and *phosphates* of the *alkali earth metals,* Ca, Sr, Ba:

C. In the case of *rock* or *mineral* analysis, *silicic acid.* This case is dealt with separately in §§ 189–192:

D. Occasionally *sulphates* of the alkali earth metals, particularly if hydrogen sulphide had been passed into the solution containing ions of these metals.

Precipitation of the substances included under D above may take place without addition of ammonia, and often does occur after the filtrate from the

second group has been heated with nitric acid, since these substances are scarcely more soluble in acids than in water. As a rule, however, the precipitate is not observed until ammonia has been added. In case it is noted in the filtrate from the second group, it is filtered and tested for sulphates of the alkali earth metals according to § 183.

When substances included under C and D are present, the third group precipitate is not completely soluble in dilute hydrochloric acid (§ 55).

The precipitation of the hydroxides included under A is incomplete in the presence of certain organic compounds (consult §§ 63, 70, 116). In § 193 will be found a description of phenomena which indicate that this precipitation may be incomplete, and also the means of overcoming the difficulty.

The conditions under which *cobalt hydroxide* precipitates in the third group are not yet thoroughly established, nor are the exact conditions known under which *this* may be *prevented*. Since cobalt belongs to the fourth group components, and hence is generally tested for only in the precipitate obtained with *ammonium sulphide*, it sometimes happens that this metal escapes detection unless the third group precipitate is also tested for cobalt (compare § 56).

52. **Oxidation and reduction.** As has already been mentioned in § 39, by heating a salt with sodium carbonate on charcoal in the luminous flame *reduction* to the metal takes place. Conversely the formation of a metallic oxide from a metal constitutes *oxidation*. These terms oxidation and reduction (§ 51) are also used for reactions in which no phenomena are observed similar to those characterizing the above cases. Hence it becomes expedient to study this question somewhat more closely with the object of determining why different reactions are classified under this one head.

Originally *reduction* was *restricted* to the *transformation* of *metallic oxides* into the *metals*, while *oxidation* meant exactly the *reverse reactions*. The connection between oxidation and reduction will become clearer by consideration of a particular case, for example, the reduction of cupric oxide with hydrogen. The equation

$$CuO + H_2 = H_2O + Cu$$

states that cupric oxide + hydrogen become copper + water under certain conditions. On applying the above definition it is at once evident that the *one substance*, cupric oxide, is *reduced*, while the *other*, hydrogen, is *oxidized*. In every other case it is found that *oxidation is always accompanied by reduction*, and vice versa. Hence to call a *reaction* an oxidation or reduction is not justifiable. The terms are, however, always applicable to definite substances which undergo oxidation or reduction in any given reaction, thus in the above case to hydrogen and cupric oxide respectively. In the case of several reactions which have been classed under these heads, viz. in § 38, (*3*); § 40, (*1*) and § 43, (*1*), the above-mentioned characteristics of oxidation and reduction are not observed, for the reason that these terms are now used in a broader sense. *Oxidation* accordingly denotes changes by means of which a *substance takes up oxygen or a non-metal, or gives up hydrogen*. In the following pairs of substances the relation of the left hand member to

the right hand member is that of an *oxidizable* substance to an *oxidized* product: thus, $Cu_2O - CuO$; $Cu_2S - CuS$; $H_3AsO_3 - H_3AsO_4$; $HgCl -$ $HgCl_2$; $SnCl_2 - SnCl_4$; $H_2O - H_2O_2$; $HCl - Cl$; $H_2S - S$; etc.

The *reverse* changes are classified as *reduction*, viz. transformation of $SnCl_4$ into $SnCl_2$, etc.

According to the above a substance is oxidized by combining with another substance which can give oxygen or some other non-metal up to it, or can remove hydrogen from it. Substances having this property are called *oxidizing agents.* Substances which produce the reverse changes, i. e. which take away oxygen or other non-metal, or give up hydrogen, are *reducing agents.* Hence hydrogen is a reducing agent towards cupric oxide, while the latter is an oxidizing agent with respect to hydrogen. According to their mutual relations in this regard, substances may be arranged in a single series in such a way that each member acts as a reducing agent towards each succeeding member and as an oxidizing agent towards every preceding member. Typical reducing agents, as sodium, iron, stannous chloride, hydrogen, arsenous acid, cuprous chloride, ferrous sulphate, sulphurous acid, etc., stand at the beginning of the series. At the other end are oxidizing agents, viz. ferric salts, nitric acid, chromic acid, oxygen, chlorine, permanganic acid. The order of the substances in this series, like that in the electric potential series,[1] is not fixed. It is found that the strength of an oxidizing or reducing agent increases or decreases with a decrease or increase respectively of its concentration. Hence even for the more commonly occurring concentrations the mutual oxidizing or reducing relations of the members are *only approximately indicated* unless their concentrations are exactly defined.

Evidently a given substance may be an oxidizing agent towards one substance and a reducing agent towards another, as the following examples will show:

$$\underline{Zn} + SnCl_2 = \underline{Sn} + ZnCl_2 \qquad (1),$$

and

$$2FeCl_3 + SnCl_2 = 2FeCl_2 + SnCl_4 \qquad (2).$$

In the first case *stannous chloride* is an *oxidizing agent* towards *zinc* which is oxidized (i. e. combines with chlorine), while in the latter case it is a *reducing agent* towards *ferric chloride.* Hence it is not quite exact to call a substance an oxidizing agent, except with reference to some other particular substance.

53. **The relation between oxidation and reduction and changes in the electric charges of ions.** To show the role of the individual substances in the case of reactions in solutions, in which ions take part, the following rule is verv serviceable. In the reactions just quoted, by abbreviating the equations so that only the essential changes are indicated (§ 14), i. e. by leaving out the anion, Cl′, the equations assume the following form:

$$\underline{Zn} + Sn^{\cdot\cdot} = \underline{Sn} + Zn^{\cdot\cdot} \qquad (1b),$$

and

$$Sn^{\cdot\cdot} + 2Fe^{\cdot\cdot\cdot} = Sn^{\cdot\cdot\cdot\cdot} + 2Fe^{\cdot\cdot} \qquad (2b).$$

It is directly evident that the *oxidizing agents, stannous-* and *ferric ion* respectively, *have lost positive electric charges,* while the *reducing agents, zinc* and *stannous ion, have gained positive charges.*

[1] The electric potential series constitutes a portion of this series. With metals the formation of cathions is an oxidation, while transformation from the ionic condition to the metallic is a reduction (see § 53).

The interpretation of different reactions in this way affords an excellent exercise to the student. Amongst other cases the reduction of ferric salts by hydrogen sulphide (§ 69 (*3*)) should be undertaken.

The separation of iodine from potassium iodide solution by means of chlorine water is easily recognized from the brown red color of the free iodine. Ordinarily the reaction is expressed by the equation:

$$2KI + Cl_2 = 2KCl + I_2.$$

Taking into account the dissociation of the two salts, KCl and KI, it is evident that the essential changes may be formulated as follows in this case:

$$Cl + I' = Cl' + I.$$

From its position in the oxidation series chlorine is a stronger oxidizing agent than iodine, that is, it will oxidize substances which follow iodine in the series, but stand in front of chlorine. Hence in the case of an exchange of negative charges the *oxidizing agent gains,* the *reducing agent loses.* The following rule results from a combination of the two cases. *Oxidation consists in an increase of positive charges or a decrease of negative. Reduction consists in a decrease of positive charges or an increase of negative.*

54. **The oxidation of ferrous ion to ferric ion with nitric acid.** The oxidation of ferrous- to ferric ion by means of nitric acid is accompanied by a simultaneous reduction leading to the production of lower oxidation forms of nitrogen, as follows:

$$2Fe(NO_3)_2 + 4HNO_3 = 2Fe(NO_3)_3 + 2H_2O + 2NO_2 \qquad (1),$$

and

$$3Fe(NO_3)_2 + 4HNO_3 = 3Fe(NO_3)_3 + 2H_2O + NO \qquad (2).$$

Apparently the above rule is not applicable to this case since the oxidizing agent, NO_3', has a negative charge which is given up. Assuming, however, that nitric acid dissociates in several ways according to the conditions existing in the solution, yielding nitrate ion, NO_3', nitro-ion, $NO_2^{\cdot}$, and nitroso-ion, $NO^{\cdot\cdot\cdot}$, respectively, it becomes possible to apply the rule also to this case. The dissociation equations for nitric acid have then the following forms:

$$HNO_3 \rightleftarrows NO_3' + H^{\cdot}$$

$$HNO_3 \rightleftarrows NO_2^{\cdot} + OH' \qquad (3),$$ [1]

$$HNO_3 + H_2O \rightleftarrows NO^{\cdot\cdot\cdot} + 3OH' \qquad (4).$$

Taking these into account the above equations (*1*) and (*2*) assume the forms:

$$2Fe^{\cdot\cdot} + \underbrace{2NO_2^{\cdot} + 2OH'} + \underbrace{2H^{\cdot} + 2NO_3'} = 2Fe^{\cdot\cdot\cdot} + 2H_2O + 2NO_2 + 2NO_3' \qquad (1b),$$ [2]

and

$$3Fe^{\cdot\cdot} + \underbrace{NO^{\cdot\cdot\cdot} + 3OH'} + \underbrace{3H' + 3NO_3'} = 3Fe^{\cdot\cdot\cdot} + 3H_2O + NO + 3NO_3' \qquad (2b).$$

[1] Ostwald, Grundriss der allgemeinen Chemie, 3d edition, p. *440*.

[2] For the sake of clearness the anion of the ferrous salt has been omitted.

The oxidizing action of nitrates depends upon the presence of *hydrogen ion.* By making the assumption that nitric acid dissociates in the above ways this dependence can be read from the equation. It is evident that dissociation according to (*3*) or (*4*), upon which the oxidizing effect of the nitrate ion depends, can take place to a considerable extent only when the concentration of hydroxyl ion is very slight. The more hydrogen ion in the solution the greater will be this dissociation. Hence it is clear that nitrates will not have the same oxidizing action in neutral solution that they exert in acid solution.

55. Separation of the hydroxides precipitated by ammonia. For the sake of simplicity it may be assumed that the *substances mentioned in* § 51 *under B, C and D are not present.* The precipitate is first washed and is then brought into solution by pouring *warm dilute hydrochloric acid* over it on the filter. To this solution alkali hydroxide is added, reprecipitating all of the hydroxides at first. When an excess of alkali hydroxide has been added, however, the hydroxides of *aluminum* and *chromium* go into solution again, according to the equation:

$$\underline{Al(OH)_3} + 3OH' = AlO_3''' + 3H_2O \qquad (1).$$

Only when a *sufficient excess* of alkali hydroxide is present does this latter reaction take place to such an extent that all of the precipitated aluminum- and chromium hydroxides are dissolved. Enough must be added to make the solution feel soapy (Principles, p. *447*). If a residue is left after this treatment it is filtered, and the filtrate or the alkali hydroxide solution, in case no residue was left, is then *diluted with water* and boiled for some time. Chromium and aluminum are separated by this operation, since the ion CrO_3''' is decomposed by water to far greater extent than AlO_3''':

$$CrO_3''' + 3H_2O = \underline{Cr(OH)_3} + 3OH' \qquad (2).$$

Hence *aluminum hydroxide* remains in solution, while *chromium hydroxide* separates as a *characteristic green precipitate.*

After removing chromium hydroxide the filtrate is tested for *aluminum* by adding *ammonium chloride,* either in solution or in the form of powder. This strongly alkaline solution is then boiled for some minutes. The ion AlO_3''' yields a precipitate of *aluminum hydroxide* under these conditions.

The precipitate, $Al(OH)_3$, forms slowly. By warming the solution, however, the precipitation is hastened and at the same time a

coagulation of the gelatinous mass is effected, which makes it more readily visible.

The action of the ammonium salt consists in a reaction with alkali hydroxide by which hydroxyl ion is used up:

$$NH_4^{\cdot} + OH' = H_2O + NH_3 \qquad (3).$$

It is evident that the separation of aluminum hydroxide is effected by this consumption of hydroxyl ion, since the *solubility* of aluminum hydroxide is due in the first place to the presence of an excess of hydroxyl ion. Any means of removing this excess will cause a reprecipitation of the hydroxide. Thus, for instance, the addition of any acid will also lead to the formation of a precipitate.

To remove hydroxyl ion by neutralizing the solution with an acid (§ 50) might give rise to errors, however, since aluminum hydroxide is easily soluble in acids. The least excess of acid (see § 63) would cause the precipitate to vanish completely, and so the precipitation might be overlooked.

As alkali hydroxide frequently contains aluminum a blank test must always be made by heating it with an ammonium salt. Only in case the alkali hydroxide gives *no* test for aluminum does the formation of a precipitate with ammonium chloride constitute a criterion of the presence of aluminum in the third group precipitate.

[The following modification affords a very distinct identification of chromium and aluminum and in addition leaves the residue, insoluble in an excess of alkali hydroxide, in a condition which facilitates its filtration. In this case the third group precipitate is transferred to a beaker or casserole and an excess of alkali hydroxide together with hydrogen peroxide (about 15 cc. of a 2 per cent solution) are added (see §§ 57 and 64). The mixture is boiled until the evolution of gas ceases. Aluminum hydroxide goes into solution, as before, without any essential changes. Chromium, however, is found in the solution as the ion CrO_4'', due to an oxidation of CrO_3''' in alkaline solution by hydrogen peroxide. Solutions of chromates have a characteristic yellow color.

The residue insoluble in alkali hydroxide filters better when hydrogen peroxide is used and a more complete separation of iron and chromium is effected.

A yellow filtrate indicates chromium. It is divided into two parts. In one of these the presence of chromium is confirmed by any of the reactions characterizing the ions CrO_4'' or Cr_2O_7'' (§ 119). The other portion is tested for aluminum with ammonium chloride or ammonium carbonate.]

56. Testing the residue insoluble in alkali hydroxide for hydroxides of iron and manganese. The *residue* of *hydroxides insoluble* in *alkali hydroxide* may contain *hydroxides* of *iron* and *manganese* (§ 78). To test for manganese a small portion of the residue is fused on a platinum foil with a few grams of sodium carbonate and potassium nitrate. Under these conditions *manganic hydroxide* (or manganese dioxide) yields an alkali manganate of a characteristic blue green color which is very readily perceptible. It is thus possible to detect *very minute quantities* of *manganese compounds* by this test. The fused mass dissolves in water and forms a green solution which turns red upon addition of a few drops of acetic acid, owing to the formation of permanganate:

$$3Na_2MnO_4 + 3H_2O = 2NaMnO_4 + MnO(OH)_2 + 4NaOH.$$

This equation shows that the acetic acid acts merely by neutralizing the alkali hydroxide formed and thus establishes the conditions for the transformation of the manganate into permanganate.

To test for iron the rest of the precipitate is dissolved in hydrochloric acid. On addition of *potassium sulphocyanate* a dark blood-red color is a characteristic indication of iron.

Another method consists in separating the iron as basic acetate. To effect this sodium carbonate is added to the hydrochloric acid solution until the precipitate that is formed about each drop of sodium carbonate as it is added, will remain permanently on shaking the solution. This precipitate is then dissolved in acetic acid or in a few drops of dilute hydrochloric acid, and a few cc. of sodium acetate solution are added. In the presence of ferric ion a dark red color results, owing to the formation of basic ferric acetate. A red precipitate of basic acetate (compare § 69) is obtained on heating this solution.

It must be borne in mind that the *potassium sulphocyanate* test for iron is extremely delicate. Chemically pure acids frequently give a weak iron test with it. Hence a red color on adding potassium sulphocyanate to a given solution indicates iron *only in case* it is deeper than the color obtained in a parallel experiment made with the acid that has been used to prepare the solution (§ 55, also § 69).

To detect *cobalt hydroxide* in the third group precipitate manganese and iron are first tested for as above given. Cobalt ion

remains in solution after precipitating ferric ion as basic ferric acetate. On adding ammonium sulphide to the filtrate from the basic acetate precipitation a black precipitate indicates cobalt. In case a very small amount of cobalt is present the solution assumes a reddish brown color. The precipitation of any iron that is present becomes practically complete if the acetate solution has been *diluted* with several times its own volume of water and is then boiled sufficiently long (several minutes). Hence, if cobalt is absent, ammonium sulphide does not color the filtrate under these conditions. The formation of the color is therefore a sufficient criterion of the presence of cobalt. Since cobalt is only partially precipitated in the third group, the precipitate obtained with ammonium sulphide after separating iron may be added to the fourth group precipitate. The whole is then tested according to § 73. Manganese does not interfere with the test for cobalt, as manganese ion yields a flesh-colored precipitate with ammonium sulphide.

57. Incompleteness of the separation of iron- and chromium hydroxides. When the ammonia precipitate contains *both iron-* and *chromium hydroxides, sodium hydroxide dissolves* the latter (§ 55) only incompletely unless hydrogen peroxide has been added. On the other hand some *ferric hydroxide* always goes into *solution* with the *chromium hydroxide* and precipitates again with it on boiling the diluted alkaline solution. Therefore the *precipitate* of *chromium hydroxide* must always be tested for iron and the precipitate which is tested for iron and manganese according to § 56 must also be tested for chromium.

To separate ferric- and chromium hydroxides the mixture of hydroxides is fused in a crucible with sodium carbonate and potassium nitrate. This oxidizes chromium hydroxide to sodium chromate (§ 64) which is soluble in water, while ferric hydroxide remains unchanged. The iron is then tested for as above given.

The hydroxides of *chromium, iron* and *manganese* are separated by means of sodium acetate (see § 57). The basic acetates of chromium (see § 64) and iron precipitate together. These are then separated by fusion as above given. Manganous ion is identified in the filtrate from the basic acetate separation by means of ammonia and ammonium sulphide, yielding flesh-colored manganous sulphide.

58. To test the third group precipitate for phosphates, oxalates, etc. When the third group precipitate contains the different salts mentioned in § 51, *B*, in addition to the hydroxides, it is first treated as described in § 55. The further treatment of the *insoluble* residue depends, however, upon the nature of the anions that are present. It is especially important to ascertain if it contains phosphates (a), or oxalates and fluorides (b), or mixtures of phosphates, oxalates and fluorides (c).

In testing for phosphates a part of the precipitate is dissolved in nitric acid. To this solution several times its own volume of *ammonium molybdate* solution is then added. Phosphates yield a *yellow precipitate* of *ammonium phospho-molybdate* (§ 127) after standing for some time, more quickly on gently warming the solution. It is essential to this test that a sufficient quantity of nitric acid is present: too large an excess is, however, harmful.

The other anions are tested for by boiling another portion of the precipitate with sodium carbonate solution for several minutes. The following reaction takes place, assuming that barium oxalate is present:

$$\underline{BaC_2O_4} + CO_3'' = C_2O_4'' + \underline{BaCO_3}.$$

Oxalates are decomposed and the oxalate ion is brought into solution, while a corresponding amount of insoluble carbonate is formed simultaneously. The precipitated carbonate is filtered and the *filtrate,* which contains the anions CO_3''[1] and C_2O_4'' in the present case, is acidified with *acetic acid.* By this means the carbonate ion is decomposed:

$$CO_3'' + 2H^{\cdot} = CO_2 + H_2O.$$

A soluble calcium salt (calcium chloride or -nitrate) is then added to the solution. With *oxalate ion this yields a white precipitate. Hydrochloric* or *nitric* acid must not be used in place of acetic acid because they would prevent the precipitation of calcium oxalate (§ 116). Fluorides behave like oxalates and are treated in exactly the same way.

Oxalates and *fluorides* do not affect the test for phosphates, and conversely phosphates have no influence upon the test for oxalates and fluorides.

[1] From the excess of carbonate.

59. The decomposition of phosphates. In case any of the substances mentioned in § 58 are present in the third group precipitate the *residue insoluble* in sodium hydroxide (§ 55) is treated with substances which will *decompose the salts in question.* In order to *decompose phosphates* (a) the residue is dissolved in concentrated nitric acid. This solution is then heated with tin foil. Tin reacts with nitric acid, evolving brown vapors of nitrogen peroxide and forming insoluble stannic acid:

$$\underline{Sn} + 4NO_2^{\cdot} + 4OH' = \underline{Sn(OH)_4} + 4NO_2 \qquad (1).^{1}$$

In case the nitric acid solution contains phosphate ion, this reacts with stannic acid, yielding insoluble stannic phosphate:

$$\underline{3Sn(OH)_4} + 4H_3PO_4 = 12H_2O + \underline{Sn_3(PO_4)_4} \qquad (2).$$

The presence of a *sufficient amount* of *stannic acid thus ensures the complete precipitation of phosphate ion.* To see if all phosphate ion has been removed, a small portion of the solution is filtered from the precipitate. In case this solution still yields a yellow precipitate with ammonium molybdate, more tin must be added to the original mixture and the boiling is continued until a further test with ammonium molybdate no longer gives any indication of phosphate. The precipitate is then separated from the solution by filtration and is washed, first with dilute *hydrochloric acid,* afterwards with water. The wash water and filtrate are mixed. Since the filtrate may contain some stannic acid, it is evaporated on the water-bath to dryness, or at any rate until nearly dry. By this means the stannic acid is converted into the less soluble modification (§ 48). The residue from evaporation is next extracted with slightly acidulated water and any insoluble stannic acid is filtered. With this solution which now contains no phosphate ion the regular tests for the components of groups III,[2] IV and V, and also magnesium ion, are carried out.

60. The decomposition of oxalates and fluorides (§ 58, *b*). In this case the residue from the *treatment* with *alkali hydroxide* is likewise dissolved in *nitric acid. Sodium carbonate* is added to this

[1] Compare § 54 on the dissociation of nitric acid: $4NO_2^{\cdot} + 4OH'$.

[2] The solution must be tested for components of group III because it may contain ferric ion. Tin foil is sometimes contaminated with copper, which will also be found in the solution. It is best removed with hydrogen sulphide.

solution in small portions until the solution has an *alkaline reaction.* The excess of nitric acid at first decomposes a portion of the carbonate ion present (see § 58). Hence the solution must be heated to completely remove all carbon dioxide. The cathions that were contained in the residue as oxalates and fluorides are then precipitated as *difficultly soluble carbonates* by the excess of CO_3''. This precipitate is filtered, washed and dissolved in hydrochloric or nitric acid:

$$\underline{MCO_3} + 2H^{\cdot} = M^{\cdot\cdot} + CO_2 + H_2O.$$

By this means a solution is obtained which contains the same cathions that were present in the insoluble residue, but has no oxalate or fluoride ions.

61. The decomposition of phosphates, oxalates and fluorides (§ 58, *c*) depends upon a combination of the methods described in §§ 59 and 60. Phosphate ion is first precipitated by means of tin. The cathions which form difficultly soluble carbonates are then separated by the addition of sodium carbonate to the filtrate. Oxalate and fluoride ions remain in solution.

[Phosphates, oxalates, etc., may also be tested for in the filtrate from the second group after hydrogen sulphide has been expelled. If detected they are destroyed as above given, and the regular analysis of the third and succeeding groups is then carried out. It must be noted that arsenic acid also yields a yellow precipitate (§ 133) with ammonium molybdate. No confusion will result, however, if the regular course of analysis is properly followed, since arsenic is precipitated as sulphide in the second group. Only in case arsenic has not been entirely removed in the second group can confusion result from this circumstance.]

62. Theory of precipitating the salts dissolved in acids by means of ammonia. The dissolving of barium oxalate by hydrochloric acid, which will serve to exemplify this question, is represented by the equation:

$$\underline{BaC_2O_4} + 2H^{\cdot} \rightleftarrows Ba^{\cdot\cdot} + C_2O_4' H_2^{\cdot\cdot}.$$ [1]

By leaving a definite amount of hydrochloric acid of any suitable concentration in contact with the solid oxalate for some time, and then filtering, a

[1] Consult § 50. The equation shows that oxalic acid results on dissolving the oxalate in an acid. It forms the ions C_2O_4'' and $2H^{\cdot}$, but its dissociation is very slight in comparison with that of the acid used to dissolve the oxalate (HCl or HNO_3). This must always be taken into account in the quantitative study of these relations.

saturated solution of oxalate is obtained. This solution has the property of yielding a precipitate of barium oxalate on addition of a little ammonia. Now in the saturated solution an equilibrium exists between the different substances taking part in the reaction. Hence, according to the law of mass action, the relation

$$\frac{C_{Ba^{\cdot\cdot}} \cdot C_{C_2H_2O_4}}{C^2_{H^{\cdot}} \cdot C_{BaC_2O_4}} = k$$

holds. In estimating the effect of the added substances account must be taken of the purely physical change of concentration brought about by the increase of volume, as well as of the change due to ammonia. Assuming that the concentration of each constituent is lowered to the *n*'th part of its former value, it follows that a change must take place in the system in consequence of the dilution alone, irrespective of the action of ammonia, for

$$\frac{\frac{C_{Ba^{\cdot\cdot}}}{n} \cdot \frac{C_{C_2H_2O_4}}{n}}{\frac{C^2_{H^{\cdot}}}{n^2} \cdot \frac{C_{BaC_2O_4}}{n}} = n \cdot \frac{C_{Ba^{\cdot\cdot}} \cdot C_{C_2H_2O_4}}{C^2_{H^{\cdot}} \cdot C_{BaC_2O_4}} > k.$$

Hence the concentrations of barium ion and oxalic acid must decrease, those of barium oxalate and hydrogen ion must increase. On diluting a saturated solution of barium oxalate with water, however, the formation of a precipitate is not observed. Nevertheless this does not disprove the conclusion that the concentrations of the substances are altered in the sense of the formation of more barium oxalate, but it merely shows that this reaction does not take place to sufficient extent to effect a separation of the solid salt. For the exact treatment of these relations other conditions must be taken into account, which cannot be considered here.

By the addition of hydroxyl ion (ammonia) the concentration of the hydrogen ion, $C_{H^{\cdot}}$, is decreased. Hence, in order that the system remain in equilibrium, the values of $C_{H^{\cdot}}$ and $C_{BaC_2O_4}$ must increase at the expense of $C_{Ba^{\cdot\cdot}}$ and $C_{C_2H_2O_4}$, and this leads to the precipitation of the solid salt as soon as its saturation concentration has been exceeded, in consequence of the increased formation of barium oxalate.

Conversely the concentration of hydrogen ion is increased by adding hydrochloric acid to the saturated solution in contact with solid oxalate, whence

$$\frac{C_{Ba^{\cdot\cdot}} \cdot C_{C_2H_2O_4}}{C^2_{H^{\cdot}} \cdot C_{BaC_2O_4}} < k.$$

To reëstablish the equilibrium in this case the concentrations of barium ion and oxalic acid must increase, those of hydrogen ion and barium oxalate must decrease. Since solid barium oxalate is present, equilibrium can exist in the system only when $C_{BaC_2O_4}$ has a definite value corresponding with its saturation concentration (§ 30). Therefore solid oxalate must go into solution and the ultimate result of adding hydrochloric acid consists in a decrease of the amount of solid, which is readily perceptible.

In similar cases analogous considerations hold for the other salts, e. g. fluorides, phosphates, etc.

REACTIONS OF THE IONS OF THE THIRD GROUP COMPONENTS.

63. Aluminum. Aluminum forms only *trivalent cathions* in which, however, the *cathion characteristics are not at all pronounced,* as is evidenced by the *hydrolysis* (§ 71) of aluminum salts and by the *amphoteric* (§ 42) character of the hydroxide.[1]

The test solution contains $\frac{1}{80}$ formula weight of *alum,* $KAl(SO_4)_2 \cdot 12H_2O$, or *aluminum chloride* per liter. The latter salt is easily prepared by dissolving *aluminum* in hydrochloric acid.

Alkali hydroxides and *ammonia* precipitate *aluminum hydroxide.* Like arsenous acid (§ 42) it is soluble in *acids* and in an *excess* of *alkali hydroxide.* It is also *noticeably* soluble in an excess of ammonia, but to much less extent. From its alkaline solutions the hydroxide is precipitated by ammonium salts, according to § 55, and also by passing carbon dioxide into the solution. This action of carbon dioxide is due to the formation of carbonic acid ($H_2O + CO_2 = H_2CO_3$). While carbonic acid is extremely little dissociated, nevertheless its solutions contain enough hydrogen ion to effect a decrease of the concentration of hydroxyl ion in consequence of the reaction $H^{\cdot} + OH' = H_2O$. Hence they give rise to the condition under which the reaction:

$$AlO_3''' + 3H_2O \rightleftarrows Al(OH)_3 + 3OH' \qquad (1),$$

will proceed in the direction $\rightarrow$.

Aluminum ion also yields a precipitate of hydroxide with *alkali carbonate solution* and with barium carbonate in suspension. These carbonate solutions contain *hydroxyl ion* in consequence of the reaction:

$$\underbrace{2Na^{\cdot} + CO_3''}_{} + 2H_2O \rightleftarrows 2Na^{\cdot} + 2OH' + H_2CO_3 \qquad (2).$$

Hence they have the action of weak bases.

Ammonium sulphide likewise precipitates *aluminum hydroxide,* since it contains OH′ and moreover because *aluminum sulphide* is not stable in an *aqueous solution.*

Towards *alkali acetate* the ions of aluminum and of trivalent iron

[1] Thus aluminum hydroxide dissociates as follows:

$$Al(OH)_3 \rightleftarrows Al^{\cdots} + 3OH',$$

and

$$Al(OH)_3 \rightleftarrows AlO_3''' + 3H^{\cdot}.$$

behave alike (§ 56). On adding *alkali acetate* to neutral solutions[1] of these ions no visible change is apparent at first. When the solutions are respectively *boiled* or *warmed,* however, separation of *basic acetate* takes place, for instance, $Al(OH)_2C_2H_3O_2$. The reason for this difference in behavior at room temperature and at higher temperatures is by no means simple. The following reaction takes place on mixing a solution which contains aluminum ion with an alkali acetate solution:

$$Al^{\cdots} + 3C_2H_3O_2' + 2H_2O \rightleftarrows Al(OH)_2C_2H_3O_2 + 2C_2H_4O_2 \quad (3).$$

This reaction proceeds extremely slowly at ordinary temperature. Hence a precipitate will form only on long standing. By increasing the temperature the velocity of the reaction is increased and the equilibrium is also shifted in the sense that more basic acetate forms at higher temperatures. In order to precipitate aluminum ion as completely as possible from a solution by means of this reaction, the solution must be diluted with several times its own volume of water before it is heated.

Aluminum phosphate is soluble in *alkali hydroxide.* This is due to the fact that aluminum phosphate solution contains enough aluminum ion to form the ion AlO_3''' with hydroxyl ion, according to the equation:

$$Al^{\cdots} + 6OH' = AlO_3''' + 3H_2O \quad (4).$$

This equation assumes the following form with aluminum phosphate:

$$AlPO_4 + 6OH' \rightleftarrows AlO_3''' + PO_4''' + 3H_2O \quad (5).$$

Solid aluminum compounds yield difficultly fusible aluminum oxide when heated with sodium carbonate on charcoal. If this is moistened with *cobalt nitrate* solution and is then strongly heated, a blue infusible mass, Thenard's blue, results.

With many *organic anions aluminum ion* forms complex ions. Hence in the presence of these organic anions aluminum ion is only incompletely precipitated by ammonia (compare § 53). The means of *overcoming* this difficulty will be found in § 193 on the method of procedure when organic compounds are present.

64. Chromium. In its relations to other elements chromium shows greater diversity than aluminum. It forms blue *chromous*

[1] Acid solutions are neutralized with sodium carbonate according to § 56.

ion, $Cr^{\cdot\cdot}$, and *violet chromic ion*, $Cr^{\cdot\cdot\cdot}$, corresponding to *chromous* and *chromic* salts. The properties of the latter are of greater analytical importance. Besides these chromium forms still other compounds in which it is trivalent, hexavalent and heptavalent. Chromium has no longer the character of a *cathion* in solutions of these latter substances. It forms different anions with oxygen, viz. CrO_3''' in which chromium is trivalent, CrO_4'' and Cr_2O_7'' in which it is hexavalent [and CrO_4'? in which it is heptavalent. The latter compounds, perchromates, are very unstable, existing only in acid solutions at low temperatures.] CrO_4'' and Cr_2O_7'' are the most important. Their properties will be taken up in detail in part II, in which anions are considered (§§ 119 and 120).

Chromous ion is so *unstable* that it *decomposes water* by reducing *hydrogen ion* that is present in extremely small concentration, to hydrogen (see § 53), being itself oxidized to *chromic ion:*

$$2Cr^{\cdot\cdot} + 2H^{\cdot}OH' = 2Cr^{\cdot\cdot\cdot} + 2OH' + H_2 \qquad (1).$$

With *hydroxyl ion* the solution of a *chromous salt* gives a precipitate of yellow *chromous hydroxide,* which quickly oxidizes in contact with air.

A *chromous salt solution* may be prepared by *reducing* the *corresponding chromic salt solution* with *zinc.*

When solutions which contain *chromic ion* are kept for any considerable length of time they exhibit striking *changes of color,* varying from *green* to *violet.* These color changes are related to changes in the condition of the chromic salt (consult Principles, p. *603*).

To prepare the test solution chrome alum solution is precipitated with ammonia and the chromic hydroxide, after washing, is dissolved in hydrochloric acid. In its behavior towards other substances this solution of chromic ion is closely related to an aluminum salt solution. With respect to sodium acetate a *difference* is noted. A solution containing only *chromic ion* yields *no precipitate.* If, however, *aluminum-* or *ferric ion* is *also* present, *basic chromic acetate* is precipitated with basic aluminum- or ferric acetate. Under certain conditions, especially when a large quantity of chromic ion is present, the *precipitation* of *basic aluminum-* or *ferric acetate* may be *incomplete.* The phenomena call to mind what has been mentioned in § 57 concerning *chromic hydroxide.*

To *separate chromic ion* from the closely *related* ions use is made of its ability to form ions of higher valence. This takes place, for example, on fusing the hydroxide with sodium carbonate and sodium nitrate (§ 57). Omitting certain complications the reaction may be formulated as follows:

$$2Cr(OH)_3 + 2Na_2CO_3 + 3NaNO_3 = 2Na_2CrO_4 + 2CO_2 + 3H_2O + 3NaNO_2 \quad (2).$$

In *solutions chromic ion* may be transformed into *chromate ion* by means of *hydrogen peroxide* (compare § 55), *bromine* or *lead peroxide.* The reaction is represented by the *general* equations:

$$CrO_3''' + H_2O + 3(^{\cdot}) \rightleftarrows CrO_4'' + 2H^{\cdot} \quad (3a),$$

or

$$CrO_3''' + H_2O \rightleftarrows CrO_4'' + 2H^{\cdot} + 3(') \quad (3b).$$

These equations mean that a cathion which reacts with the ion CrO_3''' must give up three positive charges, while an anion or substance that forms anions must take up three negative charges in order to oxidize the chromite ion, CrO_3''', that exists in alkaline solutions, into chromate ion, CrO_4''. Bromine is a representative of the latter class, as is shown by the equation:

$$CrO_3''' + 3Br + H_2O = CrO_4'' + 3Br' + 2H^{\cdot} \quad (4).$$

When heated with sodium carbonate on charcoal *solid chromium compounds* form a *green infusible* oxide.

65. Colored bead tests. In addition to the blowpipe tests (§ 39) *colored bead* tests are very widely used for a rapid determination of the components of a given mixture. The beads are formed of molten sodium metaphosphate or sodium borate, which are fused in a loop of platinum wire. They are prepared by heating the platinum loop and then dipping it in *phosphor salt* (sodium ammonium hydrophosphate) or in *borax* until enough salt is collected to form a drop filling out the loop when fused. The salts are then heated in a *small flame. Phosphor salt,* $NaNH_4HPO_4$, loses water and ammonia, yielding *sodium metaphosphate,* $NaPO_3$; *borax* loses its water of crystallization under these conditions. Sodium metaphosphate and anhydrous borax have the property of dissolving different salts, assuming colors which are characteristic of the metallic com-

ponents dissolved. These colors vary also for the different oxidation forms of a metal. Hence by using the *reducing* and *oxidizing* blowpipe flames to heat the beads distinctions are obtained which lead to the detection of the metals (compare § 173).

To dissolve the salts the *heated bead* is brought in contact with a little of the solid material. On further heating the salt becomes uniformly distributed through the bead.

66. Double salts. Aluminum and chromic salts have the property of entering into combination with other salts in simple stöchiometric proportions. The resulting compounds are called *double salts.* To this class belong, in particular, the alums, which have the general formula $M_I M_{III} (SO_4)_2.12H_2O$. Their composition is quite simple, although many different metallic components may be present. In the formula M_I stands for a monovalent metal, M_{III} for a trivalent. Now another class of salts besides double salts are built up of simpler components united in simple stöchiometric proportions. They are called complex salts because they yield complex ions, for instance, $KAg(CN)_2$ (§ 13). A knowledge of the general properties of the two groups of salts is therefore very important.

In double salts the solid components are united in simple stöchiometric proportions and the *properties of their solutions are made up of the sum of the properties of the simple ions into which the components characteristically dissociate.* Thus ammonium chrome alum solutions exhibit the properties of the ions $NH_4^{\cdot}$, $Cr^{\cdot\cdot\cdot}$ and SO_4''. *Complex salts,* however, are characterized by the *variations* they show with respect to this last point. *Their solutions have as a rule other properties than would be expected from their composition.* Nevertheless they always show, to a very slight extent at least, the properties of the simple ionic components (compare § 13).

It should also be noted that the conception of a double salt brings to mind characteristics with respect to the composition in the *solid* state, while in complex salts certain properties exhibited by their *solutions* are emphasized (Principles, p. *534*).

67. Iron. Like chromium *iron* also forms *divalent* and *trivalent, ferrous* and *ferric* ions. Since ferrous ion is much more stable than chromous ion, its properties will be considered in more detail. While ferrate ion, corresponding to chromate ion, is not wholly unknown, still it has very little analytical importance and so needs only this brief reference. With cyanogen iron forms *complex ions* of very pronounced *anion character.* These will be dealt with in detail in §§ 147 and 148.

68. Ferrous ion. A *solution entirely free from ferric ion* is an impossibility. By dissolving pure metallic iron in dilute hydrochloric or sulphuric acid a solution is obtained which contains minimal amounts of ferric ion. Owing to impurities always present in

commercial irons the gas which is evolved in this reaction is composed of hydrogen and various hydrogen compounds (hydrocarbons, etc.). Ferrous ion is almost colorless and the above solutions will have a pale green color. Stock solutions of ferrous ion usually have a darker color, due to the presence of ferric ion which is formed from ferrous ion in neutral solutions by the action of air. Hence the properties of such stock solutions will vary more or less from those of ferrous ion. They are therefore not suited for the following tests.

Alkali hydroxide gives a *white* precipitate of *ferrous hydroxide,* which quickly changes color and passes into ferric hydroxide in contact with air:

$$2Fe(OH)_2 + H_2O + O = 2Fe(OH)_3 \qquad (1).$$

Ammonia also precipitates the hydroxide, but the precipitation is always *incomplete.* This is due to the formation of ammonium salts, which may prevent the precipitation entirely in the case of the hydroxides of cobalt, nickel, manganese, zinc and magnesium (see § 78).

Alkali carbonates yield a white precipitate of *ferrous carbonate.* The precipitate dissolves on passing carbon dioxide into the solution:

$$\underline{FeCO_3} + CO_2 + H_2O \rightleftarrows Fe^{\cdot\cdot} + 2HCO_3' \qquad (2).$$

A separation of ferric hydroxide takes place when air comes in contact with the solution.

Hydrogen sulphide forms no precipitate in an acid solution of ferrous ion. With a neutral solution it may yield an incomplete precipitation of *black* ferrous sulphide, which will vanish on addition of acetic acid. This difference in the behavior of ferrous sulphide, as compared with that of the sulphides already mentioned in § 28, is due to the fact that the solution of ferrous sulphide contains enough sulphide ion to form a considerable amount of hydrogen sulphide with the hydrogen ion from acetic acid:

$$2H^{\cdot} + FeS \rightleftarrows H_2S + Fe^{\cdot\cdot} \qquad (3).$$

Thus conditions favoring the dissolving of more ferrous sulphide result.

On passing hydrogen sulphide into neutral ferrous solutions a transformation does take place, however, in the sense ←, i. e. a pre-

cipitate forms. This shows that the dissolving of ferrous sulphide does not occur beyond a certain lower limit in the concentration of hydrogen ion. By adding sodium acetate the precipitation becomes more complete. According to equation (*1*), § 38, the addition of sodium acetate decreases the concentration of hydrogen ion (see also § 89). Hence the transformation ← can take place to greater extent under these conditions.

With *alkali cyanide* ferrous ion gives a red precipitate of *ferrous cyanide* (*4*), which dissolves in an excess of alkali cyanide (*5*) to form yellow ferrocyanide ion (§ 147):

$$Fe^{\cdot\cdot} + 2CN' = \underline{Fe(CN)_2} \qquad (4),$$

and

$$\underline{Fe(CN)_2} + 4CN' = Fe(CN)_6'''' \qquad (5).$$

Occasionally a precipitation of basic salts accompanies the above-mentioned change of color which ferrous salt solutions undergo in contact with air. By adding an acid the ferrous salt solution becomes more stable. The equation:

$$2Fe^{\cdot\cdot} + 5H_2O + O = \underline{2Fe(OH)_3} + 4H^{\cdot} \qquad (6),$$

offers a possible explanation for the fact that the presence of hydrogen ion tends to prevent the formation of ferric compounds in solutions of ferrous ion, for one of the products of the reaction has been initially added to the system and hence the reaction leading to the formation of this substance is retarded. It is worthy of note in this connection that ferric ion is a stronger oxidizing agent in acid solutions than in neutral, i. e. it is less stable in acid solutions.

69. Ferric ion. The reddish color of ferric salt solutions is not characteristic of ferric ion, but is due to soluble ferric hydroxide resulting from hydrolysis (§ 71). Hence anything that lessens the hydrolysis will cause the color to fade.

Assuming that hydrolysis[1] takes place according to the equation:

$$Fe^{\cdot\cdot\cdot} + 3H^{\cdot}OH' \rightleftarrows Fe(OH)_3 + 3H^{\cdot} \qquad (1),$$

it is evident that an increase in the concentration of the hydrogen ion will lessen the hydrolysis. As a matter of fact acidifying a neutral ferric salt solution changes its color from red to yellow and sometimes even renders the solution colorless. The same color change, however, is not always produced by different acids. For

[1] It would not introduce any essential change to assume that an ion containing hydroxyl, e. g. $Fe(OH)_2^{\cdot}$ or $Fe(OH)^{\cdot\cdot}$, is formed and that this has a characteristic red color.

example, *hydrochloric acid* gives a yellowish *red solution,* the color being due to *undissociated ferric chloride.* Analogous cases occur with other acids.

An increase of hydrolysis is effected by heating the solution and a deeper color results.

Heating *ferrous ion* with nitric acid transforms it into *ferric ion* (compare § **51**).

Alkali hydroxide and *ammonia* precipitate reddish brown ferric hydroxide, $Fe(OH)_3$. According to *Haber* ferric hydroxide is soluble in alkali hydroxide, but to so slight an extent, as compared with aluminum- and chromium hydroxides, that the separation according to § **55** is practically complete.

Alkali carbonate precipitates *basic carbonate,* which is decomposed by water—very quickly at higher temperatures—into ferric hydroxide and carbonic acid.

Ferric ion is also precipitated as *hydroxide* by *barium carbonate,* which is used in form of a suspension on account of its insolubility.

With *sodium phosphate* (Na_2HPO_4) ferric ion gives a white precipitate of *ferric phosphate, insoluble in acetic acid, soluble* in *stronger* acids, as hydrochloric or nitric. Ferric phosphate is very slightly soluble in water. Hence the saturated solution contains so little phosphate ion that the reaction:

$$\underline{FePO_4} + 3H^{\cdot} \rightleftarrows Fe^{\cdot\cdot\cdot} + H_3PO_4 \qquad (2),$$

does not take place to a noticeable extent with acetic acid. The solvent action of hydrochloric acid (compare § **128**) is due to the fact that a hydrochloric acid solution contains approximately 80 times as much hydrogen ion per unit volume as an acetic acid solution of equivalent concentration.

Towards *sodium acetate* ferric ion behaves like aluminum ion (§ **63**). In order to precipitate *ferric ion completely* by the basic acetate method, the solution must be diluted with *water several* fold (10 cc. to 100 cc.) after the addition of sodium acetate, and it is then *boiled* for several minutes. Strong acids must be absent, since they prevent the complete precipitation of ferric ion owing to the fact that they dissolve basic ferric acetate.

Potassium sulphocyanate gives a dark blood-red solution with ferric ion. The color is due to undissociated *ferric sulphocyanate,* as is shown by the fact that ether dissolves the colored component, yielding a red solution.

This conclusion is based upon the general experience that in solvents such as ether, and in vapors, ions are either *not at all present or else only in extremely minute quantities,* and that, therefore, only the undissociated part is taken up by a solvent of this nature.

Nevertheless ions are present in ferric sulphocyanate solutions to a noticeable degree. Many reagents which react with very small amounts of ferric- or sulphocyanate ion, and therefore afford delicate tests for them, will cause the red color of ferric sulphocyanate solutions to fade, or will prevent its formation. *Ammonia, sodium acetate* and *sodium sulphate* react with the ferric ion. In the latter case ferric sulphate forms, which is less dissociated than ferric sulphocyanate. *Mercuric chloride* also decolorizes the red solution, because sulphocyanate ion and mercuric ion form slightly dissociated mercuric sulphocyanate:

$$2CNS' + Hg^{\cdot\cdot} = Hg(CNS)_2.$$

Mercuric nitrate has the same effect as the chloride, but in this case a precipitate of mercuric sulphocyanate forms (compare § 34).

These points must be taken into consideration on testing for ferric ion. In case substances are present which hinder the test, they must be removed or else rendered inactive. Thus hydrochloric acid nullifies the effect of acetate, for the system $HCl + NaC_2H_3O_2$ goes more or less completely into $NaCl + C_2H_4O_2$ and the latter contains much less acetate ion. This limits the formation of ferric acetate and the red color of ferric sulphocyanate will again appear. A large amount of hydrochloric acid, however, weakens the sulphocyanate test itself.

Hydrogen sulphide reduces ferric ion to ferrous ion (§ 51):

$$2Fe^{\cdot\cdot\cdot} + H_2S = 2Fe^{\cdot\cdot} + \underline{S} + 2H^{\cdot} \qquad (3).$$

Ammonium sulphide yields a *black precipitate* composed of ferrous sulphide and sulphur. Probably ferric sulphide forms at first and then decomposes rapidly into *ferrous sulphide* and *sulphur:*

$$Fe_2S_3 = 2FeS + S \qquad (4).$$

Alkali cyanide gives a brown red precipitate of *ferric cyanide,* which dissolves in an excess of the precipitant to form the complex ferricyanide ion, $Fe(CN)_6'''$ (§ 148).

When heated in the reducing flame on charcoal, dry mixtures of *iron salts* with sodium carbonate yield *metallic iron,* which is recognized by its magnetic properties.

The colors imparted to phosphor salt and borax beads depend upon the amount of iron salt present. In the *oxidizing flame* a *yellow* to *brown* bead is formed, the color becoming lighter as the bead cools. A *greenish* bead is formed in the *reducing* flame.

70. The complex ions of iron. Besides the complex cyanogen ions iron also forms complexes with the anions of many organic acids (compare aluminum, § 63). These organic complexes have strikingly variable properties. Their analytical detection will be considered later (§ 193).

Ferrous- and ferric oxalate, which may be prepared by adding alkali oxalate to ferrous and ferric solutions respectively, are soluble in an excess of alkali oxalate, owing to the formation of the complex ions $Fe(C_2O_4)_2''$ and $Fe(C_2O_4)_3'''$ (consult Principles, p. 582). The reactions are:

$$\underline{FeC_2O_4} + C_2O_4'' = Fe(C_2O_4)_2' \qquad (5),$$

and

$$\underline{Fe_2(C_2O_4)_3} + 3C_2O_4'' = 2Fe(C_2O_4)_3''' \qquad (6).$$

71. Hydrolysis. As has already been pointed out, a salt is very often decomposed by water (consult §§ 22, 37, 69), while in some instances no indications of such a decomposition are obtained. The question therefore arises: what connection has this phenomenon with the composition of the salt? It is seen from the equations for the reactions above quoted that a *hydroxyl compound* and *hydrogen ion* are formed.

The following example will show that an analogous phenomenon is met with in the case of another type of salt. A solution of pure[1] potassium cyanide does not react like a *neutral* salt towards litmus (compare NaCl), but it has a distinct alkaline reaction, resembling that of dilute potassium hydroxide.

It has been found that this is due to a reaction between potassium cyanide and water, according to the equation:

$$K^{\cdot} + CN' + H^{\cdot}OH' \rightleftarrows K^{\cdot} + OH' + H^{\cdot}CN' \qquad (1).$$ [2]

Although the salts KCl, KNO_3, NaCl, $BaCl_2$ have a composition analogous to that of KCN, nevertheless their aqueous solutions react *neutral* to litmus. The striking difference between potassium chloride and potassium cyanide, for example, is closely related to the

[1] In this case particular stress is laid on the *purity* of the salt, to make sure that the observed alkaline reaction is not due to an impurity.

[2] Compare § 50 for the significance of the formulae $H^{\cdot}CN'$ and $NH_4^{\cdot}OH'$. In order to indicate the preponderance of hydrogen ion, i. e. acid reaction, the dissociation of ammonium hydroxide has been assumed to be negligible in the latter case.

properties of the acids, hydrochloric and hydrocyanic. The specific acid properties of the latter (§ 146) are much weaker than those of hydrochloric acid, which indicates that hydrocyanic acid is much less dissociated than hydrochloric acid. In fact it belongs to the class of *weak acids* (§ 49). Therefore when cyanide ions are produced by dissolving a cyanide, there is a tendency to form undissociated hydrocyanic acid. This results from a reaction with water, as indicated by equation (*1*).

On the other hand a solution of *ammonium chloride,* not too dilute, will color blue litmus red, in consequence of the reaction:

$$NH_4^{\cdot} + Cl' + H^{\cdot}OH' \rightleftarrows NH_4^{\cdot}OH' + H^{\cdot} + Cl' \qquad (2).$$

On comparing the salts KCl and NH_4Cl it is seen that the *hydroxyl compounds* formed by their cathions, KOH and NH_4OH, are bases of very different strength, ammonium hydroxide being the weaker.

The study of these cases leads to the conclusion that a *salt* is decomposed by water (hydrolyzed), irrespective of its electrolytic dissociation, when the *hydroxyl compound* of the *cathion* is a *weak base,* or the *hydrogen compound* of the *anion* a *weak acid.* In the *former* case the *solution* has an acid reaction, in the *latter* an *alkaline.*

Conversely the character of the hydrogen and hydroxyl compounds of the respective anions and cathions may be estimated from the behavior of a given salt solution towards litmus.

A special case is met with when the *acid* and *base* from which a salt is prepared by mixing equivalent quantities of each and evaporating the solution, are *both weak. Hydrolysis* is *very pronounced* on dissolving such a salt, i. e. a greater fraction of the salt reacts with water, because both *undissociated acid* and *undissociated base* form from the *ions* of the *salt* and *water. Aluminum acetate* affords an example of this *special* case. The reaction exerted by such a salt solution towards litmus will depend upon the relative strengths of the acid and base [and furthermore upon the concentration of the solution]. If the former is stronger than the latter, the salt solution will redden blue litmus paper, and in the contrary case it will turn red litmus blue. When the acid and base have the same strength, the solution reacts neutral, although it contains appreciable quantities of free acid and free base.

Strictly speaking, all salts react in this way with water. For instance, potassium chloride solution contains the ions $K^{\cdot}$, Cl', $H^{\cdot}$, OH', the non-dissociated components KCl and H_2O, and also finite, but extremely small, quantities of KOH and HCl which are formed by combination of the ions. Similar reactions hold for the solutions of all *neutral* salts. The solution of potassium chloride represents the same condition that would result if equivalent quantities of hydrochloric acid and potassium hydroxide are mixed. In this latter case hydrogen and hydroxyl ions combine to form water up to the limits of its dissociation (§ 50).

That litmus does not detect the hydrogen and hydroxyl ions present is due to the very small amounts and also to the fact that hydrochloric acid and potassium hydroxide are equally strong.

The mixing of equivalent quantities of acid and base to prepare a salt is known as *neutralization.* The reaction is represented by the general equation:

$$H^{\cdot} + A' + M^{\cdot} + OH' \rightleftarrows H_2O + M^{\cdot} + A' \qquad (3).$$

Exact measurements have shown that the two methods of preparing a salt solution, i. e. by dissolving solid salt or by neutralization, give rise to the same condition. Hence it is evident from the variable behavior of different salt solutions towards litmus that the reaction of neutralization is only more or less complete in different cases. With salts of the type of alkali chlorides or nitrates the reaction (3) → is practically complete. It is least complete in cases where both the acid, HA, and the base, MOH, are weak electrolytes. Between these extremes lie the cases where either acid or base is weak. In this sense hydrolysis is the inverse of neutralization.

GROUP IV.

72. **The fourth group reagent is ammonium sulphide,** which precipitates the ions of *cobalt, nickel, manganese* and *zinc.* The precipitation of these ions with ammonia in the third group is prevented by the addition of *ammonium salts.* Manganese and cobalt, however, are not always entirely retained in solution by this means and hence may sometimes be found in the third group precipitate. Ammonia precipitates manganese in the presence of ammonium salts owing to the formation of manganic hydroxide through the action

of atmospheric oxygen on manganous hydroxide. The latter is contained in the ammoniacal solution when manganese is present:

$$2Mn(OH)_2 + O + H_2O = \underline{2Mn(OH)_3}.$$

Since this reaction takes place slowly the ammoniacal *filtrate from the third group* will often yield a brown precipitate on *standing.* This may be analyzed with the third group precipitate or may be dissolved in warm hydrochloric acid and added to the *filtrate from the third group.* In the majority of cases, however, manganese and cobalt ions are precipitated almost entirely in the fourth group. Still, to avoid the possibility of error, it is advisable to test for them in the third group precipitate (compare § 56).

73. **Precipitation with ammonium sulphide.** A *preliminary test* for the ions of this group should always be made with a small *portion* of the *filtrate from the third group.* In case no fourth group constituents are found, a good deal of time will be saved, owing to the necessity of removing ammonium sulphide before testing for the remaining groups. *Zinc sulphide* is frequently *missed,* particularly in the absence of the other ions of this group, for it often precipitates in the form of a film which is not easily seen. The precipitate coagulates when heated, however, and then it is more readily perceptible.

Zinc ion may also be precipitated in the second group if the solution into which hydrogen sulphide is passed has been only slightly acidulated. In this case the nitric acid solution of the sulphides (§ 22) is treated with ammonium sulphide after lead ion has been removed by dilute sulphuric acid. The precipitate so obtained is washed, dissolved in dilute hydrochloric acid and tested with sodium hydroxide and hydrogen sulphide according to § 75.

On *filtering* the *precipitate* of this *group* it is occasionally observed that the *filtrate* and also the hydrogen sulphide *solution used for washing run through brown.* This always indicates the presence of *nickel sulphide* in the precipitate. To *overcome* the *difficulty* some acetic acid is added to the brown solution. It forms ammonium acetate. Care must be taken not to add an excess of acid (the solution must still react alkaline). Coagulation takes place after a few minutes' warming and the solution may then be filtered without any further difficulty (through the same filter). The diffi-

culty can be avoided in a still more effective manner if freshly prepared ammonium sulphide is used, or better, if hydrogen sulphide is passed into the warm ammoniacal filtrate from the third group.

Hydrogen sulphide solution is used for *washing.*

74. **Theory of the precipitation of the fourth group ions by ammonium sulphide.** The fourth group ions do not precipitate in the second group owing to the greater solubility of their sulphides and also to the limited amount of sulphide ion that is present in the solution. The limitation is due to the slight solubility of hydrogen sulphide on the one hand, and on the other hand to the action of the free acid always present (§ 89). It is customary to subdivide the metals into two groups, the ions of which are precipitated respectively by hydrogen sulphide and by ammonium sulphide. The subdivision is specially conditioned by the concentration of sulphide ions which can exist in an *acid solution of hydrogen sulphide, saturated under atmospheric pressure.* Since the concentration of sulphide ions varies with the total concentration of hydrogen sulphide, and the latter varies in its turn with the gas pressure according to Henry's law,[1] it follows that a change of pressure will alter the grouping of the metallic sulphides, other conditions remaining the same. Thus zinc sulphide would precipitate from an acid solution in the second group by increasing the concentration of hydrogen sulphide, and therefore that of the sulphide ion. Conversely by passing hydrogen sulphide into a solution under very low pressure its concentration is lessened and antimony and tin sulphides do not precipitate or are only incompletely precipitated under these conditions.

Another means of altering the groups without changing the concentration of hydrogen sulphide consists in *varying* the *concentration* of *hydrogen ion* (see § 28). In this case the secondary *decrease* of the *concentration* of *sulphide ion* (§ 39) adds its effect to that of the hydrogen ion.

For practical reasons it is advisable to retain the customary grouping. Since the effect of an increase of pressure is not very pronounced and the method presents manipulative difficulties, another means is employed to increase the concentration of sulphide ion. It takes advantage of the fact that salts, even of weak acids, are dissociated to a considerable extent, and therefore ammonium sulphide is used.

75. **Separation of the sulphides.** The precipitate is first treated on the filter with *very dilute hydrochloric acid* (about one volume of reagent acid to ten volumes of water). This dissolves zinc and manganous sulphides and leaves nickel and cobalt sulphides behind on the filter. Any dissolved hydrogen sulphide is expelled by heat-

[1] According to Henry's law the saturation concentration of a gas in a liquid is proportional to the pressure under which the gas stands. Since the volume of a given quantity of gas is inversely proportional to the pressure, and the quantity of gas filling a given volume is directly proportional to the pressure, the law can be stated in another form, viz. the same volume of a given gas will always saturate equal amounts of a given liquid, independent of the pressure (compare Principles, p. *271*).

ing the hydrochloric acid solution, and then an excess of alkali hydroxide is added. *Manganous hydroxide* separates as a *white precipitate,* which rapidly goes over into *brown manganic hydroxide* in contact with air. *White zinc hydroxide* also separates at first, but the precipitate redissolves in the *excess* of *alkali hydroxide. Hydrogen sulphide* gives a white precipitate of *zinc sulphide* with the *alkaline filtrate,* in case zinc is present. The precipitate is not easily seen.

$$ZnO_2'' + H_2S = \underline{ZnS} + 2OH'.$$

Heating the solution in contact with the precipitate of zinc sulphide effects a coagulation and thus renders the precipitate more readily visible. Quite often a dark precipitate is observed. The color is generally due to ferrous sulphide that is present as an impurity. Since ferrous sulphide is much more soluble in acetic acid than zinc sulphide, acetic acid is added until the dark color just disappears on shaking the solution. The *white flocculent* precipitate of zinc sulphide is then easily recognized. Inexperienced observers, however, frequently mistake the white milky precipitation of *sulphur* which results from acidifying ammonium sulphide solutions (compare § 24, (*3*)), for a precipitate of zinc sulphide. A little practice will enable the observer to distinguish with certainty the flocculent zinc sulphide from a milky sulphur precipitate.

The *black residue* left after extracting the fourth group precipitate with dilute hydrochloric acid is dissolved in a hot *mixture* of *hydrochloric* and *nitric* acids. In order to remove most of the *excess* of *acid* the solution is next evaporated nearly to dryness on the *water-bath* or over an *Ostwald burner.* The *residue* is then *dissolved* in *water* and alkali carbonate is added *drop by drop* until the solution becomes permanently clouded. This removes any excess of strong acids remaining. *Acetic acid* is now added until the cloudy precipitate disappears and the *solution has an acid reaction.* Finally this liquid is mixed with its own volume of *concentrated potassium nitrite solution.* In case cobalt ion is present a *yellow precipitate* of *potassium cobaltinitrite,* $K_3Co(NO_2)_6$ will form after a short time or at any rate a change of color will take place from reddish to yellowish (compare § 83). The complete precipitation takes a longer time and for this reason it is advisable to let the mixture stand over night. The filtrate should be again tested with potas-

sium nitrite for cobalt, in order to ensure its complete removal. Then the test for nickel is carried out (see below).

For detecting cobalt use may also be made of the behavior of cobaltous ion towards sulphocyanate ion (§ 83).

After cobalt has been completely removed, *alkali hydroxide* is added to the filtrate. An *apple green precipitate* of *nickelous hydroxide* forms if *nickelous ion* was present. A hot solution favors the coagulation of the precipitate. In case the color leaves the presence of nickel in doubt, the precipitate is filtered and tested further in the phosphor salt bead for nickel.

Instead of first removing cobalt ion with potassium nitrite and then detecting nickel in the filtrate that is thus freed from cobalt, the solution containing cobalt and nickel may be tested with potassium cyanide, potassium hydroxide and bromine (§ 84). By this means cobaltous ion is transformed into a complex cobalticyanide ion which gives no distinctive reaction with alkali hydroxide and bromine, and moreover does not interfere with their reaction on nickelous cyanide ion.

[Cobalt and nickel are distinguished simultaneously by the following 'persulphate' method[1] and moreover small amounts of each are readily detected. The method depends upon a difference in the behavior of their ammoniacal solutions towards persulphates. To the solution of cobalt and nickel sulphides a little ammonium chloride (2 cc.), and a considerable excess of ammonia are added. The solution is diluted with water to at least 25 cc. and solid potassium persulphate, $K_2S_2O_8$, added (0.3 g.). The mixture is heated with constant stirring until the persulphate is all dissolved (note that potassium persulphate is only slightly soluble), and most of the ammonia expelled. If a precipitate forms at this point it is dissolved in hydrochloric acid and the above operation is repeated, omitting the addition of ammonium chloride. Under these conditions the nickel ammonia complex retains its blue color, but the cobalt complex changes color (to red or pink). Oxidation is ensured by adding more persulphate and boiling. Then the solution is cooled under the water tap and alkali hydroxide is added. A dark brown or black precipitate of nickelic hydroxide, $Ni(OH)_3$, forms slowly on shaking. It is filtered after standing for fifteen minutes. If

[1] This method has been worked out by Mr. H. H. Willard of the University of Michigan.

cobalt is present the filtrate will be pink or red. The amount of alkali hydroxide needed to precipitate nickel depends upon the quantity of ammonium chloride present. When the solution is warm cobaltic hydroxide also precipitates with alkali hydroxide, but none forms in a cold solution.

According to Tschugaeff[1] α-dimethylglyoxime affords an extremely delicate test for nickel. The nickel solution is rendered alkaline with alkali hydroxide, ammonia or sodium acetate, and then some of the powdered dioxime is added. A red precipitate forms on boiling. Cobalt yields no precipitate.]

76. Cobalt- and nickel sulphide. To separate these sulphides from those of manganese and zinc advantage is taken of their insolubility[2] in dilute hydrochloric acid. Comparing this phenomenon with the fact that hydrogen sulphide does not precipitate cobalt and nickel sulphides from acidulated solutions, it becomes evident that the general equation:

$$M^{\cdot\cdot} + H_2S \rightleftarrows \underline{MS} + 2H^{\cdot},$$

does not satisfy the present case. For the non-precipitation from acid solutions would indicate solubility in acids, and their actual insolubility in dilute hydrochloric acid leads to the conclusion that they should be precipitated by hydrogen sulphide. The observed variance between the actual relations and the demands of the theory may be explained by assuming that precipitated nickel and cobalt sulphides have certain properties which other sulphides do not possess. Apparently these properties are due to a change which the solid sulphide undergoes. It may be of physical or chemical nature. As a consequence less soluble sulphides are formed from more soluble modifications. Our present knowledge of the properties of these sulphides is not extensive enough, however, to clear up these relations. The above-mentioned phenomenon that nickel sulphide occasionally does not separate in the solid form, even when the conditions are favorable to its precipitation, seems to support the assumption of a physical change which leads to the formation of larger particles having a smaller total surface.

77. Colloids. In a number of cases the artifice of heating certain precipitates (hydroxides or sulphides) has been employed to render them more readily visible. There are common reasons for doing this in the different cases. The solubility relations of these substances are influenced not only by temperature, but also by their state of subdivision to a considerable extent.

[1] L. Tschugaeff, Ber. d. d. chem. Gesellschaft, 38, *2520*, 1905. This method has been recently tested by Mr. H. H. Willard. Apparently it affords a quantitative precipitation of nickel.

[2] As a matter of fact a certain amount is always dissolved, particularly on long standing in contact with hydrochloric acid. It is therefore always advisable to further test the precipitate obtained by adding an excess of alkali hydroxide to the hydrochloric acid solution, for manganese hydroxide by fusing it with sodium carbonate and potassium nitrate (§ 56), in order to avoid mistaking hydroxides of cobalt or nickel for manganese hydroxide.

Thus at a given temperature the solubility varies with the fineness of the particles. The more finely divided they are, the greater will be their total surface and also the solubility. Hence practical methods of separating such precipitates aim to give them coarser particles. This is effected by *heating* or by the *addition* of *certain other substances.*

Concerning the action of the latter it is noted that chemical transformations, which often produce a regular change in the solubility of a substance, for instance, of oxalates in acids (compare § 62), do not come into question, or more correctly speaking, very seldom come into question. In the majority of cases it has to do with effects peculiar to salts and acids, independent of their chemical nature. As yet these are not satisfactorily explained. They apply to the separation and also to the dissolving of solids. According to recent investigations a distinct relation exists between the effects produced and the valence of the ions. Compare Ostwald, Scientific Foundations of Analytical Chemistry, Translation by McGowan, p. *25*.

In washing such precipitates, for instance, arsenic trisulphide, the wash water runs through perfectly clear, but as soon as it comes in contact with the filtrate, which contains salts, a cloudiness is produced. The salts precipitate the colloid dissolved in the water. Substances which undergo so considerable a change of condition in solutions through increase of temperature, and particularly by means of certain added substances, are called *colloids.* They form *colloidal* solutions. The means of overcoming their disturbing effects consists in adding salts to the water used for washing. In selecting salts to be added, those must be chosen which do not interfere in any way with the analysis. Substances which are stable in the amorphous condition are especially liable to form colloidal solutions. Another particular characteristic of colloids is their inability to diffuse through animal membranes (consult Principles, p. *421*).

Crystalloids form a counterpart to colloids. They also show a relation between the solubility and the state of subdivision.[1] The striking influence of salts and acids upon the solubility is not observed, however, apart from certain regular effects of another kind, which are mentioned later (§ 90).

Reactions of Fourth Group Ions.

78. Manganese shows still greater diversity than chromium in its combining ratios. It exists in di-, tri-, tetra-, hexa- and heptavalent forms (see § 167). *For a beginner the knowledge of the properties of manganous ion and its relations to the other oxidation forms is sufficient.* The highest oxidation forms comprise only *oxygen anions,* the lowest only *elementary cathions,* while the intermediary form exists as the *tetravalent* cathion, $Mn^{····}$, and an *oxygen anion* (probably MnO_3''). [From analogy an ion, MnO_2', derived from trivalent manganese, would also be expected.]

[1] A very striking example of the relations between the properties of a salt and its state of subdivision, has been mentioned by Ostwald. It has been shown in the case of mercuric oxide that the yellow modification consists of smaller particles than the red. It is also more readily soluble. Consult § 96.

The chloride ($MnCl_2.4H_2O$) or the sulphate ($MnSO_4.4H_2O$) is used for the test solutions which contain $\frac{1}{20}$ formula weight of the salt per liter. Manganous sulphate forms several hydrates, hence more water of crystallization may have to be taken into consideration in some cases. Solutions of manganous ion are pale pink.

Alkali hydroxide precipitates white *manganous hydroxide*, which quickly becomes brown by taking up oxygen from air.

Ammonia gives only an *incomplete* precipitation of *hydroxide* because of the low concentration of hydroxyl ion in ammoniacal solutions (see § **49**). The reaction:

$$Mn^{\cdot\cdot} + 2OH' = \underline{Mn(OH)_2} \qquad (1a),$$

does not proceed far enough to leave only a very little manganous ion in the solution. The transformation is, moreover, limited in consequence of the resulting ammonium ion, the concentration of which may become very considerable in the solution:

$$Mn^{\cdot\cdot} + 2NH_4^{\cdot}OH' = \underline{Mn(OH)_2} + 2NH_4^{\cdot} \qquad (1b).$$

Since $NH_4^{\cdot}$ is a dissociation product of ammonia, according to:

$$NH_3 + H_2O \rightleftarrows NH_4^{\cdot} + OH' \qquad (2),$$

it lessens the dissociation (§ **89**), and hence limits the *formation* of *hydroxyl ion*, which is necessary for further precipitation of manganous ion according to (*1a*). Thus the incompleteness of the precipitation is due not only to the *original slight* dissociation of the dissolved ammonia, but also to a *secondary decrease* of the *latter* through the *increase* in the *concentration* of *ammonium ion*. If this is very much increased by dissolving solid ammonium salt in the ammonia, no precipitate forms unless the solution has stood for some time in contact with air (§ **72**). The addition of ammonium salts is thus essential to keep fourth group ions from precipitating in the third group.

Alkali carbonate yields a white precipitate of manganous carbonate.

Sodium phosphate and *ammonium chloride* produce a white precipitate of *manganous ammonium phosphate,* which passes over into pink scales on long heating. The precipitate is easily soluble in acids. Since hydrogen ion results from the reaction:

$$HPO_4'' + NH_4^{\cdot} + Mn^{\cdot\cdot} \rightleftarrows \underline{MnNH_4PO_4} + H^{\cdot} \qquad (3),$$[1]

it must be *neutralized* for a *complete separation.* Ammonia best accomplishes this.

Of the fourth group sulphides *manganous sulphide* is most easily soluble and is even decomposed by acetic acid (§ **68**, (*3*)). It is flesh colored, but on standing in contact with ammonium sulphide becomes green.[2]

A method very often used to detect manganous ion (Volhard) consists in transforming it into *permanganate ion* by heating the manganous salt solution with lead peroxide and *concentrated* nitric acid. The color changes to a deep red or purple, and thus gives a sharp indication of the reaction:

$$2MnSO_4 + 5PbO_2 + 6HNO_3 =$$
$$2PbSO_4 + 3Pb(NO_3)_2 + 2H_2O + 2HMnO_4 \quad (5).$$

Small quantities of manganous ion may be detected by this test. It must be borne in mind, however, that concentrated nitric acid and chloride ion (from any chlorides present) react with permanganate ion and hence lessen the trustworthiness and delicacy of the test. [Furthermore commercial lead peroxide is rarely free from manganese. Therefore a blank test must always be carried out with it. Red lead, Pb_3O_4, is more likely to be free from manganese and serves equally well because it reacts with nitric acid to form lead peroxide (§ **15**).]

In dry mixture with sodium carbonate manganese compounds yield a gray metallic mass in the reducing flame on charcoal.

The *borax* and *phosphor salt* beads are *violet* in the *oxidizing* flame and *colorless* in the *reducing* flame.

79. The relations of manganous ion to the higher oxidation forms can merely be pointed out, because they are complicated owing to the large number of oxidation forms.

[1] Ordinary sodium phosphate is *disodium hydrogen phosphate*, Na_2HPO_4. Essentially it dissociates according to the equation:

$$Na_2HPO \rightleftarrows 2Na^{\cdot} + HPO_4'' \qquad (4).$$

The ion HPO_4'' dissociates but very little into $H^{\cdot} + PO_4'''$. Thus the above equation (*3*) expresses the essential dissociation (compare § 129).

[2] For the relations between the different modifications of manganous sulphide see *J. C. Olsen* and *H. S. Rapalje*, J. Amer. Chem. Soc., **26**, *1615*, 1904.

Neutral solutions of *manganous ion* are unstable in contact with air. A brown precipitate separates and the solutions assume acid properties. Acid solutions are more stable. The following reaction affords a possible explanation of these relations:

$$2Mn^{\cdot\cdot} + O + 5H_2O = \underline{2Mn(OH)_3} + 4H^{\cdot} \qquad (1).$$

The stability of acidified solutions indicates that the reaction is reversible. This assumption of a reversible reaction also explains the fact that oxidation by means of atmospheric oxygen occurs much more readily in the presence of alkali hydroxide. Neutralization of the hydroxide conditions the increased transformation in this case.

In the presence of a substance which unites with hydrogen ion, e. g. alkali hydroxide, other oxidizing agents, as chlorine, bromine or hydrogen peroxide (§ **55**) give rise to manganese dioxide. From this it is evident that the reaction follows an equation similar to the above.

The method given in § **56** for detecting manganese compounds leads to the formation of the highest oxidation products (hexavalent and heptavalent). Solutions of permanganates will remain unchanged for a considerable time. They undergo decomposition in the presence of many substances, both organic and inorganic, however, since permanganate acts as an oxidizing agent towards most substances (see § **52**). In fact it is a *typical* oxidizing agent.

The decomposition of permanganate will *vary* according as the solution is *acid* or *alkaline*. *Manganous ion* forms in the presence of *hydrogen ion,* while from *alkaline solutions manganese dioxide* separates. For the *reduction* of *permanganate ion* in *acid solution* the *general* equations are:

$$MnO_4' + 8H^{\cdot} = Mn^{\cdot\cdot} + 4H_2O + 5(^{\cdot})$$

or $$MnO_4' + 8H^{\cdot} + 5(') = 4H_2O + Mn^{\cdot\cdot} \qquad (2a),$$

and $$2MnO_4' + 6H^{\cdot} + 2H_2O = 2Mn^{\cdot\cdot} + 10(OH) \qquad (2b).$$

In *alkaline solution* permanganate ion undergoes the following changes:

$$MnO_4' + 2H_2O = \underline{MnO_2} + 4(OH)' + 3(^{\cdot})$$

or $$MnO_4' + 2H_2O + 3(') = \underline{MnO_2} + 4(OH)' \qquad (3a),$$

and $$MnO_4' + H^{\cdot} + H_2C = \underline{MnO_2} + 3(OH) \qquad (3b).$$

Equations (*2a*) and (*3a*) state that one formula weight of MnO_4' can oxidize *five formula weights of a substance which takes up a positive electric charge or gives up a negative,* in an *acid solution,* while in *alkaline solution three* are oxidized. Equations (*2b*) and (*3b*) show that one formula weight of MnO_4' can give up *five* formula weights of hydroxyl in an *acid solution,* in *alkaline solution three,* to a substance which combines with hydroxyl.

80. **The formulation of reactions.** It has already been pointed out in § 9 that always the same number of positive or negative electric charges are gained or lost in a given reaction. Hence an exact equation (i. e. one corresponding to the actual relations) must have *as many positive as negative* charges *on each side,* but not necessarily the *same number* of, say *positive charges,* on *either side.*

The apparent discrepancy in the above equations is due to the fact that those ions which undergo no noticeable changes of concentration have been *omitted.* The complete formulation of a given reaction will at once show the truth of the above statement in all cases.

An equation which takes this into account must always satisfy the condition that the *same excess* of positive or negative charges is present on *each side.* Thus in equations (*2a*) there is an excess of seven and two positive charges respectively on the left side, i. e. $8H^{\cdot} - MnO_4'$ and $8H^{\cdot} - MnO_4' - 5(')$, and on the right side also an excess of seven and two respectively, i. e. $Mn^{\cdot\cdot} + 5(^{\cdot})$ and $Mn^{\cdot\cdot}$ These electric charges are by no means set free. They become *available* only when a substance is present which can take up *positive* or give up *negative* charges.

For the sake of learning the application of these equations the student should practice formulating a number of oxidation reactions, for instance, of ferrous- to ferric ion, oxalic acid to carbon dioxide and water, nitrite to nitrate, sulphite to sulphate, hydriodic acid to iodine, all in acid solution, and of ethyl alcohol (C_2H_6O) to aldehyde (C_2H_4O) or acetic acid ($C_2H_4O_2$) in alkaline solution.

The oxidation of oxalic acid will be considered here by way of illustrating this exercise in the case of organic compounds, the relations of which are less clear than for the oxidation of ions.

One formula weight of oxalic acid ($C_2H_2O_4$) in its reaction with permanganate gives two formula weights of carbon dioxide and one formula weight of water. Since

$$2CO_2 + H_2O - C_2H_2O_4 = O \qquad (4),$$

it follows that one formula weight of oxalic acid uses one combining weight of oxygen. Now one combining weight of oxygen corresponds to two formula weights of hydroxyl, according to the equation $O + H_2O = 2(OH)$. Hence from (*2b*) two formula weights MnO_4' oxidize five formula weights of oxalic acid. Thus the equation follows:

$$2MnO_4' + 6H^{\cdot} + 5C_2H_2O_4 = 2Mn^{\cdot\cdot} + 8H_2O + 10CO_2 \qquad (5).$$

81. **Zinc.** In its combining ratios zinc is the antithesis of manganese. Its salts contain only a divalent zinc ion, which is colorless.

For the tests $\frac{1}{20}$ formula weight of zinc sulphate, $ZnSO_4.7H_2O$, is dissolved to a liter of solution. Consult § 82 for the preparation of this salt.

Alkali hydroxide and *ammonia* each precipitate *zinc ion* as zinc hydroxide, $Zn(OH)_2$, soluble in an excess of either precipitant. Zinc hydroxide acts as an acid, hence it dissolves in alkali hydroxide.

$$Zn(OH)_2 \rightleftarrows ZnO_2'' + 2H^{\cdot} \text{ (compare § 42, (1))} \qquad (1).^1$$

The solubility in ammonia is due to the formation of zinc ammonia ion, $Zn(NH_3)_n^{\cdot\cdot}$,[2] as is shown by the solubility of zinc oxide in ammonia even without the presence of ammonium salts. Ammonium ion which results from the precipitation of zinc hydroxide with ammonia apparently increases the solubility of zinc oxide in ammonia however.

Alkali carbonate precipitates *basic carbonate.*

Potassium cyanide precipitates white *zinc cyanide,* $Zn(CN)_2$, which dissolves in an excess of the precipitant by the formation of a complex ion, $Zn(CN)_4''$:

$$\underline{Zn(CN)_2} + 2CN' = Zn(CN)_4'' \qquad (2).$$

Hence this solution does not give a test for zinc ion, or at most only a very slight indication of its presence.

Zinc sulphide is only partially precipitated by passing *hydrogen sulphide* into zinc sulphate solution (compare § 28), but it is practically completely precipitated on adding sodium acetate to the solution.

On charcoal *solid zinc compounds* easily reduce to metallic zinc, which volatilizes quickly at the reducing temperature. The vapors react with oxygen of air and form zinc oxide which is very much less volatile. Hence a white sublimate of zinc oxide deposits around the heated portion of the charcoal.

Zinc salts give colorless beads with *phosphor salt* and *borax.*

[1] Recent investigations indicate, however, that still other relations must be taken into account to explain the solubility of zinc hydroxide in alkali hydroxide and also the stability of this solution. Therefore the explanation of the solubility in bases by reaction (*1*) is only an approximation.

[2] 'n' is an integer which varies with the concentration of the ammonia (compare Principles, p. *621*).

82. Preparation of zinc sulphate. This salt is made by dissolving zinc in sulphuric acid. Hydrogen is evolved in the reaction:

$$Zn + 2H^{\cdot} = Zn^{\cdot\cdot} + H_2 \qquad (1).$$

To carry out the experiment dilute sulphuric acid is poured over granulated zinc in a flask. As the acid is used up and the evolution of hydrogen becomes less vigorous heat may be applied to hasten the reaction. Explosions occur if a flame is brought near the mouth of the flask, since hydrogen forms an explosive mixture with air. At a short distance, however, the intermixture of hydrogen with air is already so far advanced[1] that there is much less danger of an explosion. Hence a water-bath may always be used for heating the flask as soon as the gas begins to come off slowly.

The solution of zinc sulphate is next filtered from any undissolved zinc and is evaporated on a water-bath until a skin of crystals forms over its surface. This indicates that the solution has become saturated with zinc sulphate at the temperature of the water-bath. It is then set aside and cooled slowly. Most of the dissolved salt crystallizes, since it is much less soluble at room temperature than at higher temperatures. The crystals are collected over a filter plate in a funnel. Acid is removed by pouring a little water on the funnel several times and quickly draining it off by means of a filter pump. Finally the crystals are dried between sheets of filter paper.

The purity of this preparation must be tested before using it for the above reactions. The test is carried out by dissolving a little of the salt in some water and adding hydrogen sulphide to one portion (§ 19) and ammonia to another. Any impurities so detected are then removed from the entire mass of the crystals by these reagents.

83. Cobalt. Cobalt forms numerous complex ions and also *di-* and *trivalent cathions*. For analytical purposes the divalent cathion is most important. Still the properties of cobaltic ion are important for the separation of cobalt and nickel salts.

[1] The danger of an explosion is determined by the ratio of oxygen and hydrogen present. A definite lower limit of hydrogen is necessary to give an explosion from local heating, i. e. in order that the reaction taking place at a certain definite point, which results in the formation of water, spreads instantaneously through the whole mixture.

The test solution may be prepared from the chloride, $CoCl_2.6H_2O$, nitrate, $Co(NO_3)_2.6H_2O$ or sulphate, $CoSO_4.7H_2O$, by dissolving $\frac{1}{20}$ formula weight per liter of solution.

Alkali hydroxide gives a rose red precipitate of *cobaltous hydroxide.* When only a small amount of alkali hydroxide is present, blue basic salts form. These go over into hydroxide on heating or by adding a moderate excess of the precipitant.

Atmospheric oxygen and other oxidizing agents transform the red hydroxide into brown cobaltic hydroxide. Freshly prepared *cobaltous hydroxide* dissolves in *very concentrated solutions* of *alkali hydroxide* to a deep blue solution.[1] The property of forming anions of the composition CoO_2'' is only slightly developed, however. Hence very concentrated solutions of alkali hydroxide are required to dissolve cobaltous hydroxide. On diluting the blue solution with water, hydroxide separates in various shades.

The solution likewise gives a precipitate of basic salt with *ammonia.* This dissolves to a red solution on adding ammonium salts, and with an excess of ammonia it gives a yellowish green solution. The first reaction is related to the decreased dissociation of ammonium hydroxide through the presence of ammonium salts (compare §§ 78, (*2*) and 89), the latter to the formation of complex *cobaltous ammonia cathions* (Principles, p. *613*). For the most part these two influences are active at the same time. It has already been mentioned in § 51 that precipitation in the third group is sometimes not prevented by ammonium salts.

The precipitation of cobaltous ion as *potassium cobaltinitrite* (see § 75) always occurs when a given solution contains potassium ion, cobaltous ion, nitrite ion and hydrogen ion simultaneously, the latter in small amount. The reaction betweeen these ions is expressed by the equations:

$$Co^{\cdot\cdot} + NO_2' + 2H^{\cdot} = Co^{\cdot\cdot\cdot} + H_2O + NO \qquad (1a),$$

and

$$Co^{\cdot\cdot\cdot} + 6NO_2' + 3K^{\cdot} = \underline{K_3Co(NO_2)_6} \qquad (1b).$$

Essentially *cobaltous ion* is oxidized to *cobaltic ion* by this reaction. The necessary positive electric charge comes from hydrogen ion, in case nitrite ion is also present, owing to the formation of $NO^{\cdot}$ +

[1] This fact has been referred to by Donath, Zeitschrift für analytische Chemie, 40, *437*, 1901.

OH′. Cobaltic ion then yields a difficultly soluble characteristic precipitate of *potassium cobaltinitrite*, $K_3Co(NO_2)_6$, with potassium- and nitrite ions. The solution of this salt contains the complex cobaltinitrite ion, $Co(NO_2)_6'''$, besides the ions $K^{\cdot}$, NO_2' and $Co^{\cdot\cdot}$. The formation of this complex ion may be represented by the equation (*1b*) above.

From the fact that *one* formula weight of *nitrite ion* is needed to *form* one formula weight of cobaltic ion, and *six* are required to precipitate it according to the formula of the precipitate, it is evident that a *concentrated solution* of potassium nitrite must be used for this test. No precipitate forms on using sodium nitrite, since sodium cobaltinitrite is easily soluble.

A solution of cobaltous ion which has stood a short time in contact with sodium nitrite and acetic acid, however, gives a precipitation *immediately* on addition of any potassium salt solution. This shows that the oxidation of cobaltous ion to cobaltic ion and the *formation* of the complex cobaltinitrite ion takes place independently of the presence of the *cathion* which is added with the nitrite ion.

Since to increase the concentration of substances that are used up in a reaction always leads to the formation of greater amounts of the reaction products, it would seem advisable to increase the concentration of hydrogen ion as much as possible for the sake of a larger yield of cobaltic ion. While the inference is logical, the operation itself does not have the desired effect. This is due to the weakness of nitrous acid. Under the above conditions a great deal of undissociated nitrous acid would form and this, moreover, undergoes a decomposition according to § 156 (*2*). Hence the result would be an actual decrease of the concentration of nitrite ion, which more than compensates the advantages aimed at. Owing to this reason acetic acid is added. It does not follow, however, that the most favorable conditions are obtained with acetic acid.

According to Vogel cobaltous and nickelous ions exhibit a striking difference in their behavior towards sulphocyanates. Cobaltous ion yields a pink solution, turning blue on addition of ordinary (ethyl) alcohol.

A similar phenomenon is observed with other alcohols, for instance, methyl alcohol. In the case of amyl alcohol, which is not miscible with water in all proportions and therefore forms two layers with it, the alcohol layer is colored deep blue. Since nickel salts do not give this reaction the phenomenon affords a convenient means of distinguishing cobaltous ion in the presence of nickel. Mercuric ion interferes with the test for the same reason that was stated in §

69. The test is rendered more delicate by the addition of an acid, as hydrochloric or nitric.

The use of amyl alcohol may be avoided by evaporating to dryness a small quantity of the solution under examination for nickel and cobalt. Then alcohol is added, and finally ammonium sulphocyanate solution, drop by drop. In case no blue color is observed it is advisable to compensate any excess of water which may prevent the formation of the color, by increasing the amount of alcohol.

A closer study[1] of the chemical changes involved in this test indicates that the blue color is characteristic of a double salt of the formula $K_2Co(CNS)_4$ or $(NH_4)_2Co(CNS)_4$. It ionizes in aqueous solution and the color vanishes.

This very striking behavior of cobalt salts is also utilized as a means of detecting alcohol.[2]

Alkali cyanide forms a red brown precipitate of *cobaltous cyanide.* The precipitate dissolves in an excess of alkali cyanide to a brown solution containing the complex cobaltocyanide ion, $Co(CN)_6''''$. This is, however, very unstable and oxidizes readily, for example, in contact with atmospheric oxygen, to form the lighter colored cobalticyanide ion, $Co(CN)_6'''$:

$$2Co(CN)_6'''' + O + H_2O = 2Co(CN)_6''' + 2(OH)' \quad (2).$$

Cobalticyanide ion is much more stable than cobaltocyanide ion, as is shown by the fact that the yellow solution yields no precipitate when acidified, or at most gives only a slight cloudiness on long standing. Cobaltocyanides are decomposed by acids. Addition of *bromine* and *potassium hydroxide* causes *no precipitation* of *cobaltic hydroxide* in the cobalticyanide solution. This fact also shows that the solution contains practically no cobaltous ion. In case the oxidation was not complete, however, it would contain cobaltous ion and would then yield a precipitate of cobaltic hydroxide on the addition of bromine.

Cobaltous sulphide is only partially precipitated by hydrogen sulphide from acid or neutral solutions of cobaltous salts, even when sodium acetate is present. It is soluble in *concentrated* nitric acid (consult § 75).

[1] Rosenhain and Cohn, Ber. d. d. chem. Gesellschaft, 33, *1111*, 1900. Treadwell, Zeitschrift für anorganische Chemie, 26, *108*, 1901.

[2] F. Morrel, Zeitschrift für analytische Chemie, 16, *251*, 1877.

Cobalt salts are reduced on *charcoal* to the *gray metal.* The *phosphor salt* and *borax beads* are colored a *deep blue.* After a long reduction the color becomes gray owing to the separation of metallic cobalt.

84. Nickel. Nickel has likewise two oxidation forms. It is, however, much more difficult to oxidize nickelous ion to nickelic, i. e. stronger oxidizing agents are needed, than is the case with cobalt. The following tests relate to the properties of nickelous ion.

For the test solution $\frac{1}{20}$ formula weight of nickel chloride, $NiCl_2.6H_2O$, or nitrate, $Ni(NO_3)_2.6H_2O$ or sulphate, $NiSO_4.7H_2O$ is dissolved to a liter. The solutions have a deep green color, characteristic of nickelous ion.

Alkali hydroxide precipitates *apple green nickelous hydroxide.* The precipitate is transformed into black nickelic hydroxide by bromine and alkali hydroxide. How is this reaction formulated?

Ammonia also precipitates green hydroxide, soluble in an excess of ammonia. The solution has a characteristic blue color, due to a complex nickel ammonia cathion, $Ni(NH_3)_4^{\cdot\cdot}$ or $Ni(NH_3)_6^{\cdot\cdot}$ (Principles, p. *616*).

Nickelous ion is not precipitated by ammonia if ammonium salts are present in considerable quantity. In this case the decrease of the dissociation of ammonium hydroxide by ammonium salts (§ 78) adds its effect to that of the complex formation.

Alkali carbonate and *ammonium carbonate* precipitate green *nickelous carbonate,* soluble in an excess of ammonium carbonate.

Alkali cyanide gives a green precipitate of *nickelous cyanide.* It dissolves in an excess of alkali cyanide to a yellow solution containing chiefly the ion $Ni(CN)_4''$ (compare § 81). The solution contains practically no nickelous ion. Hence it forms no precipitate with *potassium hydroxide* or with *ammonium sulphide.* When heated with *bromine* and *alkali hydroxide,* however, it yields a *black* precipitate of *nickelic hydroxide* (distinction from cobalticyanide ion).

A method of detecting nickelous ion in the presence of cobalt is based upon this latter property. To carry out the test potassium cyanide is added to the solution until the precipitate which forms at first is redissolved. A small excess over the amount just necessary to dissolve the precipitate favors the formation of the more stable anion $Co(CN)_6'''$ (§ 83), but a large excess tends to prevent

the formation of $Ni(OH)_3$ in case nickel is present. If too much cyanide is added, a large amount of bromine must be used. The cyanide solution is left standing for some time and is occasionally shaken. Finally it is poured into a hot solution of bromine in alkali hydroxide. Very small quantities of nickel give rise to a brown coloration instead of forming a precipitate.

Potassium sulphocarbonate affords a particularly delicate test for nickelous ion. With an ammoniacal solution of a nickel salt it gives a brown red solution, appearing almost black in incident light. Very small amounts of nickel give a rose red coloration.

According to Braun (Zeitschrift für analytische Chemie, 7, *346*, 1868) the solution of potassium sulphocarbonate is made as follows: A definite volume of potassium hydroxide containing one formula weight of KOH per liter is divided into two equal parts. The one part is saturated with hydrogen sulphide. It is then mixed with the other half and the mixture is gently heated with $\frac{1}{25}$ its volume of carbon disulphide, being frequently shaken during this operation. The dark orange red solution is then poured off from any remaining carbon disulphide and is kept tightly stoppered.

[The reaction of Tschugaeff (§ 75) is an exceedingly delicate test for nickel, detecting one part in four hundred thousand. None of the commoner cathions interfere with this reaction. Cobalt also shows a characteristic reaction with this reagent, but it does not interfere with the detection of nickel, as it gives no precipitate.]

Nickel sulphide resembles cobalt sulphide in its behavior. Like the latter it also dissolves in *concentrated* nitric acid (compare § 75).

Solid nickel compounds are reduced to the gray metal on *charcoal.*

In the oxidizing flame the phosphor salt and borax beads are *brown red when hot,* almost colorless when cold. The borax bead becomes gray in the reducing flame in consequence of the separation of metallic nickel.

GROUP V.

85. **Removal of ammonium sulphide** from the filtrate from the fourth group. Before precipitating with the fifth group reagent, *ammonium carbonate,* the filtrate from the fourth group must be freed from ammonium sulphide. To accomplish this acetic acid is added and the mixture is heated for some time. Ammonia and hydrogen sulphide are driven off. Any sulphur separated is filtered before adding the group reagent.

Heating the solution causes the ions $NH_4^{\cdot}$ and S'' to react with water, giving rise to ammonium hydroxide and hydrogen sulphide:

$$2NH_4^{\cdot} + S'' + 2H_2O = 2NH_4OH + H_2S.$$

The products of this reaction vaporize at the elevated temperature, either directly (H_2S) or in consequence of decomposition ($NH_4OH = NH_3 + H_2O$). Hence they escape from the solution as fast as they are formed and all the ammonium sulphide present will be thus driven off. Addition of acetic acid acts in the same sense by neutralizing (§ 50) the ammonia.

In case the solution contains yellow ammonium sulphide, the hydrogen compound H_2S_n is formed, corresponding to the ion S_n''. This separates sulphur on evaporating the solution.

86. Precipitation with ammonium carbonate.[1] The filtrate treated according to § 85, or in case the preliminary test indicated no fourth group ions, the filtrate from some preceding group, is first made *alkaline* with *ammonia* and *boiled.* Then *ammonium carbonate* solution is added *drop* by *drop* until the resulting precipitate no longer increases in bulk. *Barium, calcium* and *strontium ions* are precipitated with this reagent:

$$Ba^{\cdot\cdot} + CO_3'' \rightleftarrows BaCO_3.$$

The precipitate should be *stirred* for *several minutes* in *contact* with the *liquid* before filtering, to avoid an *incomplete separation.* Otherwise errors may arise owing to the fact that the above reaction is reversible. The latter point is readily shown by heating a small quantity of any of these carbonates with an ammonium salt solution. Provided sufficient ammonium salt is taken all will finally go into solution.

The explanation of this is as follows. A certain definite amount of ammonia and carbon dioxide will always be formed in a solution which contains the dissociation products of ammonium carbonate:

$$2NH_4^{\cdot} + CO_3'' = 2NH_3 + H_2O + CO_2.$$

These products volatilize when the solution is heated. Hence the reaction proceeds until practically all carbonate ion is used up.

[1] Concerning the composition and dissociation of ammonium carbonate, which in reality is somewhat more complicated than has been assumed here, consult Principles, pp. *394* and *504*.

This decrease in the concentration of carbonate ion disturbs the equilibrium with respect to solid carbonate (from which the carbonate ion has been formed). Accordingly more of the solid carbonate will go into solution. The dissolving continues as long as sufficient ammonium ion is present. Hence the greater the concentration of ammonium ion, the more carbonate ion will disappear from the solution, i. e. a considerable quantity of solid carbonate redissolves. If, therefore, too great an amount of ammonium ion has been introduced anywhere in the course of the analysis, the solution must be evaporated to dryness. The residue is then heated in a porcelain evaporating dish over an asbestos wire gauze (not over the direct flame of a bunsen burner). The ammonium salts are volatilized in this way. The residue is next dissolved in water, acid being added if necessary to bring it into solution, and *the solution is rendered ammoniacal. If ammonia produces a cloudiness, ammonium chloride is also added drop by drop, but only in this case.* The cloudiness indicates that *magnesium ion* is present (§ 102) and the ammonium chloride prevents its precipitation as carbonate.

Long continued heating also favors the dissolving of the fifth group precipitate. Still the heating should never be entirely omitted because these carbonates precipitate in the amorphous[1] state. By remaining in contact with the solution they go over into the crystalline, less soluble form. Experience has shown that this change takes place more rapidly when the precipitate is in contact with a hot solution. For most cases it suffices, however, *to heat the solution immediately before the precipitation.* Errors which might arise from this operation are easily avoided by observing whether a precipitate is formed as a consequence of the heating alone or not.

87. The separation of barium-, calcium- and strontium carbonates. The precipitate of carbonates is first washed and then is dissolved in a little *acetic acid. Hydrochloric or nitric acid must not*

[1] The rate of formation of the precipitate bears the following relation to its condition. The more difficultly soluble modification results when the precipitate forms slowly. This may be shown by precipitating calcium ion with ammonium carbonate in one case, in another by adding ammonium chloride and -carbonate simultaneously. In this last case the precipitate is crystalline and forms much more slowly, while in the other it is amorphous, more bulky, and forms at once. It passes over into the less soluble modification after standing for a short time (compare Principles, p. *207*).

be used in place of acetic, since this would give rise to errors which will be explained later (§ 98). *A small portion* of the acetic acid solution is first tested with *potassium dichromate. In case a precipitate forms* the reagent is added to the *entire* solution until the *reddish yellow color* of the *filtrate* indicates that an *excess* of the precipitant is present. The yellow precipitate is *barium chromate,* $BaCrO_4$. The precipitation is more complete if sodium acetate is added to the solution (consult § 98).

The *reddish yellow filtrate* is then made slightly ammoniacal. It is heated and again tested with *ammonium carbonate* for *strontium-* and *calcium ion.* Any precipitated carbonates are filtered and *washed.* They are next dissolved in a very little *acetic acid.*[1] To a small portion of this solution, or if no barium ion had been precipitated, to a small portion of the original acetic acid solution, a few cc. of *saturated calcium sulphate solution* are added. A precipitate of *strontium sulphate* forms if strontium ion is present. The precipitate separates slowly, however. Hence the solution must be set aside for *several hours* before it is examined for a precipitate.

If this test indicates strontium ion, most of the remaining solution is precipitated with dilute sulphuric acid or a solution of sodium- or ammonium sulphate. Both strontium and calcium ion are hereby precipitated up to the saturation limits of strontium sulphate and calcium sulphate corresponding to the excess of SO_4'' (according to § 90). It must be specially noted that the separation of the sulphates is *not completed as soon as the precipitate becomes visible.* Accordingly it is not filtered immediately, but is shaken up frequently and left 20–30 minutes[2] in view of any further tests to be made with the filtrate. To detect calcium ion the filtrate is made alkaline with ammonia and a few drops of *ammonium oxalate* are added. *Calcium ion* yields a white precipitate of *calcium oxalate.*

When calcium sulphate solution gives a negative test for stron-

[1] It is advisable to use acetic acid instead of hydrochloric acid (which is commonly used), because the precipitation of strontium sulphate by the addition of gypsum solution is apparently accelerated by acetic acid. Another reason is that strontium sulphate is noticeably soluble even in dilute hydrochloric acid (according to 'qualitative' experiments). Hence the delicacy of this test is lessened if hydrochloric acid is used.

[2] The *incomplete* separation of these sulphates may give rise to errors with respect to the detection of calcium ion, since strontium ion also gives a precipitate with oxalates which might be mistaken for calcium oxalate.

tium it may happen that no precipitate forms on adding sulphate ion. It is evident in this case that the portion of the solution to which no calcium sulphate has been added is taken for the calcium test.

Another method of separating $Ba^{\cdot\cdot}$, $Ca^{\cdot\cdot}$ and $Sr^{\cdot\cdot}$ is based upon the behavior of their solid *chlorides* towards *absolute alcohol.* Strontium and calcium chlorides dissolve, while barium chloride is practically insoluble. The alcoholic solution is then evaporated to dryness and any residue of chlorides is transformed into *nitrates* by repeated evaporation with concentrated nitric acid. Calcium nitrate dissolves in absolute alcohol and strontium nitrate is left behind. The method makes greater demands upon the experience and skill of the student than is usually acquired in an ordinary course of qualitative analysis. Therefore the detailed description is omitted.

88. Application of the law of mass action to electrolytes. The application of the mass action law to the dissociation of salts is particularly valuable. It leads up to a conception which affords a more exact representation of many phenomena.

It has already been stated in § 49 that the ions and the non-dissociated part of an electrolyte are mutually transformable, as is expressed by the following equation for the dissociation of sodium chloride:

$$NaCl \rightleftarrows Na^{\cdot} + Cl'.$$

From § 29 (*2*) the following relation exists between the concentrations of the ions and the non-dissociated salt. Attention should be called to the fact that this expression holds for substances like sodium chloride only approximately (qualitatively). This objection can of course only be made if *quantitative* conclusions are drawn. As this is not the case the inferences drawn from this equation are not affected by this statement.

$$\frac{C_{Na^{\cdot}} \quad C_{Cl'}}{C_{NaCl}} = k \qquad (1).$$

The first question to be considered is that of the effect produced upon the system by diluting the solution when equilibrium has been reached, i. e. what changes result from diluting, say ten times. Assuming that the dilution will not occasion a transformation in either direction, it follows that the individual concentrations are decreased 'n' times through the n-fold increase of volume. Thus they become $\frac{C_{Na^{\cdot}}}{n}$, $\frac{C_{Cl'}}{n}$ and $\frac{C_{NaCl}}{n}$ respectively. The mass action law should also be applicable after the dilution, whence

$$\frac{\frac{C_{Na^{\cdot}}}{n} \cdot \frac{C_{Cl'}}{n}}{\frac{C_{NaCl}}{n}} = k \qquad (2).$$

Now equations (*1*) and (*2*) cannot both hold simultaneously. Accordingly the condition assumed to exist after diluting the system is not the equilibrium

condition. A transformation must take place to bring the system to a stable state. The direction of this transformation is indicated by equation (*2*). In order that the quotient $\frac{C_{Na^{\cdot}} \cdot C_{Cl'}}{nC_{NaCl}}$ becomes equal to 'k,' the values of $\frac{C_{Na^{\cdot}}}{n}$ and $\frac{C_{Cl}}{n}$ must increase, that of $\frac{C_{NaCl}}{n}$ must decrease. In other words the *undissociated portion* of the salt undergoes further dissociation. This deduction agrees with the actual relations, for all electrolytes are *proportionately* more dissociated in dilute solutions than in concentrated. When the solution is already extremely dilute a proportional increase of the dissociation with the dilution is no longer observed because the electrolyte is practically completely dissociated under these conditions. The increase of the per cent dissociation with the *dilution*, i. e. the ratio of the *dissociated parts* to the *total amount* of electrolyte in the solution, *cannot be shown by qualitative experiments.* A detailed description of the quantitative methods used to test these ratios is found in Ostwald, Grundriss der allgemeinen Chemie, 3d edition, pp. *390.* . .

89. **The influence of foreign substances** upon the dissociation of a given electrolyte may also be deduced by means of the mass action law. The independent variation of the concentration of each substance which reacts in any way with one of the dissociation products or non-dissociated salt, disturbs the equilibrium and occasions a transformation. Thus, for example, the *effect* of *ammonium salts in decreasing the dissociation of ammonium hydroxide* and the *similar influence of acetates upon the dissociation of acetic acid,* which have already been referred to in § 55, can be explained more exactly by applying the above considerations. The following relation holds for an ammonia solution of given concentration, which contains ammonia, NH_3, and in addition ammonium hydroxide, NH_4OH, and the ions $NH_4^{\cdot}$ and OH':

$$\frac{C_{NH_4^{\cdot}} \cdot C_{OH'}}{C_{NH_4OH}} = k$$

The values of $C_{NH_4^{\cdot}}$ and $C_{OH'}$ are small in accordance with the slight dissociation of ammonium hydroxide. Hence 'k' is also small. On adding ammonium salt to the solution the concentration of ammonium ion becomes the sum of the concentrations due to dissociation of ammonium hydroxide and the ammonium salt together. The quotient would then be too large if no change resulted from the introduction of ammonium salt. In consequence of the addition of ammonium salt, however, ammonium hydroxide will form from its dissociation products until the above relation holds. Since ammonium salts are relatively much dissociated into their ions it follows that the addition of a small amount will cause a proportionately great increase of $C_{NH_4^{\cdot}}$. Hence they will have a very pronounced effect upon the dissociation of ammonium hydroxide. This is in fact the case.

Just as an ammonium salt influences the equilibrium between ammonium hydroxide and its ions, so the reciprocal relation holds conversely. Ammonia decreases the dissociation of ammonium salts to a much less extent, however, in accordance with its proportionately slight dissociation. Hence its effect may be neglected in practice.

The explanation of the influence of acetates upon the dissociation of acetic acid is left to the student.

This action of a salt upon the dissociation of an acid or base with a common anion or cathion is by no means limited to the cases cited. With all medium strong and weak acids and bases it is pronounced. A case that has been already mentioned (§ 74) is that of the decrease in the dissociation of hydrogen sulphide (in itself but little dissociated) by adding stronger acids, as hydrochloric, sulphuric or nitric.

90. The influence of foreign substances upon solubility. The equilibrium existing in the *saturated* solution of a salt is very markedly influenced by introducing a substance with a common ion.

Thus, for example, on passing hydrochloric acid gas into a saturated solution of potassium chloride a separation of the solid salt takes place.

The same result may also be obtained by adding concentrated hydrochloric acid to the potassium chloride solution. In this case, however, the conditions are less favorable, since an amount of potassium chloride corresponding to the volume increase must first be formed before supersaturation and separation of the solid takes place. Other experiments illustrating the same relations are afforded by precipitating barium chloride from a concentrated solution through the addition of hydrochloric acid, or barium nitrate with nitric acid. Potassium chlorate affords a particularly instructive example. Its *saturated* solution yields a precipitate of potassium chlorate on addition of *saturated* solutions of *potassium chloride* or *sodium chlorate.* This is due in the first case to the increase in the concentration of $K^{\cdot}$-ion, in the second to the increase in the concentration of ClO_3'-ion.

The phenomenon is a direct consequence of the alteration in the dissociation of potassium chloride. Increasing the concentration of chloride ion by the dissolved hydrochloric acid causes more undissociated salt to form. The solution was saturated to begin with, however, i. e. it was in equilibrium with solid potassium chloride. The formation of non-dissociated salt leads to a precipitation,[1] for in the saturated solution of a salt the concentration of the non-dissociated part has a definite value. To exceed this limit conditions a reaction which leads again to a state of equilibrium. In the present case a separation of solid salt takes place. Hence it follows that *the total quantity of a salt contained in unit volume of its saturated solution is often much decreased by salts with a common ion.*

This fact shows that the value assigned as the solubility of a substance at a definite temperature is quite indefinite if it is taken to be the total amount contained in unit volume, which is made up of the dissociated and non-dissociated parts. For a given substance, however, the *solubility* of the *non-disso-*

[1] Supersaturation in itself need not give rise to a precipitation. The separation of solid salt is determined by certain other conditions which will be considered in detail in connection with some specially striking examples (compare § 100).

ciated part is *fixed* within certain definite limits. The variability of the *total solubility* is hence due to the decrease in the concentration of the ions which are formed from the salt itself. This is caused by the presence of an excess of one of the ions, arising, however, from another salt.

The following considerations will make these relations clearer. Let a_0 and b_0 represent the concentrations[1] of the ions (Cl', $K^{\cdot}$) in the saturated solution of a binary salt, KCl, i. e. one decomposing into two ions, and c_0 that of the undissociated part. With certain limitations the following relation:

$$a_0.b_0 = k.c_0 \qquad (1),$$

holds, according to the law of mass action, where $a_0 = b_0$.[2] The total solubility of the salt, L_0, is expressed as follows:

$$L_0 = c_0 + a_0 \qquad (2),$$

where c_0 is the constant solubility of the undissociated part.

In the presence of a second substance, say HCl, with the common ion, Cl', let the concentrations corresponding to the equilibrium condition be a_1 and b_1 for the ions arising from the first substance, KCl, and a_2 for the common ion formed from the second substance. From the mass action law the following relation holds, *at least approximately:*

$$(a_1 + a_2).b_1 = k.c_0 \qquad (3).$$

Now as long as a_2 has a positive value, a_1 (and of course also b_1) must be

[1] The number of formula weights in unit volume, one liter.

[2] Sometimes difficulties are experienced in understanding this fact. Hence the present case may be treated in a little more detail, Assume that 'm' grams of potassium chloride are left on evaporating one liter of a solution saturated at 18° C. In formula weights the solubility is accordingly $\frac{m}{74.60}$. Assume further that 70 per cent of the salt is dissociated (an assumption approximating closely the actual relations). Then the undissociated potassium chloride amounts to $0.30 \times m$ grams, the dissociated portion to $0.70 \times m$ grams. The latter is made up of $0.70 \times m \times \frac{39.15}{74.60}$ g. potassium ion and $0.70 \times m \times \frac{35.45}{74.60}$ g. chloride ion. To arrive at the concentrations a_0, b_0 and c_0, these quantities in grams must be divided respectively by the formula weights, whence

$$a_0 = \frac{0.70 \times m \times 39.15}{74.60 \times 39.15}$$

$$b_0 = \frac{0.70 \times m \times 35.45}{74.60 \times 35.45}$$

$$c_0 = \frac{0.30 \times m}{74.60}$$

From this it follows that

$$a_0 = b_0 = \frac{0.70 \times m}{74.60} \qquad (1),$$

$$\text{and } L_0 = \frac{m}{74.60} = a_0 + c_0 = b_0 + c_0 \qquad (2).$$

smaller than a_0 and b_0. From this it follows that the total solubility, L_1, in the presence of a substance with a common ion, is also smaller, as

$$L_1 = a_1 + c_0 \qquad (4).$$

From (*2*) and (*4*) it follows that the *difference in the total solubility* caused by the presence of a substance with a common ion is equal to the *difference in the concentrations of the ions formed from the original salt:*

$$L_0 - L_1 = a_0 - a_1 = b_0 - b_1 \qquad (5).$$

Strictly speaking this is not actually the case, but it approximates the reality fairly closely. The mass action law may be applied to electrolytes like potassium chloride and hydrochloric acid only with considerable limitations.

91. Solubility product.[1] From the above explanations it is evident that in the case of the solubility of a salt in a solution which contains another salt with a *common* ion, the concentration of the latter has a great influence upon that of the former. The *concentrations of the ions formed from the one salt decrease as the concentration of the common ion arising from the other salt increases.* Accordingly the *decrease* of the *solubility* of the *first salt* becomes greater as the *concentration* of the *other increases.*

These deductions are based upon the invariability of kc_0 at constant temperature. The expression is termed the "*solubility product*" and denotes the *value* of the *product* of the *ionic concentrations* in the *saturated* solution of a salt, according to § 90, (*1*).

Hence *saturated* solutions of electrolytes or salts may be defined to be those in which the product of the ionic concentrations, "the *concentration product,*" *is equal to the solubility product.* In *unsaturated* (§ 6) solutions the *concentration product* is *smaller* than the *solubility product,* in *supersaturated solutions* it is *greater.*

In the case of difficultly soluble substances a knowledge of the solubility product is particularly important. In their saturated solutions (as in almost all very dilute solutions, consult § 88), the non-dissociated part, c_0, is so small in comparison with a_0 that the solubility may be placed equal to the concentration of one of the ions, thus:

$$L_0 = a_0 = b_0 \qquad (6).$$

Hence not only the *difference* between the solubility in pure water and that in the presence of a salt with a common ion, but also the *quantity dissolved in unit volume* may be estimated from the value of a_2.

A study of equation (*3*) shows this in particular. With very difficultly soluble salts and larger excess of the salt with the common ion, a_1 may be neglected in comparison with a_2. Hence equation (*3*) assumes the form:

[1] Consult Ostwald, Scientific Foundations of Analytical Chemistry, Translation by McGowan, p. *75*.

$$a_2.b_1 = kc_0 \qquad (7).$$

Thus the concentration of the ions formed from the difficultly soluble salt ($a_1 = b_1$) is inversely proportional to the concentration of the ion present in excess (through the presence of the second salt).

This leads to a conclusion very important for the precipitation of difficultly soluble salts. By using an *excess* of the salt serving as precipitant, for example, of hydrochloric acid for $Ag^{\cdot}$ or of sulphuric acid for $Sr^{\cdot\cdot}$, the *precipitation is more complete* (of $Ag^{\cdot}$ and $Sr^{\cdot\cdot}$ respectively in the above cases). The increase in the concentration of hydrochloric and sulphuric acids respectively carries with it also an increase in the concentration of the chloride and sulphate ions, which in turn lessens the solubility of silver chloride and of strontium sulphate.

In certain cases, a very striking example of which is afforded by silver cyanide (§ 11), an *increase* in the concentration of the precipitant renders the precipitation less complete. It has already been stated, however, that the cause of this phenomenon is a new reaction between cyanide ion, serving as precipitant, and the precipitated silver cyanide. Hence *variations* from the above-mentioned decrease of solubility frequently lead to the discovery of new reactions. The significance of the solubility product is clearly illustrated by the determination carried out for potassium perchlorate in Principles, p. 442.

As a simple reflection will show, the solubility product has the significance of a concentration to the second power for binary electrolytes, and for electrolytes which dissociate into 'n' ions it has the significance of a concentration to the n'th power. This may be readily shown, for example, by applying the mass action law to the dissociation of barium chloride, $BaCl_2 \rightleftarrows Ba^{\cdot\cdot} + 2Cl'$. The deduction is left to the reader for the sake of practice. All necessary information has already been given.

92. **The separation of strontium- and calcium ions.** The test for $Sr^{\cdot\cdot}$ mentioned in § 87, by adding saturated calcium sulphate solution to a solution which may contain both strontium and calcium ions, has the disadvantage that the precipitation of strontium sulphate takes place very slowly. This is due to the fact that the concentration of the sulphate ions in a saturated solution of calcium sulphate is rather low. Therefore the concentration product of $Sr^{\cdot\cdot}$ and SO_4'' will exceed the solubility product of strontium sulphate to only a limited extent under ordinary conditions. To use a solution containing more SO_4'', say of sodium sulphate or sulphuric acid, would avoid the difficulty. However, this would also give rise to the conditions for precipitating calcium sulphate, which might be mistaken for strontium sulphate, and hence may not be used.

As a matter of fact conditions for forming a precipitate of calcium sulphate are often met with. This is the case when a *saturated* calcium sulphate solution is added to a solution containing calcium ion in ordinary concentrations. The liquid will become supersaturated with respect to calcium sulphate. In the *precipitant the solubility product* is already reached, as the solution is saturated. The decrease of concentration through mixing it with the solution under investigation is usually less than the increase arising from the calcium ions in the solution. Often no precipitate will form, however, probably because the *excess over the solubility product of calcium sulphate* is only

slight. Under these conditions a transformation leading to equilibrium may be put off for some time. A sure means of doing away with the supersaturation consists in introducing a small quantity of the substance with respect to which the solution is supersaturated (compare § 104). This means is purposely avoided in the present instance.

Reactions of the Fifth Group Ions.

93. *Calcium, strontium* and *barium* form only divalent colorless cathions. Complex ions are not formed in detectible amounts under the ordinary conditions of analysis. Consequently the properties of the ions of these metals are readily kept in mind.

The strong development of cathion characteristics is particularly shown in the hydroxide solutions, which have specific basic properties. A limitation is caused, however, by the slight solubility of the hydroxides.

94. **Calcium.** The test solution is most easily prepared by dissolving $\frac{1}{20}$ formula weight of calcium chloride, $CaCl_2.6H_2O$,[1] or of calcium nitrate, $Ca(NO_3)_2.4H_2O$, in a liter of water.

Potassium hydroxide forms a white precipitate of calcium hydroxide, $Ca(OH)_2$. Ammonia yields *no precipitate*. This is due to the fact that much less hydroxyl ion is contained in an equimolecular ammonia solution. Hence the solubility product of calcium hydroxide is not reached.

Ammoniacal solutions of calcium salts often yield a precipitate on standing for some time in contact with air. The precipitate is not calcium hydroxide, however, but calcium carbonate. This may be readily shown by filtering and pouring dilute hydrochloric acid over the precipitate on the filter. It decomposes with effervescence of carbon dioxide. If the amount of precipitate is very minute an effervescence may not be detected. The carbon dioxide which gives rise to this precipitate comes from the air which, especially in laboratories, always contains considerable quantities of this gas.

Alkali carbonates give a precipitation of difficultly soluble *calcium carbonate*. The precipitate dissolves on passing carbon dioxide into the liquid with which it is in contact. The CO_3''-ion formed from dissolved calcium carbonate is used up through the following reaction:

$$CO_3'' + CO_2 + H_2O \rightleftarrows 2HCO_3'.$$

[1] There are several hydrates of calcium chloride. Under ordinary conditions the most stable one is that containing $6H_2O$ (Principles, p. 515). For the preparation consult § 95.

Consequently solid carbonate goes into solution as long as any carbon dioxide is present. The salt $Ca(HCO_3)_2$ is not known in the solid form. All attempts to isolate it by evaporating the solution result in the precipitation of calcium carbonate in consequence of a reversal of the above reaction.

With solutions of *sulphates* and with dilute *sulphuric acid* calcium ion yields a white precipitate of *calcium sulphate.*

Disodium hydrogen phosphate solution precipitates white calcium phosphate (compare § 78, footnote). The precipitate dissolves in acetic acid.

From the solution *ammonium oxalate* precipitates *calcium oxalate,* soluble in dilute hydrochloric or nitric acid, practically insoluble in acetic acid (the precipitate will dissolve in a very large quantity of acetic acid).

A drop of calcium salt solution, evaporated on a platinum wire and heated to redness, imparts an orange color to the bunsen flame.

When heated with sodium carbonate on charcoal, solid calcium compounds yield white calcium oxide, infusible at the temperatures reached with a blast flame.

95. The preparation of carbon dioxide and calcium chloride. For laboratory purposes carbon dioxide is prepared from natural calcium carbonate (chalk, marble or limestone). Decomposition is effected either by heating the *dry* salt or by means of an acid.

The following reaction takes place on heating calcium carbonate:

$$CaCO_3 = CaO + CO_2.$$

The only volatile product is carbon dioxide, calcium oxide being practically non-volatile. The reaction is reversible, $\leftarrow$, as is shown by the fact that calcium oxide goes over into carbonate when left in an atmosphere containing carbon dioxide. In case all three substances, $CaCO_3$, CaO and CO_2 are present in a system, it depends upon the pressure exerted by the carbon dioxide how the reaction will proceed.

For every temperature there is a fixed value of the pressure under which equilibrium exists, i. e. no transformation occurs. When the pressure is decreased more carbonate decomposes, when increased more carbonate forms. In preparing carbon dioxide by heating calcium carbonate the gas is allowed to escape. In this way the condition for the complete decomposition of a definite amount of calcium carbonate is established.

The carbonate may be heated in a tube similar to the one used for the experiment in § 35. The escaping gas is passed into *baryta* solution, a solution of *barium hydroxide.* It precipitates *barium carbonate.*[1] The residue left in the tube is shaken into a porcelain evaporating dish. When moistened with water it forms calcium hydroxide, developing a considerable amount of heat (slaking). The liquid in contact with the calcium hydroxide that is formed in this experiment colors red litmus paper blue (compare § 93 on the properties of the hydroxide).

The preparation of carbon dioxide by decomposing a carbonate with an acid is usually carried out in an apparatus which permits the generation of gas to be regulated and stopped at any point desired. Of the various contrivances serving this purpose the Kipp apparatus, illustrated in Fig. 5, is most used. The method of operation is readily seen on using a Kipp. The middle bulb is filled with the solid material (marble), and acid is poured into the upper bulb through a funnel, the stopcock "h" being open, until it remains partly filled when the stopcock is closed.

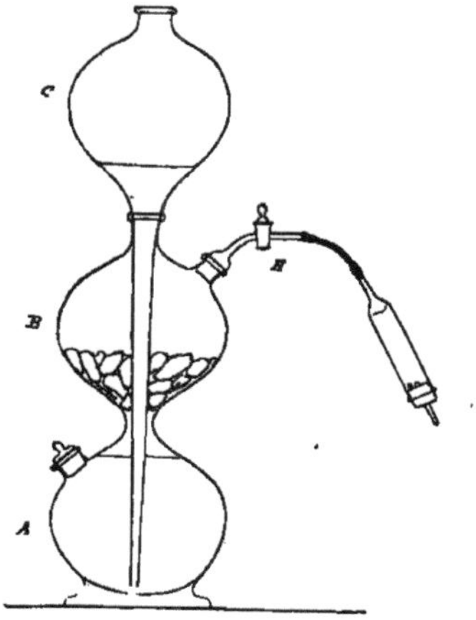

FIG. 5.

The middle and lower bulbs are connected in such a way that the acid can pass into the middle bulb on opening the stopcock and is driven out of it when the stopcock is closed. The opening in the lower bulb is closed with a glass or rubber stopper. It serves to draw off the liquid after the acid has been used up, when the solution has become inactive through the accumulation of *one of the reaction products,* $Ca^{\cdot\cdot}$, and through the disappearance of hydrogen ion. The solution may then be used to prepare calcium chloride. Even when pure hydrochloric acid was used in decomposing the carbonate, the solution will generally have a yellow color due to small quantities of ferric chloride. It is evaporated on a water-bath until a skin of crystals

[1] If a solution of *barium chloride* or *barium nitrate* is used instead of baryta water, *no separation* of barium carbonate takes place, even when the chloride or nitrate solution is more concentrated, i. e. contains more barium ion, than the baryta solution. Though this is quite self evident, nevertheless it often surprises the student. How is it to be explained? (compare § 137).

forms over the surface, and then under constant *stirring* it is evaporated further over an asbestos wire gauze until it becomes thick with small crystals. When cooled this becomes solid. The composition is somewhat variable in consequence of a partial dehydration. In calculating the amount to be used in preparing the test solution (§ 94) the formula may be assumed to be $CaCl_2.4H_2O$.

When commercial hydrochloric acid is used, the solution invariably contains other ions which interfere with the preparation of calcium chloride by leaving other chlorides mixed with the calcium chloride on evaporating the solution. To obtain a pure product in this case the group reagents are systematically added to the solution. The precipitate obtained with ammonium carbonate is then dissolved in dilute hydrochloric acid and the solution evaporated on a water-bath.

On a commercial scale this method would be too expensive. When the solution no longer reacts with calcium carbonate it is heated with a suspension of slaked lime. By this means the most common impurities, as $Fe^{\cdot\cdot\cdot}$, $Mn^{\cdot\cdot}$ and $Mg^{\cdot\cdot}$, are precipitated and then they are filtered. The filtrate is next evaporated in any convenient way and towards the end of the operation some concentrated hydrochloric acid is added to limit the formation of basic salts. For the same reason the evaporation is also carried out more slowly from this point on. The product is anhydrous calcium chloride.

96. Theory of the decomposition of carbonates by means of acids. The general equation for this reaction is the following:

$$\underline{MCO_3} + 2H^{\cdot} = M^{\cdot\cdot} + H_2O + CO_2,$$

where M is the symbol of a divalent metal. The law of mass action may be applied to this reaction just the same as it was applied to the precipitation of metallic ions by hydrogen sulphide and the reverse reaction of the dissolving of sulphides in acids.

In the present case only the dissolving of the precipitate has practical importance. The reverse reaction which takes place under certain conditions when carbon dioxide is passed into a solution containing the ions $M^{\cdot\cdot}$, ← is represented by the equation:

$$CO_3'' + CO_2 + H_2O \rightarrow 2HCO_3'.$$

For an acid to effect a complete decomposition of a carbonate the carbon dioxide which, along with water, results from the carbonate- and hydrogen ions, must escape as fast as formed. The fact that a weak acid, as acetic

acid, can decompose a carbonate, indicates that carbonic acid, which first forms from the ions:

$$CO_3'' + 2H^{\cdot} = H_2CO_3,$$

is much less dissociated than acetic acid.

In the case of the decomposition of calcium carbonate the rules stated in § 28, concerning the influence of the concentration of substances taking part in a reaction, are not so evident so far as the end condition comes into question. Equivalent amounts of hydrochloric and acetic acids will completely decompose a given quantity of calcium carbonate, provided carbon dioxide is allowed to escape. There is a difference, however, in the rate of action of various acids. Acetic acid will take several times as long as hydrochloric or nitric acids. A primitive experiment serves to illustrate this difference. In each of three 150 cc. flasks the same quantity, 1 g., of calcium carbonate is placed, and equal amounts of acetic, hydrochloric and nitric acids are poured into the respective flasks. The quantity of acid must be a little in excess of the amount[1] just needed to effect complete decomposition. It is observed that hydrochloric or nitric acid dissolve the carbonate in a short time, while acetic acid takes considerably longer. For class demonstration the dissolving of zinc is better adapted, however (Principles, p. *241*).

The slight dissociation of acetic acid in comparison with the others affords an explanation of this phenomenon. It is seen that the *rate* of *reaction* is *influenced* by the *concentration* of *hydrogen ion*, which is immediately concerned in the reaction. The final result of the transformation indicates no difference in the behavior of the various acids employed.

The quantitative study of these relations has led to the following law. In most cases the reaction velocity, which is measured by the quantity of substances transformed in unit time, is proportional to the n'th power of the concentrations of the components, when 'n' formula weights of the components take part. However, in reactions where 'n' has a high value complications are very often met with.

It has already been stated in § 30 that the concentration of solid substances may be regarded as constant, independent of form and quantity. Only in certain cases have exceptions to this law been noted, for example, in the case of mercuric oxide (§ 77, footnote).

The influence of the fineness of the particles makes itself apparent in studying the velocity of a reaction. This is shown by comparing the action of hydrochloric acid upon equal quantities of calcium carbonate, in one case coarsely powdered, in another very finely divided.

[1] The acid solutions contain two formula weights of acetic, hydrochloric or nitric acid per liter. Since one formula weight of any of these acids will give one combining weight of hydrogen ion, *equal volumes of the solutions can neutralize equal volumes of a given base.*

The quantity of acid required to decompose 1 g. of calcium carbonate is calculated as follows. One formula weight of calcium carbonate (100 g.) is decomposed by two combining weights of hydrogen ion. These come from two formula weights of acid, i. e. from one liter of the solution. Hence 1 g. of calcium carbonate will take 10 cc. acid. For each experiment, however, 12 cc. should be used. It may be measured in a graduated cylinder or test-tube.

97. **Strontium.** The ions of strontium, barium and calcium resemble one another so closely that differences in their properties are observed only in quantitative experiments. With the *chromates, sulphates* and *sulphites* the solubility products *decrease,* with the *hydroxides* and *oxalates* they *increase* in the order: calcium, strontium, barium. The difference in the properties of the chromates is most striking of all. Calcium chromate is rather easily soluble, while barium chromate is one of the least soluble barium salts.

Volatile strontium salts impart a carmine red color to the non-luminous bunsen flame. The chloride serves best for this test.

98. **Barium.** Barium ion is separated from strontium- and calcium ion by precipitation as barium chromate (§ 87). To effect a complete separation the fifth group carbonates must be dissolved in acetic acid. Only in the presence of acetate ion or the anion of a similar weak acid is the precipitation practically complete. This may be readily shown by adding a slight excess of *potassium dichromate* to a solution of *barium chloride,* filtering and then adding sodium acetate to the yellow filtrate.

A second precipitation is observed under these conditions. The precipitate redissolves upon addition of hydrochloric acid. By considering the reaction:

$$K_2Cr_2O_7 + 2BaCl_2 + H_2O \rightleftarrows 2BaCrO_4 + 2KCl + 2HCl \quad (1a),$$

from the standpoint of the mass action law (§ 29) and remembering that acetic acid is but little dissociated and hence always forms when acetate ion and hydrogen ion are present in equivalent amounts in the same solution, it is easy to see why the addition of apparently indifferent sodium acetate causes a further precipitation.

The relations become still clearer when the dissociation of the various substances is taken into account, and those ions which undergo no essential change of concentration are left out of the equation. The equation then reads:

$$Cr_2O_7'' + 2Ba^{\cdot\cdot} + H_2O \rightleftarrows \underline{2BaCrO_4} + 2H^{\cdot} \quad (1b).$$

The study of this equation shows that dichromate ion, Cr_2O_7'', reacts with water to form $2CrO_4''$ and $2H^{\cdot}$. Barium ion yields a precipitate with chromate ion, CrO_4'', thus removing it from the system as fast as it forms. In this way the conditions for a complete transformation are established.

The equation:

$$Cr_2O_7'' + H_2O \rightleftarrows 2CrO_4'' + 2H^{\cdot} \quad (2),$$

indicates the relations existing between dichromate ion and chromate ion. They are clearly shown by the following experiments.

The yellow color of potassium chromate solution changes to red upon addition of acetic acid, or even by passing carbon dioxide into the solution. On the other hand ammonia or alkali hydroxide will turn dichromate solution yellow. Evidently hydrogen ion reacts with chromate ion, giving a transformation in the direction ←. Since the concentration of hydrogen ion is very low in alkaline solutions (§ 50), they will contain practically only chromate ion.

The fact that dichromate solutions have an acid reaction, while chromate solutions are alkaline, agrees with these relations (Principles, pp. *606–7*).

Apparently potassium chromate would furnish a more efficient means of separating barium ion, since its solutions contain only chromate ion. It would also precipitate strontium ion, however, and hence is not used as a reagent for Ba¨.

Volatile barium salts impart an apple green color to the non-luminous bunsen flame.

GROUP VI.

99. The filtrate from the fifth group may contain the ions Mg¨, K˙ and Na˙. Whenever a large quantity of ammonium salts have been introduced in the course of the analysis, the precipitation of fifth group ions with ammonium carbonate is incomplete. Hence it is always essential to make sure of a complete separation of the fifth group before proceeding further. To detect any Ba¨ or Sr¨ that may be left, dilute sulphuric acid is added to a small portion of the filtrate, while a separate portion is tested with ammonium oxalate for Ca¨.

In case a cloudiness is observed within a few minutes after adding sulphuric acid, the entire solution is boiled. To the hot solution *dilute sulphuric acid* or *ammonium sulphate* solution is then added drop by drop. In many cases it will be necessary to lessen the volume by evaporating the solution before adding sulphate ion, since the precipitation of these sulphates from dilute solutions always requires considerable time (§ 87). If the system is frequently shaken and it set aside *20–30* minutes it is safe to assume that equilibrium has been reached (i. e. that the ions precipitated by SO_4'' are practically completely removed. The precipitate is then filtered. A portion of the filtrate or acidulated solution is made *ammoniacal* and is tested for *calcium ion* with *ammonium oxalate.* In case a precipitate forms the entire solution is treated with oxalate.

The filtrate is now ready to be tested for sixth group ions. *Magnesium ion* is detected by the insolubility of *magnesium ammonium*

phosphate,[1] $MgNH_4PO_4$. Only *part* of the solution is used for this test. Since it already contains some ammonium ion, just enough ammonia is added to give it an alkaline[2] reaction, and then disodium hydrogen phosphate is added. Under these conditions a clear crystalline precipitate is obtained if magnesium ion is present (compare § 102).

The above method cannot be used to *remove* magnesium ion from the rest of the solution since it introduces sodium ion, which is to be tested for along with potassium ion. Hence another means must be employed. This will be described after the simpler case has been considered, viz. that a solution containing no magnesium ion is to be tested for $K^{\cdot}$ and $Na^{\cdot}$.

100. Detection of potassium- and sodium ion. The solution which contains no *magnesium ion* is evaporated *to dryness* on a water-bath or over an *Ostwald burner*. The residue is shaken into a porcelain crucible or evaporating dish and is heated over an asbestos wire gauze, at first gently and finally at the full heat of a bunsen burner. This removes ammonium salts. If the temperature is raised too high, however, alkali salts (chiefly the chlorides) volatilize to a noticeable extent. *After cooling the residue is dissolved in a small amount of water.* If necessary *this solution* is filtered and is then used for the tests.

To a portion of the *neutral* solution an equal volume of *sodium hydrogen tartrate* solution is added. *Potassium ion* is indicated by the precipitation of *potassium hydrogen tartrate* (cream of tartar) after a few minutes' standing. The test is made much more certain by introducing an extremely minute (invisible) quantity of cream of tartar[3] into the solution on the end of a platinum wire. The reasons for this addition are stated in § 104.

Another portion of the solution is tested for *sodium ion* by means of a solution of dipotassium dihydrogen pyroantimonate, $K_2H_2Sb_2O_7$, which is prepared just before use by heating a small quantity

[1] This salt has a great tendency to form supersaturated solution. Hence the test-tube should be vigorously shaken after adding the reagents. It is frequently recommended to rub the walls with a glass rod instead of shaking. Errors are often caused by this means, however (compare § 100, last footnote).

[2] The above simple condition is very frequently neglected.

[3] In the form of a dry mixture containing 1 per cent cream of tartar, 99 per cent sodium nitrate, which must be free from *potassium* nitrate.

of the salt with water. The solution must be cooled before it is used in the test for sodium ion. A granular crystalline precipitate, $Na_2H_2Sb_2O_7$, is formed if sodium ion was present.

When the solution contains a little *hydrogen ion* or *ammonium salts* an amorphous precipitate of antimonic acid is obtained, which interferes with the recognition of disodium dihydrogen pyroantimonate. Hence inexperienced observers verv often *draw* wrong conclusions from the test. However the difference between the densities of antimonic acid and the pyroantimonate serves to separate and identify the precipitates. The contents of the test-tube are transferred to a watch glass and the liquid is then slowly poured away from the precipitate. The salt has a much greater density than antimonic acid, hence it settles to the bottom quickly and remains behind on the watch glass. The acid, however, follows the movement of the liquid and is poured away with it. By placing some more water on the watch glass and then pouring the liquid off again, even minute quantities[1] of antimonates can be separated from antimonic acid.

The precipitate of pyroantimonate also has the characteristic property of sticking tight to the glass after standing for a short time. In consequence it is more readily perceived on pouring off the rest of the liquid and the antimonic acid. This is indicative of the sodium salt only when the solution of potassium pyroantimonate which is used as a reagent is not supersaturated. Hence the reagent solution must be cooled and shaken from time to time before it is used. Otherwise the potassium salt might separate during the test and of course it would be mistaken for the sodium salt.

In carrying out these tests for $K^{\cdot}$ and $Na^{\cdot}$ the reagents must always be allowed to react for a long time (say *30* minutes).

101. **The removal of magnesium ion** is effected by precipitating with baryta water. Since this precipitation is incomplete if *ammonium salts are present,* the solution is first evaporated to dryness and the residue heated over an asbestos wire gauze until ammonia vapors cease coming off (compare § 99). The *ignition residue* is next dissolved in water (a little dilute hydrochloric acid being added if neces-

[1] Rubbing the walls of the test-tube with a glass rod is sometimes recommended to effect the separation of the precipitate. Glass may, however, be rubbed off by this means and is then mistaken for antimonate. Hence it is best to avoid this operation entirely. Moreover by shaking the test-tube vigorously the precipitate can always be made to form.

sary). The hot solution is then *precipitated*[1] with *baryta water*[2] until the *filtrate will no longer yield any precipitate with baryta.* The solution now contains barium ion in consequence of the removal of magnesium ion with barium hydroxide. This is precipitated from the filtrate by means of ammonium carbonate, or better by sulphuric acid.

To remove the excess of sulphuric acid or ammonium salts the filtrate from the barium precipitation is heated on a water-bath until water vapor no longer escapes, and finally in an air-bath or on a sand-bath. The residue is then tested for $K^{\cdot}$ and $Na^{\cdot}$ according to § 100.

Errors often arise from boiling the above alkaline solution in porcelain or glass vessels. They are caused by a reaction between the glass or porcelain and baryta water, in consequence of which ions of the alkali metals go into solution. To avoid this as far as possible Jena glass vessels should be used, and in place of porcelain, platinum vessels wherever practicable.

Reactions of the Sixth Group Ions.

102. Magnesium. Magnesium forms only divalent cathions which are analogous in properties with the ions of the fifth group metals. Magnesium salts are considerably more soluble, however, except the hydroxide.

The test solution contains $\frac{1}{20}$ formula weight per liter of the sulphate, $MgSO_4.7H_2O$, or the chloride, $MgCl_2.6H_2O$. The latter salt may be easily prepared by dissolving magnesium in hydrochloric acid.[3]

Alkali hydroxide and *baryta solution* precipitate *magnesium hydroxide.* The precipitate redissolves upon addition of an *ammonium salt* or *ammonia.* Hydroxyl ion from magnesium hydroxide

[1] The formation of a precipitate with baryta water is *not an absolute criterion of the presence of magnesium ion,* since sulphate ion (§ 112) may also be contained in the solution and it gives a precipitate with baryta. A wrong conclusion is very often made with regard to this point.

[2] [For the use of HgO in place of baryta consult Abegg, Handbuch der anorganischen Chemie, Vol. II, 2, p. 42.]

[3] On evaporating the solution to prepare the solid salt a basic chloride is frequently obtained, owing to a partial volatilization of the acid resulting from the hydrolysis:

$$\underbrace{2Cl' + Mg^{\cdot\cdot}} + H^{\cdot}OH' = \underbrace{Mg(OH)^{\cdot} + Cl'} + \underbrace{H^{\cdot} + Cl'}.$$

and ammonium ion form ammonia. This uses up hydroxyl ions and hence gives rise to the condition necessary for the dissolving of more magnesium hydroxide. The solution process continues as long as sufficient ammonium ion is present (Principles, p. 532).

Solutions of *alkali carbonates* form a white precipitate of *basic carbonate*. It is not certain whether this has a *definite composition* or whether it is simply a *mixture* of *carbonate* and *hydroxide* in *variable* proportions. It dissolves in ammonium salts.

The formation of crystalline *magnesium ammonium phosphate* (§ 99) is a characteristic test most frequently used in analysis to detect magnesium ion. The *addition* of an *ammonium salt* is necessary to effect the precipitation for two reasons. In the first place ammonium salts are much more dissociated than ammonia, hence are more active in giving the solubility product of magnesium ammonium phosphate. In the second place the precipitation of *magnesium hydroxide* is thereby prevented (compare §§ 78 and 89). A little ammonia must be added, but only enough to neutralize the acid which is formed by the precipitation of magnesium ammonium phosphate.

As was mentioned above (§ 99), hydrogen ion has a disturbing influence upon the precipitation owing to the formation of the ions HPO_4'' and H_2PO_4' (consult § 129):

$$\underline{MgNH_4PO_4} + H^{\cdot} = Mg^{\cdot\cdot} + NH_4^{\cdot} + HPO_4'',$$

and

$$\underline{MgNH_4PO_4} + 2H^{\cdot} = Mg^{\cdot\cdot} + NH_4^{\cdot} + H_2PO_4'.$$

Any magnesium compound forms a white sublimate of *magnesium oxide* when heated with sodium carbonate in the reducing flame on charcoal.

103. Potassium. The ions of the alkali metals have very few reactions which can be used to identify and distinguish them. With most anions they form soluble compounds having no other striking characteristics. The test solutions are made from the chloride, nitrate or sulphate and contain $\frac{1}{10}$ formula weight of the salt per liter.

With *chloroplatinic*[1] (or *hydrochloroplatinic*) acid, H_2PtCl_6,

[1] The term chloroplatinic acid will always be used later (Miolati). Corresponding with this hydrofluosilicic acid will be called fluosilicic acid, H_2SiF_6, i. e. silicic acid with 6F instead of 3(O).

potassium ion yields a yellow precipitate of potassium chloroplatinate, K_2PtCl_6. Formation of the precipitate is promoted by mechanical means, as rubbing the wet walls of the test-tube with a glass rod, or *better* by shaking. Addition of alcohol renders the test more delicate because the salt is less soluble in a mixture of alcohol and water. The liquid becomes brown in case iodide ion is present (compare § 198 on platinum).

Accordingly potassium iodide must be transformed into the chloride by evaporating the solution with concentrated hydrochloric acid.

Fluosilicic acid, H_2SiF_6, yields a difficultly soluble precipitate, K_2SiF_6, which is not readily perceptible, however (Principles, p. 465). The test must be carried out in an acid solution, for hydroxyl ion effects a decomposition of fluosilicate ion, giving rise to fluoride ion and insoluble silicic acid:

$$SiF_6'' + 4OH' = \underline{Si(OH)_4} + 6F' \qquad (1).$$

The same decomposition occurs in neutral solutions, but considerably more slowly.

Potassium ion also forms a difficultly soluble salt with the anion of perchloric acid, ClO_4'. [In 70 per cent alcohol potassium perchlorate is much less soluble than in water, Principles, p. 454.]

Sodium hydrogen tartrate,[1] already mentioned in § 100, is the reagent most frequently used to detect potassium ion. *Customarily* a *mixture* of *tartaric acid* and *sodium acetate* is taken in place of *sodium hydrogen tartrate.* This is less *suitable* because potassium hydrogen tartrate is much more soluble in solutions containing sodium acetate than in water. Consequently the delicacy of the reaction is considerably lessened, especially when a large *excess* of *sodium acetate* is added.

The greater solubility of cream of tartar in solutions containing acetate ion is due to the *formation* of *acetic acid* with the hydrogen ion of the tartrate. When alkali acetate is dissolved in a *saturated solution* of *cream of tartar* which is in *contact* with the *solid salt,* hydrotartrate ion and acetate ion will form acetic acid. Since this disturbs the equilibrium between the solution and solid salt, more of the latter will go into solution. The dissolving continues until the products of the reaction have accumulated in sufficient amounts to bring it to an end. This comes from the reversibility of the reaction, which is shown by the fact that a solution of potassium sodium tartrate

[1] A solution adapted to this purpose is obtained by dissolving 18.8 g. tartaric acid and 28.8 g. disodium tartrate in a liter of water.

likewise precipitates cream of tartar when acidified with acetic acid. The following equation formulates these relations:

$$HC_4H_4O_6' + C_2H_3O_2' \rightleftarrows H^{\cdot}C_2H_3O_2' + C_4H_4O_6'' \qquad (2).$$

Only when very large amounts of sodium acetate are added and the concentration of potassium ion is particularly low, is the delicacy of the above reaction lessened sufficiently to spoil the test entirely. The precipitation is incomplete, however, if too small a quantity of sodium acetate is present in a mixture containing potassium ion and tartaric acid, because the concentration of hydrogen ion is then not low enough. For the dissociation of tartaric acid and the formation of the ion $C_4H_5O_6'$, on which the delicacy of the test depends, is *increased* by small amounts of sodium acetate.

The use of sodium hydrogen tartrate as a reagent for neutral solutions of potassium ion very easily does away with all difficulties in the matter of judging correctly the required concentration of hydrogen ion with the other method.

104. To remove supersaturation. In all cases where the conditions favor the separation of a solid, and yet no precipitation takes place, it is found that the addition of a very minute quantity of the solid in question will give rise to a separation. Thus in the above example (§ 100) equilibrium was reached in this way by adding a scarcely detectible amount of cream of tartar (in the form of a mixture with sodium nitrate).

These considerations refer to cases where the state of equilibrium has been overstepped and where no spontaneous establishment of the equilibrium takes place. They are not limited to solutions supersaturated with respect to solids, but are also observed with solutions of gases in liquids, the solidification of molten substances, etc., in fact wherever a system is met with which can be separated by physical means into several spatially distinct phases (heterogeneous phases). Extreme instability is represented by the *labile state.* To realize it is an unusually difficult matter, for it passes into the stable condition of equilibrium very rapidly and without the addition of foreign substances. Systems in the *metastable state* are characterized by greater stability. Their existence is characteristically influenced by other substances, however (consult Principles, pp. *116, 127, 214, 438*).

The analyst has frequently to deal with conditions of supersaturation (§§ 87, 94) which come under the head of metastable states. Vigorous shaking and rubbing the wet walls of the containing vessel with a glass rod are generally employed to obtain a precipitation

in cases where the supersaturation is persistent. The *most effective* and *least objectionable* method, however, is the introduction of a very minute quantity of the solid itself: *method of inoculation.* For purely practical reasons the *active* substance is mixed with an *indifferent* solid in order to effect the introduction of the smallest possible amounts.

Experiments with specially dilute solutions of a potassium salt will best illustrate the activity of such an inoculating mixture (§ 100). It is advisable to keep some always in stock. The quantity of the mixture which will produce the desired result when introduced into the liquid on a platinum loop will vary from 1–3 milligrams. Hence the weight of potassium hydrogen tartrate actually used is only 0.01–0.03 milligrams. Mixtures[1] which contain a still smaller percentage of cream of tartar will also be active in many cases. The one per cent mixture is nevertheless more trustworthy and hence more advisable.

105. For sodium ion the test with *dipotassium dihydrogen pyroantimonate* (§ 100) is the most important.

Use is also made of the insolubility of potassium chloroplatinate in aqueous alcohol to separate potassium- and sodium ion. Sodium chloroplatinate dissolves in alcohol. When the solution is slowly evaporated it crystallizes in triclinic columns or plates.

106. Flame reactions. The salts of the alkali metals as well as barium, strontium and calcium salts, give off colored vapors which serve to identify these metals. To carry out flame tests one end of a platinum wire is sealed into a glass tube and the other end is bent into a small loop. The loop is heated red hot and is then brought in contact with the solid substance under investigation. The salt melts to a bead in the flame.

In case a solution is to be tested in this way a drop is taken up on the platinum loop which is slowly brought near the flame until the solvent evaporates. Care must be taken not to let the liquid boil because it will then spirt and so occasion a loss of the solute. The evaporation residue is finally treated like the bead. It is heated to redness in the hottest part of the bunsen flame, the *fusion zone,* adac, Fig. *6*, which reaches a temperature as high as 1800° C. The glass tube into which the platinum wire is sealed may be conveniently attached to the end of one of the wires of a wire filter stand.

[1] Concerning the lowest limits with which inoculation can be effected, detailed information is given in Ostwald, Grundriss der allegemeinen Chemie, 3rd Edition, p. *328*. Consult also Zeitschrift für physikalische Chemie, **22**, *289*, 1897.

A piece of rubber tubing of suitable dimensions drawn over the wire serves to increase the friction between the wire and glass tube, and thus affords a more solid support, Fig. 7.

Volatile sodium compounds (chlorides, carbonates, nitrates, sulphates) impart a deep yellow color to the bunsen flame. Interposing a crystal of potassium dichromate or a solution of this salt be-

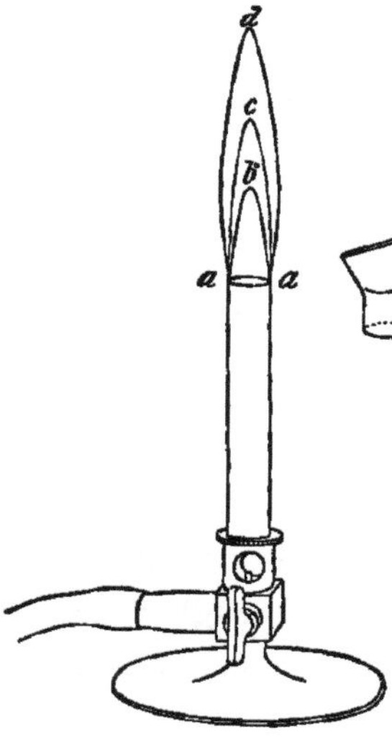

FIG. 6.

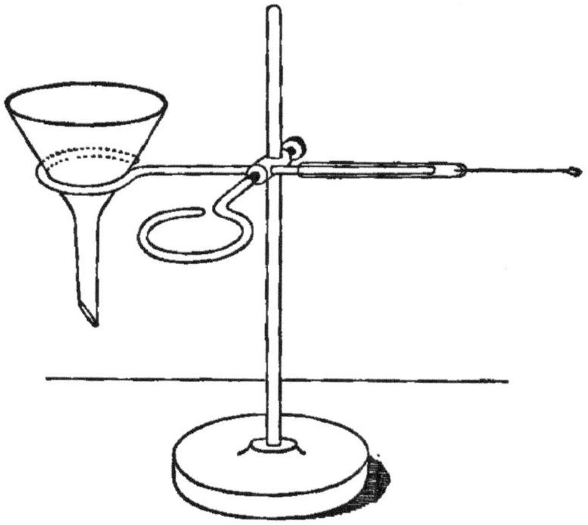

FIG. 7.

tween the eye and the flame affords a ready means of distinguishing the sodium color. The crystal or the solution appears then a bright to dark yellow.[1] In place of this a piece of paper, 1 cm. square, painted with mercuric iodide, may also be used. The paper may be attached adjustably to a wire that is riveted to the tube of a bunsen burner. By illuminating the paper, first with the light from the glowing platinum just beyond the bead and then with that sent out from the bead alone, extraordinarily minute quantities of sodium vapor may be sharply detected.

The detection of sodium compounds in this way is so delicate that

[1] The phenomenon is due to the fact that an object appears of a definite color only when it absorbs light rays. Now potassium dichromate absorbs all rays of the spectrum up to red and yellow. Therefore when a crystal of potassium dichromate is illuminated, not with white light, but with yellow which does not pass through the crystal, only that fraction of the yellow light *reflected* from the surface will meet the eye. If the crystal is simultaneously *penetrated* by *white light* it will appear a more intense yellow.

0.0004 milligrams of sodium chloride suffice to give a distinct color (Bunsen). Hence for the sake of analytical practice the chemical test is to be preferred wherever it is practicable.

Potassium compounds impart a blue color to the bunsen flame. The flame test will detect 0.0002 milligrams of potassium chloride. However many substances which make the bunsen flame luminous also give it a color similar to that imparted by potassium compounds. All organic compounds which separate carbon during combustion act in this way. Volatile or combustible substances may be removed by heating the wire to redness just before the test is made.

Sodium compounds obscure the characteristic color of the potassium flame. If a substance which absorbs the yellow rays of the sodium flame is interposed between the eye and the flame, the detection of potassium compounds will not be affected. Cobalt glass or indigo solution contained in a prism-shaped glass vessel[1] are used as an absorbent for the yellow rays. Viewed from the thinner to the thicker end of the prism the potassium flame appears sky blue, violet or carmine red. [Through cobalt glass the color appears carmine red. The non-luminous bunsen flame appears blue through cobalt glass. This is frequently regarded as the potassium color by inexperienced observers.]

The detection of potassium and sodium in non-volatile compounds, as silicates, will be considered in § 172 on the Hempel method of reduction.

The above-mentioned flame reactions of barium, calcium and strontium are not so delicate because their compounds are less volatile. The chlorides are the most volatile of their salts, hence they give the plainest flame tests. However the light sent out by the vapors of their compounds is of more complex character than that sent out by the vapors of the corresponding alkali salts. Therefore the distinctions are less sharp. To identify them with certainty recourse must be had to a special means which permits a further decomposition of light into its components.

107. Spectrum analysis. That subdivision of analytical chemistry which deals with the means of decomposing light emanating from the vapors of different substances is termed spectrum analysis.

[1] According to Bunsen, Liebig's Annalen, 111, *266*, 1859, the indigo solution contains 1 g. indigo dissolved in 8 g. fuming sulphuric acid and then diluted with water to 1500–2000 cc.

Since spectral methods are rarely used in introductory courses of qualitative analysis, a detailed description of the apparatus serving to produce the vapors and to decompose light into its components will be omitted.

The minimal quantities which can be detected by the colors of the vapors can be reduced much further with the aid of spectrum analysis.

108. **Ammonium ion.** The solution left after removal of the other cathions does not serve for the detection of ammonium ion, because ammonium salts have been introduced in precipitating certain groups. Therefore a *special* test is carried out for ammonium ion by heating a little of the original solution with alkali hydroxide and holding *moistened red litmus* or *turmeric* paper in the vapors. Alkali hydroxide reacts with ammonium salts according to the equation:

$$NH_4^{\cdot} + OH' = NH_3 + H_2O.$$

The resulting ammonia turns red litmus blue and browns turmeric paper. A strip of filter paper moistened with mercurous nitrate solution may also be used in place of litmus or turmeric. It is blackened by ammonia.

Errors often result from carelessness in carrying out the experiment. They are avoided by performing the test in a tall narrow beaker without a lip. This is covered with a watch glass on the convex surface of which is fastened the reagent paper moistened with water. Spirting of the alkaline liquid can be avoided by gently warming, best on a water-bath. Furthermore diffusion of the ammonia to the outside takes place slowly under these conditions and hence the time during which the test paper is in contact with the vapors is lengthened. In this way the test is made more delicate.

109. **The reactions of ammonium ion** resemble those of potassium ion. *Ammonium chloroplatinate* and *ammonium hydrogen tartrate* are difficultly soluble. The color imparted to the flame by potassium salts on the one hand, and the formation of ammonia from ammonium salts through the action of alkali hydroxide (§ 108) on the other hand, serve to distinguish them.

"Nessler's reagent," an alkaline solution of potassium mercuric iodide, K_2HgI_4, affords a particularly delicate test for ammonium

ion. It yields a red brown precipitate with ammonium salt solutions. Very small quantities are indicated by a yellow brown coloration. The precipitate has the composition $NHg_2I.H_2O$. It is formed according to the equation:

$$NH_4.OH + 2K_2HgI_4 + 3KOH = \underline{NHg_2I.H_2O} + 3H_2O + 7KI.$$

By leaving out the ions which undergo no changes the equation assumes the following form:

$$NH_4^{\cdot} + 2HgI_4'' + 4OH' = \underline{NHg_2I.H_2O} + 3H_2O + 7I'.$$

This reaction is specially important in testing the value of natural waters for drinking purposes. The presence of ammonium salts indicates that the water has been in contact with substances which produce ammonia through the spontaneous changes which they undergo. In particular this is true of organic substances undergoing decay. Consequently when the ammonium salts exceed a certain limit—fixed by law—the water must not be used for drinking purposes.

[Care must be taken to have the solutions dilute in carrying out the ammonia test, owing to the excess of alkali in Nessler's reagent. Otherwise hydroxides may be precipitated and inexperienced observers frequently mistake this precipitate for the ammonia reaction. By distilling the ammonia into Nessler's reagent this trouble may be avoided.]

PART II.

TESTING A SOLUTION FOR ANIONS.

110. Characteristics of the method. The procedure which has been adopted for the solution of the problem now before us differs from that described in the preceding paragraphs. Although the anions are also divided into group precipitates in accordance with their behavior towards definite reagents, the groups are not *systematically* separated. The reactions which characterize the various groups merely indicate the presence or absence of anions belonging to any one of them.

The subdivision into groups is based upon reactions with three cathions. Accordingly the grouping is defined within narrow limits. Comparatively few experiments in addition to the group test will determine, as a rule, which components are present and which are lacking in a given group. Occasionally the observations recorded with the group reagents are alone sufficient for this purpose.

Barium, lead and *silver* are the cathions which yield the information needed for subdividing anions into groups. Solutions of the nitrates of these metals are used. Barium chloride and lead acetate may also be substituted for the nitrates in the cases of $Ba^{··}$ and $Pb^{··}$ respectively.

The following are the characteristics by means of which anions belonging to individual groups are identified:

Group I: the *barium salts* are practically *insoluble* in hydrochloric acid (§ 111);

Group II: the *barium, lead* and *silver salts* are soluble in *hydrochloric* and *nitric* acids. The *barium* and *lead salts* are only *slightly soluble* in *acetic acid* (§ 114);

Group III: the *barium, lead* and *silver* salts are in part quite readily soluble in water. *All are readily dissolved by acetic acid* (§ 126);

Group IV: the barium salts are *readily soluble,* the silver salts but little soluble in water. The latter are *not* dissolved by *nitric acid,* but are dissolved by *ammonia,*

with the exception of silver iodide and silver ferrocyanide (§ 140);

Group V: the *barium, lead* and *silver* salts are *readily soluble in water.*

The above statements concerning the behavior of the precipitates towards hydrochloric and nitric acids, or towards ammonia, are wholly vague unless the quantities of salt, the behavior of which towards the added substance is to be tested, are defined, as well as the volume of the solution added. A particular note of warning must be sounded against the view that a salt, say the barium precipitate of a first group anion, does not dissolve at all in hydrochloric acid when it is stated that it is practically insoluble in hydrochloric acid: or that another salt, for instance, the barium salt of a third group anion, will dissolve completely in any volume of acetic acid, when it is claimed that the barium salts of third group anions are soluble in acetic acid. To dissolve such a precipitate a definite minimal amount of acid is always required. Hence at least this amount of acid must be used in bringing the precipitate into solution, otherwise errors will arise. With respect to the barium salts that are practically insoluble in hydrochloric acid it follows that a definite amount of the precipitate also dissolves in the definite volume of hydrochloric acid that is added. In this case, however, the quantity of hydrochloric acid which will completely dissolve a given amount of salt is *disproportionately greater* than with the barium salts of second group anions. Also the quantity of acetic acid which dissolves a definite amount of the precipitated salts is disproportionately greater for the second group precipitates than for those of the third group.

As will be seen from what follows, testing for the anions of a solution deals with the behavior of *unknown* salts in *unknown* quantities. Hence in studying the action of added substances on the precipitates an exact estimate of the amounts that are requisite to effect solution is needless. It suffices to estimate them approximately. For example, the quantities of second group barium salts obtained by precipitating 2–4 cc. of a given solution with barium chloride or -nitrate will dissolve completely and readily in 10–20 cc. of dilute hydrochloric acid. In case a slight residue remains after this treatment, it is tested again with some more acid to see if all will go into solution. The amount of a barium precipitate with first group anions which dissolves in this volume of acid can be detected only by using special precautions.

A similar relation, varying more or less from one pair of salts to another, however, is found in the behavior of barium salts of third and second group anions towards acetic acid, and also in the behavior of silver salts towards ammonia.

The tests for anions are *carried out* with a *neutral* solution, to which solutions of the three typical cathions are added. Nitric acid serves as the neutralizing agent in the majority of cases, where the given solution has an alkaline reaction. A *special portion* of the original solution, or of a solution prepared according to § 181, must then be taken to test for the *nitrate ion.* It may be neutralized with acetic or dilute sulphuric acid.

Litmus paper serves as the reagent for testing neutrality, a neutral solution changing neither red nor blue litmus. It very seldom happens that an alkaline *solution* is made *exactly neutral* by adding, say nitric acid. In most cases the point of neutrality will be exceeded. The excess of acid must then be neutralized with a base, ammonia being the best to use for this purpose. After a little practice a given solution may easily be made very nearly exactly neutral in this way.

The chief difficulty met with in carrying out the above operation lies in the addition of the *small* amounts of liquid (acid or base) that are used to effect neutrality. It is easily overcome, however, by dropping the solution from a glass tube instead of from a reagent bottle, which is not readily manipulated to deliver a drop at a time. The dropping tube is conveniently made from a piece of glass tubing, 5–8 mm. in diameter. The one end has its edges rounded, while at the other end the tube is drawn down to a point with a suitable orifice. The acid or base is filled into the tube by suction, and when full the upper end is closed with the index finger of the hand holding the tube. Then the liquid can be suitably dropped by regulating the pressure of the finger. In order to avoid frequent cleaning it is advisable to have *two* such dropping tubes, one for acids, the other for bases. They may be distinguished by giving them different dimensions or by etching with hydrofluoric acid (§ 118).

Another means of arriving easily at the neutral point consists in the use of dilute solutions for neutralizing. Then very small amounts of acid or base may be added and the danger of having either in excess is much lessened.

The precipitates are tested as follows with respect to their solubility in acids.

Acetic acid is first added to the solution containing the barium precipitate in suspension. In case a residue is left hydrochloric acid is then added, or else a fresh portion of the precipitated solution is taken for this experiment. *Complete* solution in *acetic acid* indicates that only *anions* of the *third group* are present. A *partial* dissolving is recognized by the accompanying decrease in the amount of precipitate left. Similarly anions of the second group are indicated when *hydrochloric acid completely* or *partially dissolves* the residue *insoluble* in *acetic acid*. A residue not dissolved by hydrochloric acid contains anions of the first group.

The behavior of the silver precipitate towards nitric acid serves particularly to characterize fourth group anions.

The notes given in the last column of table II on the tests for anions, together with the results of preliminary tests (§§ 170–178),

are of service in determining the anions of the fifth group and also for identifying those of the other groups.

The above-mentioned differences in the behavior of members of the various groups makes the tests for anions quite simple in case only one non-metallic component is contained in the solution. Similarly the problem is still simple when several anions from the same group are present. Difficulties arise, however, if the solution contains several anions belonging to different groups. In this case it is necessary to fall back upon differences in the solubility of analogous salts, which are not at all sharply marked.

It is at once evident that the difficulties relate to that anion, the salt of which reacts more readily, i. e. to greater extent, with the added substance (acid or ammonia). For the less soluble salt remains behind, and from the fact that a residue is left it is clear that the precipitate contains an anion which is not brought into solution by the added substance. In particular only indefinite conclusions may be drawn from the above-mentioned *indication* of the partial solubility of a barium or lead precipitate in acetic or hydrochloric acid, or of a silver precipitate in ammonia, which consists in a *decrease* of the *quantity* of the *precipitate.* It *always* seems as if the *amount* of *precipitate decreases* upon addition of acid or ammonia. This is true even when only a slight reaction takes place between the added substance and the precipitate, and in consequence the decrease in the quantity of the precipitate is also very slight. It is the case because the precipitate is distributed through a larger volume of liquid, and is noticed in particular when the precipitate remains in suspension. But even when quick settling takes place, the estimation of the amount of precipitate from the height of the layer in the test-tube is by no means accurate.

The uncertainty is still more pronounced when the anion which reacts to greater extent with the added substance by reason of the greater solubility product (§ 91) is contained in proportionately small amount as compared with the anion, the precipitate of which is less soluble.

The following procedure affords a more certain distinction in such cases. A few cc. of the solution (HCl, NH_3), the action of which on the precipitate is to be studied, are added to the liquid containing the precipitate in suspension. The mixture is allowed to react for a few minutes with frequent shaking. Then the undissolved residue is filtered. The substance that brought about the solution is next removed from the (clear) filtrate and it is noted if a precipitate results. Acids may be neutralized with ammonia,[1] and in certain

[1] If a silver precipitate is tested in this way it must be remembered that an excess of ammonia will redissolve the precipitate which would form on exactly neutralizing the acid (exceptions are AgI and $Ag_4Fe(CN)_6$. The silver precipitates do not offer any advantages over barium or lead precipitates, however.

cases with sodium acetate (compare § 114). (Why is it not permissible to use potassium or sodium hydroxide in place of ammonia?) Ammonia may be removed by evaporation or by neutralization with an acid. For this, however, no acid may be used which yields a *precipitate* with silver ion, for example, hydrochloric acid (why?).

It is clear that the delicacy of this test may be still further increased by evaporating the filtrate to a fraction of its original volume before adding the neutralizing agent.

A *clear* filtrate is obtained by precipitating from a hot solution to begin with (compare § 115). Consequently it is advisable to heat the solution which is being tested with $Ba^{\cdot\cdot}$, $Pb^{\cdot\cdot}$ or $Ag^{\cdot}$ to the boiling point, and then add the precipitant to this hot solution drop by drop.

Acid or ammonia is allowed to react only after the solution (plus precipitate) has cooled. The table giving the behavior of precipitates towards solvents relates to room temperature.

GROUP I.

111. **Sulphate-, SO_4'', and fluosilicate, SIF_6'', ions** yield with barium ion a white precipitate, practically insoluble in hydrochloric acid. To identify fluosilicate ion in the presence of sulphate ion the property possessed by the former of yielding a precipitate of potassium fluosilicate (§ 103) with potassium ion is important.

The hepar reaction described in §§ 171–172 is essential to the independent identification of sulphate ion. This test may be carried out with the barium precipitate.

Lead salts exhibit some further phenomena characterizing the sulphate ion, which may be utilized to distinguish these two ions. *Lead sulphate* is fairly readily soluble in *concentrated nitric acid*. Reprecipitation takes place, however, on diluting the solution or clear filtrate (§ 112), or better, by driving off the acid. Lead sulphate is further readily soluble in *potassium hydroxide* and in a solution of *ammonium tartrate*. From these solutions hydrogen sulphide precipitates *lead sulphide* (see § 14).

112. **Sulphate ion and sulphuric acid.** Dilute sulphuric acid or alkali sulphate solution containing one formula weight of SO_4 in 20 liters may be used for the following tests.

The two difficultly soluble sulphates of barium (§ 47) and lead

Hence in case third group ions are to be distinguished from those of the second group, the difficulty may be avoided by investigating the barium or lead precipitate instead of the silver precipitate.

(§ 14) are distinguished by the fact that the latter is quite readily soluble in acids.[1]

This is readily shown by precipitating lead sulphate and then pouring concentrated hydrochloric or nitric acid over the precipitate. Gently warming the acid hastens the reaction. On decanting the liquid from any residue that is left and diluting the clear solution with water, a precipitation of lead sulphate takes place.

The above fact may seem surprising. In view of the phenomena hitherto met with, according to which nitric and sulphuric acids are considered as equally strong acids, the reason for the present observation is not at once evident. From the facts already mentioned no reaction is known by which the concentration of the ions formed from the precipitate is altered. Nevertheless a change in concentration does take place and this, in conjunction with other phenomena, has produced a broadening of our views concerning the dissociation of sulphuric acid, whereby the solubility of lead sulphate in hydrochloric, nitric or concentrated sulphuric acid becomes comprehensible. According to this broader conception sulphuric acid and other polybasic[2] acids do not dissociate into ions in only one simple way, as is expressed by the equation:

$$H_2SO_4 \rightleftarrows 2H^{\cdot} + SO_4'' \qquad (1),$$

but they dissociate in successive stages:

$$H_2SO_4 \rightleftarrows H^{\cdot} + HSO_4' \qquad (2a),$$

and

$$HSO_4' \rightleftarrows H^{\cdot} + SO_4'' \qquad (2b).$$

Therefore the concentration of the sulphate ion, SO_4'', is not related quite simply to the concentration of the hydrogen ion and non-dissociated sulphuric acid in accordance with the law of mass action, but the "strength" (that is, the proportionate further dissociation) of the ion HSO_4' also comes into play. If this has the character of a weak acid[3] the concentration of the ion SO_4''

[1] For the reasons given in § 91 an excess of dilute acid will render the precipitation of lead ion more complete. The effect is limited, however, by the solubility of lead sulphate in dilute acids. Hence the solubility of lead sulphate in dilute sulphuric acid passes through a minimal value.

[2] Di- and polybasic acids are those which contain two or more combining weights of hydrogen per formula weight of acid. Each combining weight of hydrogen can react with a base to form water according to the equation:

$$H^{\cdot} + OH' \rightleftarrows H_2O.$$

As will be seen from later examples this latter point is very essential to the adequate application of the term (compare § 129).

[3] Indications of this are afforded by the fact that a solution of sodium sulphate shows an alkaline reaction towards phenol-phthalein, which becomes more evident on warming the solution. The reaction is:

$$SO_4'' + H^{\cdot}OH' \rightleftarrows HSO_4' + OH'.$$

It states that the concentration product of SO_4'' and $H^{\cdot}$ (from its value in water) is large enough to cause the formation of practically perceptible amounts of the ion HSO_4'.

may have a quite small value, although dilute sulphuric acid has the characteristics of a strong acid on account of the extensive dissociation according to (*2a*). If the concentration of hydrogen ion is increased by the addition of another acid, the dissociation according to (*2b*) is considerably influenced in the sense of a decrease in the concentration of the sulphate ion.

This reaction also takes place when sulphate ion is introduced into a solution which contains a corresponding amount of hydrogen ion, but in some other way, for example, by dissolving a sulphate. More lead sulphate will go into solution as soon as the concentration value of sulphate ion sinks below its value in the saturated solution of lead sulphate, in consequence of the reaction ← according to (*2b*). The reaction between a difficultly soluble sulphate and hydrogen ion follows the equation:

$$H^{\cdot} + \underline{MSO_4} \rightleftarrows M^{\cdot\cdot} + HSO_4' \qquad (3).$$

The extent to which it takes place depends accordingly upon the concentration of the ion SO_4'' (and hence upon the solubility product of the sulphate), and also upon the concentration of hydrogen ion. The increase of concentration of the metallic ion caused by the reaction produces a decrease in the amount of sulphate dissolved (compare § 90) and thus limits the dissolving action.

The fact that barium sulphate is practically insoluble in acids shows that the concentration of the *sulphate ion* is *very slight* in the saturated solution of barium sulphate.

The above explanation for the dissolving of lead sulphate by concentrated acids may be applied only to those cases where the relations justify the assumption that *hydrosulphate ion*, HSO_4', is formed. This interpretation would not be applicable, however, to the dissolving of lead sulphate by very concentrated sulphuric acid, since *anhydrous* or *nearly anhydrous sulphuric acid is not sufficiently dissociated into ions.* Therefore the dissolving could not be referred exclusively, nor even for the most part, to the formation of the *ion* HSO_4'. It stands, however, in a simple relation to the formation of *lead hydrogen sulphate,* which takes place according to the equation:

$$\underline{PbSO_4} + H_2SO_4 = Pb(HSO_4)_2 \qquad (4).$$

By diluting this system with water, lead hydrogen sulphate that remains in solution dissociates chiefly as follows:

$$Pb(HSO_4)_2 = Pb^{\cdot\cdot} + 2HSO_4'$$

Hence the explanation of the dissolving of lead sulphate in concentrated acids not only is not opposed to that holding for the dissolving in nearly anhydrous sulphuric acid, but stands in a close relation with it.

Concerning the solubility of lead sulphate in ammonium salts, particularly in ammonium acetate and ammonium tartrate, consult § 14.

113. The properties of fluosilicate ion: SiF_6''. A solution of fluosilicic acid is used for the tests, since its salts are rather difficultly soluble. Fluosilicic acid, H_2SiF_6, is a strong dibasic acid.

Barium-, potassium and *ammonium ion* form precipitates of the corresponding salts.

Fluosilicates decompose when heated with concentrated sulphuric acid, forming hydrofluoric acid, HF, and silicon tetrafluoride, SiF_4. Hence they give the reactions characteristic of fluorides (§ 117) under these conditions.

GROUP II.

114. **The anions of this group comprise:** oxalate- (C_2O_4''), fluoride- (F'), chromate- (CrO_4''), dichromate- (Cr_2O_7''), sulphite- (SO_3'') and thiosulphate ion (S_2O_3''). In *neutral* solution they yield with $Ba^{\cdot\cdot}$, $Pb^{\cdot\cdot}$ and $Ag^{\cdot}$ precipitates which are *soluble* in *dilute hydrochloric* or *nitric acid, insoluble* in *acetic acid.* The anions of this group are distinguished from those of the first group by the solubility of their barium precipitates in hydrochloric or nitric acid, from third group anions by the insolubility of the precipitated barium salts in acetic acid.

The following is the practical method of determining whether members of the second group are contained in the given solution or not. The barium precipitate is first dissolved in hydrochloric acid. To this solution sodium acetate is then added drop by drop. A precipitate forms at once or after a few minutes if any of these anions are present.[1] Hydrochloric acid must be added only *drop by drop.* In case the barium precipitate does not dissolve completely in dilute hydrochloric acid owing to the presence of first group anions, the liquid and precipitate are separated by filtration, and the *filtrate* is tested with sodium acetate. The means of identifying individual *anions of this group* will be evident from the following paragraphs.

115. **Filtration of barium sulphate.** On attempting to separate "freshly precipitated" barium sulphate from a liquid by means of filtration, it almost always happens that the filtrate runs through cloudy, i. e. that the precipitate goes through the filter. This is due to the fineness of the barium sulphate grains, which are not held back by the coarser pores of the filter. The simplest means of overcoming the difficulty (which also suffices in most cases of this nature) consists in forming larger crystals by *heating the solution before precipitating* it with $Ba^{\cdot\cdot}$, and by adding the barium salt *drop by drop* to the *hot* solution until further precipitation ceases. If necessary a double filter is used. It is made by folding two filters

[1] A separation of *barium hydrogen phosphate* may also take place. The means of distinguishing whether this is the case or not will be considered in § 127.

together so that three layers of the paper lie against each half of the funnel.

Still another way of overcoming the difficulty consists in leaving the precipitate in contact with the solution for a long time. The smaller grains of barium sulphate go over into larger crystals. The transformation is due to variations in the solubility of different sized grains. The smaller are more readily soluble and hence dissolve more quickly and more abundantly. They vanish, while the larger grains grow. The barium sulphate which has gone into solution through the dissolving of the smaller grains reprecipitates upon the larger grains.

The freshly formed precipitate is made up of grains of different size. During the first stages of the separation the precipitate forms more quickly than in the later stages. The velocity of formation of a solid influences the size of its crystals in the sense that the crystals are larger the more slowly they form.

According to what has been stated an *invariable condition* (equilibrium) is *not* reached when the formation of the precipitate *ceases* to be macroscopically *visible.* For a solution which is *saturated with respect to small grains,* i. e. is in equilibrium with them, is *supersaturated* with respect to *large crystals* (as has been mentioned). Hence the dissolved substance will continue to separate on the surfaces of contact between the large crystals and the liquid until these are in equilibrium with the solution. Since the solution is now unsaturated with respect to the small crystals, more of them will go into solution. The dissolving of the small crystals and the growth of the larger will continue until all the crystals have the same size, or at any rate differ very little in size.

Thus smaller grains will have vanished and larger grains will have formed as the result of leaving the precipitate in contact with the solution for a long time, and of agitating it.

For practical purposes it suffices to *heat* the precipitate in contact with the liquid for *20–30 minutes.* The transformation will certainly not have gone to an end in this time, but it will have proceeded far enough for the purposes of filtration. Heating and frequent stirring essentially favor the equalization of concentration differences in the neighborhood of the crystals and also favor precipitation and dissolving.

116. Reactions of oxalate ion and oxalates. The hydrogen compound, "oxalic acid," $C_2H_2O_4$, is a *dibasic* acid belonging to the *medium strong* acids. It is *stronger* than *acetic acid,* but is *weaker* than *hydrochloric* or *nitric* acid. From this is explained the behavior of different acids towards barium oxalate. The reaction:

$$\underline{BaC_2O_4} + 2H^{\cdot} \rightleftarrows Ba^{\cdot\cdot} + H_2^{\cdot\cdot}C_2O_4'',$$

takes place to a major extent only when the acid that is added to the oxalate contains enough hydrogen ion to cause the formation of non-dissociated oxalic acid in noticeable quantity.

To dissolve *lead oxalate* in nitric acid requires a greater volume of acid than is needed to dissolve *barium oxalate*. This indicates that the saturated solution of lead oxalate contains less oxalate ion per unit volume than the saturated solution of barium oxalate.

Silver oxalate dissolves in dilute nitric acid, but is not very soluble in acetic acid.

With the ions of the trivalent metals, aluminum, chromium and iron, oxalate ion forms complex ions which weaken the reactions characteristic of these ions as well as those of the oxalate ion itself. In particular chromium oxalate is very stable. Hence the precipitation of chromic ion is rendered especially incomplete by the presence of oxalate ion. Barium ion precipitates barium oxalate from this solution only *on long standing*.

Furthermore the great insolubility of *calcium oxalate* in water is often utilized to identify oxalate ion. Even very dilute *neutral* solutions of oxalates give a white precipitate of *calcium oxalate* with soluble calcium salts. With calcium sulphate solution the precipitate forms especially rapidly (compare § 125). When heated with concentrated sulphuric acid solid oxalates yield water, *carbon dioxide* and *carbon monoxide*. The latter burns with a characteristic blue flame on igniting the escaping gases.

Dry oxalates are also decomposed by merely *heating* them. At first carbon monoxide escapes and the corresponding carbonate forms. Thus, for example,

$$CaC_2O_4 = CO + CaCO_3.$$

Finally the carbonate decomposes into metallic oxide and carbon dioxide (compare § 95).

The tests for oxalate ion in a solution containing the other anions of this group are given in § 125.

117. Reactions of fluoride ion and of fluorides. *Calcium-* or *barium* fluoride is soluble in dilute hydrochloric or nitric acid, but is scarcely soluble in acetic acid. *Lead fluoride* behaves similarly.

These phenomena, together with others, lead to the conclusion that hydrofluoric acid is less dissociated than hydrochloric or nitric acid.

Fluoride ion differs from the other halide ions by forming no difficultly soluble compound with *silver ion.*

The reactions which enable fluorides to be distinguished with certainty from the salts of other anions of this group depend upon the transformations which hydrofluoric acid undergoes with glass. When calcium fluoride is heated with concentrated sulphuric acid in a test-tube, hydrofluoric acid and silicon tetrafluoride are evolved. The latter is formed by the action of hydrofluoric acid upon glass, which contains silicates. If a glass rod is moistened with water and is held in the escaping vapors, it becomes coated with a white skin of silicic acid, due to the reaction:

$$3SiF_4 + 3H_2O = 2H_2SiF_6 + H_2SiO_3.$$

Since this reaction occurs only with *fluorides* or *fluosilicates,* it has special analytical significance. Another test makes use of the changes produced by hydrofluoric acid upon a glass plate or watch glass with which the generator tube is covered. The surface coming in contact with the escaping gas mixture becomes roughened.

118. **Etching with hydrofluoric acid.** The action of hydrofluoric acid upon glass is very often utilized in chemical laboratories for making marks on glass vessels. To accomplish this the portion of the object which it is desired to etch is gently heated over a bunsen burner and is thinly coated with paraffin or beeswax. Then the marks are made with a pointed tool through the protective coating down to the glass surface. By means of a feather or brush they are painted with a solution of hydrofluoric acid or are exposed to the action of gaseous hydrofluoric acid for some time. Etches produced in the first way are clear, while by the latter method they are dull (Principles, p. *426*). All traces of hydrofluoric acid are then washed off and the protective coating is softened by heating and is wiped off. *Very great care must be taken in working with hydrofluoric acid on account of the very painful burns it causes.* The vapors furthermore attack the eyes.

119. **Reactions of chromate- and dichromate ion.** Even when present as a chromate, chromium is detected in the third group, after reduction with hydrogen sulphide, by the method described for determining cathions. It is nevertheless important to consider specially the properties of these ions since the compounds of trivalent chromium are not infrequently changed into this form (§ 57) and are identified by the properties of these ions.

For their relations to one another consult § 98. Mention of the most important reactions and of the relations to the lower states of oxidation will suffice here.

The red solutions contain chiefly dichromate ion, the yellow principally chromate ion.

With barium ion they give yellow *barium chromate,* practically insoluble in acetic acid, soluble in hydrochloric and nitric acid. With lead ion yellow *lead chromate* is formed, quite difficultly soluble in nitric acid. Silver ion yields brown red *silver chromate.* Silver dichromate precipitates at first from concentrated dichromate solutions, but it is transformed at higher temperatures spontaneously into the chromate:

$$\underline{Ag_2Cr_2O_7} + H_2O = \underline{Ag_2CrO_4} + H_2CrO_4.$$

Mercurous ion precipitates red mercurous chromate from cbromate- and dichromate solutions.

For the identification of this state of oxidation the behavior of a slightly acidified solution containing dichromate ion towards *hydrogen peroxide* is of *special importance.* A blue color forms on mixing suitable quantities. It may be made more plainly visible by shaking out the solution with some ether, since the substance which has the blue color is more soluble in ether than in water.

On account of its great delicacy this reaction is also used in testing for chromium compounds of the lower states of oxidation. The compounds are first oxidized with hydrogen peroxide in alkaline solution (see § 55) and then this solution is tested as above given.

The coloration is probably due to an unstable higher oxidation form of chromium.[1] An excess of dichromate or of hydrogen peroxide destroys the color, since a decomposition accompanied by an evolution of oxygen takes place more readily under these conditions.

120. The relations between chromate- and chromic ion. The relations expressed in equations (*3a*) and (*3b*) of § 64 between dichromate ion and chromic ion dealt with the conditions under

[1] [Investigations by H. G. Byers and E. E. Reid, Am. Chem. J., 32, *503*, 1904, make it probable that the compound formed in this test is a derivative of perchromic acid, $H_2Cr_2O_8$, stable only at low temperatures. Consult also K. A. Hofmann and H. Heindlmaier, Ber. d. d. chem. Gesellschaft, 38, *3059*, 1905, also E. H. Riesenfeld, loc. cit., *3578*.]

which the former is produced from the latter. These have now to be supplemented by the reverse transformation. This reaction takes place in the presence of hydrogen ion and a substance which can take up positive charges or give up negative. The general equation for the reaction is:

$$Cr_2O_7'' + 14H^{\cdot} \rightleftarrows 2Cr^{\cdots} + 7H_2O + 6(^{\cdot}) \qquad (1).$$

For the sake of greater clearness the ions which undergo only an *inconsiderable change of concentration* are left out of the equation. With the transformation of one formula weight of dichromate ion + hydrogen ion[1] into two formula weights of chromic ion + water, positive charges become available for six formula weights of a substance or ion which can take up one positive charge. However, there must be present another substance which *simultaneously takes up six positive charges* for each formula weight of dichromate ion, or *gives up six negative charges.* In this way the condition is fulfilled that the same excess of positive charges occurs on each side of the equation.

The explanation of the oxidation of hydrogen sulphide to sulphur on passing the gas into an acidified solution of dichromate ion, and of sulphite to sulphate through the action of water, will help in understanding this reaction. In the first case the sulphide ion of hydrogen sulphide gives up its two negative charges. Hence one formula weight of dichromate ion will oxidize three formula weights of hydrogen sulphide according to the equation:

$$Cr_2O_7'' + 14H^{\cdot} + 3S'' = 2Cr^{\cdots} + 7H_2O + 3S \qquad (2).^{2}$$

Similarly the transformation of sulphite ion + water into sulphate ion + hydrogen ion takes place when a substance is present which gives up the necessary ionic charges to the hydrogen (compare equation 7 of the following paragraph). Two positive charges are required for this, since a sulphite ion uses up one combining weight of oxygen. However six charges become available upon the transition of one formula weight of Cr_2O_7'' into $Cr^{\cdots}$. Hence three formula weights of sulphite ion are oxidized by one formula weight of dichromate ion:

$$Cr_2O_7'' + 14H^{\cdot} + 3SO_3'' + 3H_2O = 2Cr^{\cdots} + 3SO_4'' + 7H_2O + 6H^{\cdot} \qquad (3a),$$

or

$$Cr_2O_7'' + 8H^{\cdot} + 3SO_3'' = 2Cr^{\cdots} + 3SO_4'' + 4H_2O \qquad (3b).$$

[1] Since the transformation of this into the trivalent form takes place only in acid solution, and chromate ion goes over into dichromate ion under these conditions (§ 98), the following reactions are formulated with dichromate ion.

[2] Occasionally a separation of green chromic hydroxide or of brown chromium dioxide or a hydrated form of the dioxide takes place when not enough acid is present. A sufficient amount of hydrochloric acid must always be added to ensure the *desired* reduction of chromate ion with hydrogen sulphide.

The following equation illustrates the relations for the oxidation by dichromate ion of non-electrolytes and substances, the dissociation of which is not known with certainty:

$$Cr_2O_7'' + 8H^{\cdot} = H_2O + 6(OH) + 2Cr^{\cdot\cdot\cdot} \qquad (1b).$$

Thus, for example, aldehyde is formed on heating an acidified dichromate solution with alcohol, C_2H_6O. The aldehyde volatilizes and is detected by its odor. It has the composition C_2H_4O, differing from C_2H_6O by having two combining weights of hydrogen less. However no gaseous hydrogen is formed in this reaction, but water results. One combining weight of oxygen (§ 80) or two formula weights of hydroxyl are needed for the oxidation of two combining weights of hydrogen. Hence one formula weight of dichromate ion will oxidize three formula weights of alcohol in the presence of hydrogen ion. How is the equation for this reaction formulated?

121. Reactions of sulphite ion and of sulphur dioxide. The tests are carried out with a saturated solution of sulphur dioxide (see § 122), which is neutralized with potassium- or sodium hydroxide.

Barium ion precipitates white *barium sulphite,* readily soluble in dilute hydrochloric or nitric acid, but only slightly soluble in acetic acid. *Lead sulphite* behaves similarly.

With nitric acid lead sulphite of course yields a clear solution, but it forms lead chloride with hydrochloric acid. Solutions of both the lead and barium precipitates have the property of becoming clouded in contact with air. This is due to the formation of sulphate ion, and correspondingly of solid lead- or barium sulphate, through the action of atmospheric oxygen. The fact that a freshly prepared solution of alkali sulphite occasionally forms with barium ion a precipitate, which is not completely soluble in hydrochloric or nitric acid, is also explained by a change of the solid salt into sulphate. Addition of some alcohol or sugar makes sulphite solutions more stable. These substances retard the rate of oxidation by atmospheric oxygen (Bigelow).

Silver ion yields a white precipitate of *silver sulphite* which is soluble in acids, in ammonia, and also in an excess of sulphite. The latter phenomenon is probably due to the formation of a complex ion of the composition SO_3Ag'. Neither this complex ion nor solid silver sulphite are very stable. Gray metallic silver separates on warming the solution containing this complex, or by heating the precipitate of silver sulphite with water. Sulphur dioxide and sulphate ion are simultaneously formed:

$$2(SO_3Ag)' = SO_4'' + SO_2 + \underline{2Ag} \qquad (1),$$

and

$$\underline{2Ag_2SO_3} = 2Ag^{\cdot} + SO_4'' + \underline{2Ag} + SO_2 \qquad (2).$$

Acidifying a sulphite solution or pouring dilute acid over solid sulphite causes an evolution of sulphur dioxide, which is readily identified by the smell characteristic of a burning sulphur match.

The formation of sulphur dioxide is analogous to the formation of carbon dioxide by the decomposition of carbonates (§ 96). The acid which first results according to the equation:

$$2H^{\cdot} + SO_3'' \rightleftarrows H_2SO_3 \qquad (3),$$

decomposes further into sulphur dioxide and water.

Potassium iodate is a more delicate reagent for sulphur dioxide. It is reduced to *iodine*. The reagent is most conveniently used in the following form. Filter paper is cut into strips of suitable dimensions and these are dipped in a solution of potassium iodate and then in starch solution (§ 144). In contact with sulphur dioxide the moistened strips become blue. Sulphite ion and hydrogen ion which exist in the solution of sulphur dioxide reduce the iodate ion, (IO_3'), to iodine according to the equation:

$$2IO_3' + 5SO_3'' + 2H^{\cdot} = 2I + 5SO_4'' + H_2O \qquad (4).$$

On longer action or with larger amounts of sulphur dioxide a *decolorization* will take place within a short time in consequence of a further reduction of *iodine* to *iodide ion*:

$$SO_3'' + 2I + H_2O = SO_4'' + 2I' + 2H^{\cdot} \qquad (5).$$

The summation of these two reactions leads to the following:

$$2IO_3' + 6SO_3'' = 6SO_4'' + 2I' \qquad (6).$$

This apparently simpler equation expresses the quantitative relations between the initial and resulting ions. In comparison with the two equations, however, it is less complete since it does not indicate the special conditions. Thus, for example, it does not show that hydrogen ion is necessary for the transformation.

Sulphite ion + water also acts as a reducing agent towards many other substances besides iodate ion and atmospheric oxygen. It re-

duces, for example, chromate- and permanganate ion. The oxidation product of sulphite ion + water is sulphate ion + hydrogen ion. The general equations for the reaction are:

$$SO_3'' + H_2O + 2(^{\cdot}) = SO_4'' + 2H^{\cdot} \quad (7),$$

and

$$SO_3'' + H_2O = SO_4'' + 2H^{\cdot} + 2(') \quad (8).$$

Accordingly the oxidizing substance must have the property of giving up positive charges or taking up negative. Ferric and mercuric ion are examples of the first kind. They are reduced to ferrous and mercurous ion respectively. The decolorisation of iodine affords an illustration of the second sort; compare further the reduction of arsenic solutions (§ 43).

122. **Preparation of sulphur dioxide.** Copper foil is cut into strips and heated with concentrated sulphuric acid. The copper goes into solution, evolving sulphur dioxide:

$$Cu + 2H_2SO_4 = CuSO_4 + SO_2 + H_2O \quad (1).$$

The accompanying figure (8) shows how the experiment is set up. The "round bottomed" flask must be thick-walled. Rather more than twice the quantity of acid determined by the above formula is poured in through the thistle tube. The column of liquid remaining in the thistle tube forms a safety valve to let out the gas in case the delivery tube becomes stopped. The flask is heated on a sand-bath by means of a large burner. To absorb any sulphuric acid vapors that are carried over, the tube 'r' is filled with pieces of potassium sulphate. It is transformed into potassium hydrogen sulphate (potassium bisulphate):

$$K_2SO_4 + H_2SO_4 = 2KHSO_4 \quad (2).$$

The wash bottles are connected as shown in the diagram to keep the water that is used for absorbing the gas from flowing back into the three-necked Wolff bottle (A), which contains concentrated sulphuric acid. The experiment should be carried out under a hood or in a separate room, since a leakage of sulphur dioxide cannot, as a rule, be avoided by beginners.

The saturated solution of the gas has the characteristic penetrating odor of sulphur dioxide and turns blue litmus red. The solution

contains sulphurous acid, a medium strong dibasic acid. Solutions of dialkali sulphites have a distinct alkaline reaction in consequence of the following reaction between SO_3'' and water:

$$SO_3'' + H_2O = HSO_3' + OH'.$$

Copper sulphate may be recovered as a bye-product from the above reaction by cooling the flask at the end of the operation and adding carefully twice as much water as there is solution in the flask. (The

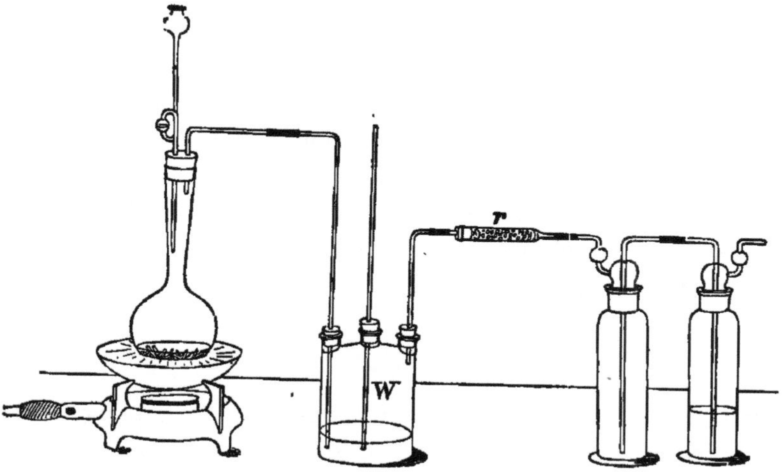

FIG. 8.

liquid will become quite hot on mixing with water.) The solution is then warmed, if necessary, and any undissolved residue is filtered The filtrate is evaporated to a smaller volume and is allowed to crystallize slowly.

123. Thiosulphate ion, S_2O_3''. For the test reactions a solution is used containing $\frac{1}{10}$ formula weight of sodium thiosulphate, $Na_2S_2O_3.5H_2O$, per liter.

Barium chloride forms a white precipitate of barium thiosulphate, scarcely soluble in acetic acid, more readily soluble in nitric acid. The nitric acid solution becomes clouded on standing, in consequence of a separation of finely divided sulphur, owing to the reaction:

$$S_2O_3'' + H^{\cdot} = HSO_3' + \underline{S} \qquad (1).$$

Even the small quantity of acid which is formed by absorption of carbon dioxide from the atmosphere is sufficient to produce an opalescence in the

solution. Hence it is difficult to obtain a clear solution by filtration. In concentrated solutions or with a large excess of hydrogen ion an evolution of sulphur dioxide also takes place:

$$S_2O_3'' + 2H^{\cdot} = H_2O + SO_2 + \underline{S} \qquad (2).$$

Barium thiosulphate has the property of readily forming supersaturated solutions. Vigorous shaking will always bring about a separation of the solid, however. It is rather readily soluble in water.

Lead acetate yields a white precipitate, soluble in an excess of thiosulphate and in nitric acid.

Silver ion also gives rise to a white precipitate which dissolves in an excess of thiosulphate, owing to the formation of a complex ion which probably has the composition AgS_2O_3'.

Solid silver thiosulphate and the complex silver thiosulphate ion are quite unstable. The *former* turns black very quickly by reacting with water. Silver sulphide results and the solution becomes acid through the formation of sulphuric acid:

$$\underline{Ag_2S_2O_3} + H_2O = \underline{Ag_2S} + 2H^{\cdot} + SO_4'' \qquad (3).$$

The *solution* of the complex becomes clouded on standing, more quickly when warmed. Separation of sulphur and silver sulphide takes place according to the equation:

$$2AgS_2O_3' = \underline{Ag_2S} + SO_4'' + SO_2 + \underline{S} \qquad (4).$$

Like sulphite ion thiosulphate ion also acts as a reducing agent towards many substances. Tetrathionate ion, S_4O_6'', is formed as the oxidation product in these reactions. The general equations are:

$$2S_2O_3'' + 2(^{\cdot}) = S_4O_6'' \qquad (5),$$

and

$$2S_2O_3'' = S_4O_6'' + 2(') \qquad (6).$$

The oxidation of thiosulphate ion to tetrathionate ion by means of another ion is accompanied by a loss of positive charges or a gain of negative. Ferric ion is an oxidizing agent of the first sort. It gives a red coloration with thiosulphate solutions. On standing the color vanishes, however, through transition of ferric- into ferrous ion. A solution of iodine in potassium iodide (§ 144) is decolorized

by thiosulphate in consequence of the formation of iodide ion:

$$2S_2O_3'' + 2I = S_4O_6'' + 2I' \quad (7).$$

This is an illustration of the second class of oxidizing agents.

124. The identification of sulphite- and thiosulphate ion. These two ions have many properties in common and it is therefore a difficult matter to *distinguish them*. In particular difficulties are met with in identifying sulphite ion when thiosulphate ion is also present, the separation of sulphur upon acidifying a thiosulphate solution affording a delicate and distinctive test for thiosulphate ion.

From equations (5), § **121**, and (7), § **123**, for the oxidation of these ions by iodine, it is evident that *hydrogen ion* results from the oxidation of *sulphite ion*, while the solution remains *neutral* when *thiosulphate ion* is *oxidized*. This difference in behavior may accordingly be used to detect sulphite ion in the presence of thiosulphate ion, since it is an easy matter to detect an increase of the concentration of hydrogen ion through the resulting acid reaction of the solution. Hence to test for sulphite ion in the presence of thiosulphates the solution is *neutralized* and a solution of iodine in potassium iodide (the latter serves merely as solvent, § 144) is added drop by drop to it. Any excess of iodine which would interfere with the test for acid reaction, is removed by means of a little thiosulphate.

The iodine solution is dropped from a tube similar to the one used for neutralizing (§ 110), by regulating the pressure on the end of the tube with the finger. Then the action of the solution towards litmus paper is tested. If it turns blue litmus red the solution contains a sulphite.

In doubtful cases the distinction is made certain by means of a parallel experiment in which only pure thiosulphate is oxidized by iodine. The change in the behavior of the solution towards litmus may be still more conveniently determined with a litmus solution by adding several drops to the original solution until it is colored pale violet.

125. The detection of oxalate ion in the presence of other anions of this group. A good many difficulties oppose the conclusive detection of *oxalate ion,* especially in mixtures of salts containing several anions. Hence errors are not infrequently made in con-

nection with this point. The following method permits a practically conclusive identification of oxalates, even when salts containing all the other anions of this group are present.

The procedure is the following. The solution to be tested for these anions is heated with alcohol and dilute hydrochloric acid. By this means dichromate ion is reduced to chromic ion, sulphite ion decomposes and evolves sulphur dioxide, while thiosulphate ion separates sulphur and also forms sulphur dioxide. The heating is continued until the alcohol is evaporated and the milky precipitate of sulphur has been changed to a form which filters readily. Chromic ion is then precipitated from the filtrate with ammonia and the chromic hydroxide is filtered. The solution thus freed from chromium is next acidified (see below) and is heated to boiling. To the hot solution barium chloride is added drop by drop. A precipitate formed at this point may contain barium sulphate, -fluosilicate, -fluoride and -oxalate. It is filtered according to the directions mentioned in § **115** and, after washing, is tested for *oxalate ion* as follows.

The method is based upon the production of carbon dioxide through a reaction between *barium oxalate, permanganate ion* and *hydrogen ion*. With baryta solution carbon dioxide gives a white precipitate of barium carbonate.

A portion of the precipitate (or all of it) is heated in a test-tube with 5–10 cc. dilute sulphuric acid. By this means barium oxalate is changed into barium sulphate and oxalic acid. Permanganate ion is next added until the red color of permanganate remains permanently (§ **145**). The test-tube 'a' (see Fig. 9) is immediately closed with a rubber stopper, through which a delivery tube of the form shown in the diagram is passed.

The long arm of the delivery tube dips below the surface of some baryta solution in the test-tube (*b*). To keep out air, which also contains carbon dioxide, a plug of cotton wool is loosely fitted around the delivery tube at the mouth of (*b*).

Since the volume of solution in (a) is rather large a considerable amount of carbon dioxide may be dissolved in it. By warming the solution, however, this is driven out. The generator tube is accordingly placed in a broader vessel (c) containing hot water. The result is therefore the formation of a white precipitate of barium carbonate in (b) after a few minutes, if the first barium precipitate

contained barium oxalate. The barium carbonate dissolves with effervescence on adding acetic acid. The stopper through which (a) is led, must not close the tube (c) *tightly,* in order to allow the water vapors generated on warming (c) to escape.

The method is not specially delicate. However, this disadvantage is at any rate partially compensated by taking a larger quantity of the original solution for the precipitation of the barium salts.

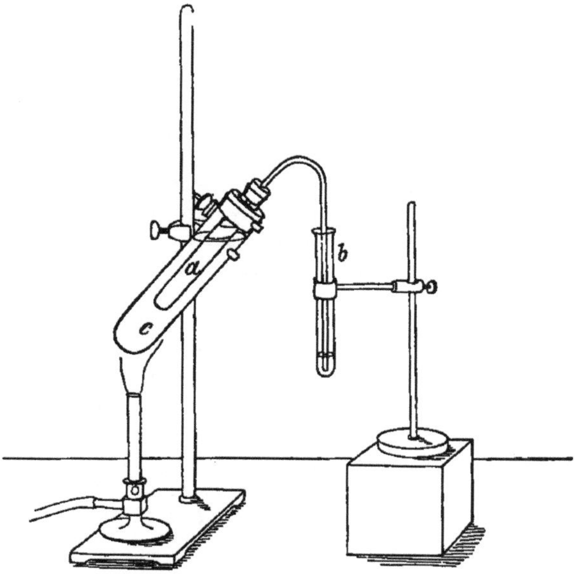

FIG. 9.

Mistakes are sometimes made in carrying out this test if the solution from which the anions are precipitated by $Ba^{..}$, contains *tartrate ion.* For tartaric acid, which is brought into the solution by decomposing the barium salt with sulphuric acid, is likewise oxidized by permanganate ion with evolution of carbon dioxide. The difficulty opposing the detection of oxalate- in the presence of tartrate ion, may be overcome readily in the following way.

A small quantity of *dilute hydrochloric acid* is added to the *barium precipitate obtained as above given.* To the *solution* in hydrochloric acid, or the *clear* filtrate (if the precipitate contains the barium salts of first group anions it is not completely soluble in hydrochloric acid) a solution of *sodium acetate* is then added in small portions at a time. This forms a precipitate of *barium oxalate at once,* even when only a small amount of *oxalate ion* is present.

Barium tartrate may also precipitate under these conditions, when the volume of sodium acetate added is *considerably in excess* of that of the hydrochloric acid.[1] Apparently barium tartrate forms supersaturated solutions very readily. Hence if *oxalate ion is not present, the precipitation either does not take place or occurs only after a much longer interval* (hours), which practically amounts to the same thing.

The delicacy of this test can moreover be increased by adding to the hydrochloric acid solution or the filtrate about an equal volume of a calcium salt solution. *Calcium oxalate* is much less soluble than *barium oxalate.* To avoid precipitating calcium tartrate, which might be mistaken for oxalate, the concentration of the calcium salt must be low. A solution containing 0.015 gram combining weights of calcium ion per liter is very serviceable. If this solution is made from calcium chloride, $CaCl_2.6H_2O$, 3.3 g. are weighed out for a liter of solution. The nitrate may also be used in place of the chloride. Calcium sulphate is not suitable since it gives a precipitate of barium sulphate with the hydrochloric acid solution of a barium precipitate.

The only errors which could possibly arise with the above method would occur when *tartrate* and *fluoride* ions are both contained in the solution to be tested for oxalate. In this case the formation of a precipitate on adding calcium chloride of the above concentration, or sodium acetate, is not *in itself* a *satisfactory criterion* of the presence of *oxalate ion*, since *calcium fluoride* is likewise scarcely soluble in acetic acid. The precipitate obtained with calcium ion is then tested with permanganate as above given for oxalate ion.

To prevent the separation of calcium tartrate as far as possible no more sodium acetate is added when enough precipitate is obtained for the further tests.

GROUP III.

126. The anions of this group give precipitates with the typical cathions (§ 110) in neutral solutions. The barium precipitates are soluble in *acetic acid.* These anions are distinguished from those of the following group by the solubility of the *silver precipitates* in acids. Use is occasionally made of this property to separate these anions from others.

The special reactions are of aid in identifying the individual members. The following anions belong here: *phosphate, arsenite, arsenate, borate, silicate, carbonate* and *tartrate* (compare the table of tests for anions).

[1] In this case it is justifiable to use volume as a measure of the quantity of hydrochloric acid and sodium acetate only when the solutions contain the same number of formula weights of hydrochloric acid and sodium acetate respectively. If the acetate is only half as concentrated as the acid, at least twice the volume of acetate would be used to transform a given volume of hydrochloric acid into acetic acid (compare also § 158).

127. **The anions of phosphoric acid.** A solution of disodium hydrogen phosphate containing $\frac{1}{20}$ formula weight of $Na_2HPO_4 \cdot 12H_2O$ per liter is used for the tests. It has a slight alkaline reaction since it is a little hydrolyzed according to the equation:

$$2Na^{\cdot} + HPO_4'' + H_2O = 2Na^{\cdot} + (OH)' + H_2PO_4'.$$

Barium ion yields a precipitate of *barium hydrogen phosphate*, $BaHPO_4$, which is very easily soluble in hydrochloric or nitric acid, but dissolves only in a large volume of acetic acid. The precipitate obtained with lead acetate is still more difficultly soluble in acetic acid.

Silver phosphate, Ag_3PO_4, has a yellow color. Towards acids it acts like barium phosphate. The precipitate is also soluble in ammonia for the reasons already mentioned in the case of silver chloride (§ 11).

Disodium phosphate gives a white precipitate with *ferric ion*. The solubility product is very small. Hence *ferric phosphate* is practically *insoluble* in *acetic acid*. When determining the cathions this property is occasionally utilized to remove phosphate ion from a solution instead of the method described in § 59.

Magnesium ion + ammonium ion + ammonia precipitate *magnesium ammonium phosphate* from phosphate solutions (compare §§ 99 and 102).

Ammonium molybdate precipitates yellow *ammonium phosphomolybdate*, $(NH_4)_3PO_4 \cdot 12MoO_3$, from the solution of a phosphate or phosphoric acid. The reaction is carried out in a strong nitric acid solution.[1] The precipitate is soluble in ammonia and alkali hydroxides (compare § 206 on molybdenum).

The transformation proceeds rather slowly and is best accelerated by warming the solution to 40° C.

This reaction has been mentioned already in § 58. It also serves particularly to test for barium phosphate in the precipitate obtained according to § 114, footnote. To make the test the precipitate produced by sodium acetate is filtered and treated with acetic acid. Not quite enough acetic acid is used to completely dissolve the precipitate. Since barium phosphate is much more readily decomposed by acetic acid than the oxalate, fluoride or chromate, the phosphate will go into solution for the major part in case these other salts

[1] Sometimes a white precipitate of molybdic acid forms. It is readily brought into solution again by adding more nitric acid. The amount of acid needed for this does not influence the other reaction.

are also present. The solution is filtered from any residue and is evaporated, if necessary, with the addition of a few drops of nitric acid to a small volume. It is then tested with ammonium molybdate and nitric acid.

128. The dissociation of phosphoric acid. Complicated dissociation relations have already been mentioned in the case of sulphuric acid (§ 112). They are still more complex with phosphoric acid. In this case three stages of dissociation may be distinguished:

$$H_3PO_4 \rightleftarrows H^{\cdot} + H_2PO_4{}' \qquad (1),$$

$$H_2PO_4{}' \rightleftarrows H^{\cdot} + HPO_4{}'' \qquad (2),$$

$$HPO_4{}'' \rightleftarrows H^{\cdot} + PO_4{}''' \qquad (3).$$

As was stated in § 69 the solubility in acids of difficultly soluble phosphates which contain the anions $HPO_4{}''$ and $PO_4{}'''$ in their aqueous solutions depends upon the degree of dissociation of the ions $H_2PO_4{}'$ and $HPO_4{}''$, according to (*2*) and (*3*). The fact that salts of the types $BaHPO_4$ and Ag_3PO_4 are soluble in acetic acid, although phosphoric acid is a *stronger* acid than *acetic acid*, indicates that the following reactions take place:

$$\underline{BaHPO_4} + H^{\cdot} \rightleftarrows Ba^{\cdot\cdot} + H_2PO_4{}'' \qquad (4),$$

$$\underline{Ag_3PO_4} + H^{\cdot} \rightleftarrows 3Ag^{\cdot} + HPO_4{}'' \qquad (5a),$$

$$\underline{Ag_3PO_4} + 2H^{\cdot} \rightleftarrows 3Ag^{\cdot} + H_2PO_4{}' \qquad (5b).$$

Now the concentration of hydrogen ion is very low in acetic acid of usual strength. Furthermore in the saturated solutions of these difficultly soluble salts the concentrations of the anions are likewise very low. Hence the fact that the above reactions take place under these conditions indicates that the dissociation of phosphoric acid in the second and third stages is very slight.

This also explains why a solution of *phosphoric acid* does *not* form precipitates with the ions $Ba^{\cdot\cdot}$, $Pb^{\cdot\cdot}$ and $Ag^{\cdot}$. The acid contains so few anions $HPO_4{}''$ and $PO_4{}'''$ that the solubility product of the salts in question is not reached. Solutions of mono- and dialkali phosphates give precipitates with these cathions. Hence they contain proportionately more of the anions $HPO_4{}''$ and $PO_4{}'''$. This is due to the smaller content of hydrogen ion, which permits a greater dissociation according to (*2*) and (*3*). The dissociation of monoalkali phosphate in the first stage is as follows:

$$NaH_2PO_4 \rightleftarrows Na^{\cdot} + H_2PO_4{}'.$$

No hydrogen ion, which would limit the dissociation according to (*2*) and (*3*) from the reasons stated in § 89, is formed in this reaction. For the same reasons disodium hydrogen phosphate contains much more of the ions $HPO_4{}''$ and $PO_4{}'''$ than phosphoric acid (consult Principles, p. *363*).

These relations are further illustrated by the fact that dialkali-, and still more trialkali phosphate, are hydrolyzed by water. Hence the explanations in § 71 likewise lead to the view that the ions $H_2PO_4{}'$ and $HPO_4{}''$ have the characteristics of a medium strong and a weak acid respectively.

129. The basicity of phosphoric acid. Phosphoric acid is regarded as tribasic on the grounds that it forms salts of the composition Ag_3PO_4, for example. According to § 112, footnote, basicity is determined by the number of formula weights of hydroxyl ion or monovalent base which are neutralized

by one formula weight of acid. From the preceding paragraph it is evident, however, that even less than two formula weights of hydroxyl are actually neutralized by one formula weight of phosphoric acid.

The discrepancy arises from the above-mentioned circumstance that the conception of basicity is used with different meanings. In one case it emphasizes the behavior of phosphoric acid towards bases and in the other the composition of certain difficultly soluble phosphates.

Ambiguity also occurs in other cases where one or more combining weights of hydrogen have very much weakened acid properties. As a consequence the transformation according to § 71, (*3*), remains incomplete with a quantity of base corresponding to one formula weight. The soluble salts therefore hydrolyze in solutions to a great extent.

Other similar cases will be dealt with in the following paragraphs.

130. Pyrophosphate ion. By heating disodium phosphate in a porcelain crucible it first melts and then solidifies again, losing water and forming pyrophosphate:

$$2Na_2HPO_4 = Na_4P_2O_7 + H_2O.$$

A freshly prepared solution of sodium pyrophosphate has a distinct alkaline reaction. With *barium, lead* and *silver ion* it yields *white precipitates,* which are but little soluble in acetic acid, readily soluble in hydrochloric and nitric acid. The difference in the color of silver ortho- and pyrophosphate is worthy of note. The former is yellow (§ 127).

The term orthophosphoric acid distinguishes ordinary phosphoric acid from meta- and pyrophosphoric acids. Correspondingly its salts are called orthophosphates.

131. Metaphosphate ion. If *sodium ammonium hydrogen phosphate,* the salt used to prepare the phosphor salt bead and hence termed phosphor salt, is heated to redness in place of disodium phosphate, *sodium metaphosphate* results:

$$NaNH_4HPO_4 = NH_3 + H_2O + NaPO_3.$$

On dissolving the fusion the solution yields *white precipitates* with *barium-* and *lead ion, insoluble* in *acetic acid,* soluble in *nitric* acid. The silver salt is likewise white, but dissolves readily. Hence it separates only when concentrated solutions are used.

Metaphosphoric acid, or a solution of metaphosphate + acetic acid, has the property of coagulating albumen solution (distinction from phosphates and pyrophosphates).

Solutions of pyro- and metaphosphoric acid are unstable. They take up water and form orthophosphoric acid.

132. The anions of trivalent arsenic (compare § 42). A solution suited for the following tests is prepared by dissolving arsenic trioxide in potassium hydroxide and neutralizing the excess of the latter with dilute nitric acid. Arsenic trioxide also dissolves in water to form a slightly acid solution. The reaction takes place very slowly, however, since arsenic trioxide is not easily wetted by water. Hence the above method is more convenient, as it takes less time.

$\frac{1}{20}$ formula weight of As_2O_3 is used for a liter of solution.

With *barium ion* only concentrated solutions yield a precipitate of *barium arsenite*. *Lead ion* forms *white lead arsenite* even with dilute solutions. The precipitate is soluble in acetic acid.

Silver nitrate gives a slight precipitation of yellowish white *silver arsenite*, Ag_3AsO_3. By adding a few drops of potassium hydroxide the amount of precipitate is considerably increased.

Neutral solutions of an arsenite contain *but little arsenite ion*. The arsenite reacts with water according to the equation:

$$AsO_3''' + 3H^{\cdot}OH' \rightleftarrows H_3AsO_3 + 3(OH)' \qquad (1).$$

By adding nitric acid to neutralize OH' almost all AsO_3''' is transformed into H_3AsO_3. Only an amount of arsenite ion corresponding to the dissociation of arsenous acid is left in the solution. Conversely an addition of OH' *increases* the *concentration* of *arsenite ion* and hence *strengthens* the properties of the solution with respect to arsenite ion.

Silver arsenite dissolves in acids (including acetic acid), in ammonia[1] and ammonium nitrate. On heating the ammoniacal solution a separation of silver takes place, while the arsenite is oxidized to arsenate. The reaction is formulated as follows:

$$AsO_3''' + 2Ag^{\cdot} + H_2O = AsO_4''' + \underline{2Ag} + 2H^{\cdot} \qquad (2).$$

From the composition of the silver salt arsenous acid is tribasic. In consequence of hydrolysis, however, a solution which contains one combining weight of sodium for one combining weight of arsenic has an alkaline reaction. This is owing to the fact that the acid is only weakly dissociated, even in the first stage. The dissociation is about equivalent to that of phosphoric acid in the second stage.

[1] In consequence of the above-mentioned behavior of bases, however, ammonia at first causes an increase of the precipitated silver arsenite.

This comparison is based on the assumption that the non-dissociated acid in the solution has the formula H_3AsO_3. As a matter of fact the relations are more complex (compare Principles, p. *708*), although not essentially different. The explanation would be less clear, however, by taking into account these special features.

133. The anions of pentavalent arsenic. The test solution is prepared according to the details given in § 43. It is evaporated down to a small volume, neutralized with alkali hydroxide and afterwards again diluted to the desired volume. One liter of the solution should contain $\frac{1}{10}$ formula weight of As_2O_5.

With *barium ion* the *neutral* solution forms a *white* precipitate, soluble in acetic acid. *Lead acetate* precipitates white *lead arsenate*, rather difficultly soluble in acetic acid, readily soluble in nitric acid.

Silver arsenate is red brown and is soluble in ammonia and acetic acid.

Magnesium- + ammonium ion + ammonia form a white precipitate of *magnesium ammonium arsenate* (see § 26), completely analogous with magnesium ammonium phosphate (§ 102). The similarity of the reactions of phosphates and arsenates extends also to the precipitation of the latter by ammonium molybdate, but the precipitate forms more slowly with arsenates.

On account of this great similarity in the properties of the anions of arsenic acid with those of phosphoric acid the following procedure must always be followed when arsenic is detected amongst the cathions. After removal of arsenic a *part of the second group filtrate* is tested with ammonium molybdate for anions of phosphoric acid. The analogy may also be extended, particularly to the dissociation relations of these acids. With respect to this question a reference to § 128 suffices.

134. Distinction of arsenite- and arsenate ion. To detect *arsenite ion* the solution is heated with an alkaline solution of copper sulphate and tartaric acid, which serves to prevent the precipitation of cupric ion by hydroxyl ion (§ 139). Arsenite ion oxidizes to arsenate, while cupric ion is reduced to cuprous ion, which yields reddish yellow cuprous oxide with hydroxyl ion. The equation for this reaction is:

$$2Cu^{\cdot\cdot} + AsO_3''' + H_2O = 2Cu^{\cdot} + AsO_4''' + 2H^{\cdot} \qquad (1).$$

This reaction is completely analogous with reaction (*2*), § 132.

It must be remembered that grape sugar, aldehyde and a few other organic compounds also give rise to the reaction which serves to distinguish arsenites from arsenates. Hence the formation of a precipitate of cuprous oxide may be used as a criterion for arsenite ion only when these substances are not present.

To identify arsenate ion in the presence of arsenite ion hydrochloric acid and potassium iodide are added to the given solution. A yellow coloration due to free iodine indicates arsenates. The reaction is expressed by the equation:

$$AsO_4''' + 2H^{\cdot} + 2I' \rightleftarrows AsO_3''' + I_2 + H_2O \qquad (2).$$

This reaction can also proceed in the reverse direction. A solution of iodine in potassium iodide is decolorized by neutral arsenous acid to which sodium hydrogen carbonate is added. Solid iodine is also transformed into a colorless solution by this means.

Since iodide ion in the presence of hydrogen ion is also oxidized by other anions besides arsenate ion (for example, by Cr_2O_7'', (NO_3'), NO_2', ClO_3', ClO', $Fe(CN)_6'''$, it is evident that a yellow coloration from the formation of iodine indicates arsenate only when these other anions are excluded.

From a mixture of these anions and arsenate ion to obtain a solution which contains practically only the latter, the given solution is neutralized and precipitated with barium chloride. Then the barium precipitate is filtered, washed and treated with acetic acid. In case the original solution contained no chromate- or dichromate ions, the barium precipitate is completely soluble in acetic acid; otherwise it is only partly soluble. In any case if a residue remains it is filtered and the filtrate is tested with hydrochloric acid and potassium iodide.

135. Reactions of the anions of boric acid. A solution of *borax* (disodium tetraborate), $Na_2B_4O_7.10H_2O$, containing $\frac{1}{4}$ formula weight per liter is used for the tests. The solution is neutralized with nitric acid.

The requisite amount of borax is first dissolved in a somewhat smaller volume than is desired. Then the solution is neutralized with nitric acid and is diluted to the volume desired in a graduated cylinder.

If the solution has a very slight acid reaction it gives *no precipitates* with $Ba^{\cdot\cdot}$ and $Ag^{\cdot}$, even when it is very concentrated. Hence it is advisable to leave it *slightly alkaline.*

The formation of a precipitate depends upon the concentration of the ion BO_2'. Its concentration is much lower in the presence of hydrogen ion than in an alkaline solution. These relations are expressed by the following equation:

$$B_4O_7'' + H_2O \rightleftarrows 4BO_2' + 2H^{\cdot} \qquad (1).$$

It is evident that the concentration of the ion BO_2' is decreased by hydrogen ion and accordingly is influenced by hydroxyl ion in the opposite direction (§ 50).

Barium chloride forms a precipitate of *barium borate,* soluble in acetic acid. It is also rather soluble in water. Hence this reaction is not specially delicate.

Lead borate is likewise white and soluble in acetic acid.

Silver borate is white, easily soluble in acids. It is decomposed by long continued heating with slightly alkaline water. Boric acid goes into solution and a separation of brown silver oxide takes place:

$$2AgBO_2 + H_2O = Ag_2O + 2HBO_2 \qquad (2).$$

By moistening boric acid or a borate with alcohol [or glycerine], adding a few drops of concentrated sulphuric acid and igniting, a characteristic green edge is imparted to the flame. The phenomenon is specially clear when air has free access. Hence the experiment is conveniently carried out by dipping a glass rod in the mixture and *igniting* the drop of liquid hanging to the end of the rod in a gas flame. Mistakes arising from salts of a green color are completely excluded in this manner. *In case the experiment does not succeed it is usually due to the fact that too little alcohol and too much sulphuric acid had been used.*

The reaction depends upon the formation of volatile *boric ethyl ester,* a compound which contains the radical ethyl, (C_2H_5), in place of the hydrogen of the acid.

Turmeric paper is not changed by dipping it into boric acid or a borate + hydrochloric acid, but on drying it assumes a brown color. Moistening with alkali hydroxide transforms the color into blue black. This reaction is not specially characteristic, however, since other acids give similar phenomena.

Commercial boric acid has the composition H_3BO_3. The composition of solid borates shows that they are derived from an acid HBO_2, which contains one formula weight of water less than H_3BO_3.

The acid character of boric acid is only very slightly developed. Solutions of the acid *do not redden litmus paper* at all *significantly* and *solutions* of the *alkali salts* are *strongly hydrolyzed.* The relations are furthermore obscured by the property of forming condensed acids according to the equation:

$$mH_3BO_3 = nH_2O + H_{3m-2n}B_mO_{3m-n} \qquad (3).$$

Borax has the composition $Na_2B_4O_7.10H_2O$. It is the salt of an acid which is derived from orthoboric acid by placing $m = 4$, $n = 5$, in equation (3).

Boric acid is volatile and colors the bunsen flame green. When moistened with sulphuric acid borates show the same characteristics.

136. The anions of silicic acid. The solution of an alkali silicate has a strong alkaline reaction. It yields precipitates with the three cathions (§ 110), which are soluble in acids. On attempting to neutralize the solutions a separation of silicic acid takes place. Occasionally the entire mass solidifies to a jelly. Silicic acid has an extremely weak acid character, which is also shown by an indefiniteness of its other properties. Thus the composition of the anions of silicic acid is indefinite, and hence it will not be considered here.

The paragraphs on the analysis of silicates will consider the behavior of silicic acid in other respects (consult also Principles, p. *420*).

137. Carbonate ion. The behavior of carbonate ion towards the cathions of barium (§ 97), silver (§ 11) and lead (§ 14) has alreadv been dealt with. In the cases cited, however, a solution of sodium carbonate was used. This has an *alkaline* reaction in consequence of hydrolysis. The precipitation is *less complete* or *does not occur* at all on using a solution *neutralized* with nitric acid. This is explained by a decrease of the total quantity of CO_3'' present, owing to an escape of carbon dioxide on neutralizing, and to the proportionately slighter dissociation of the carbonic acid still present. It is ranked as a very *weak* acid.

The formation of carbonic acid by neutralizing an alkali carbonate solution is evident from a consideration of the equation which represents the reaction between carbonate ion and water:

$$CO_3'' + 2H_2O \rightleftarrows H_2CO_3 + 2OH' \qquad (1).$$

If hydrogen ion is added to a solution of sodium carbonate which is in a state of equilibrium, the concentration of hydroxyl ion decreases according to § 50 through the reaction $H^{\cdot} + OH' = H_2O$. The consequence is an increased hydrolysis and hence an increase in the concentration of carbonic acid. Since carbonic acid is only slightly soluble in water it decomposes with an evolution of carbon dioxide:

$$H_2CO_3 = CO_2 + H_2O \qquad (2).$$

By this means a further condition for an increase of the hydrolvsis arises, viz. the decrease in the concentration of carbonate ion.

From the composition of its solid normal[1] salts carbonic acid is

[1] Normal salts contain metal for all the replaceable hydrogen of the acid (consult Principles, p. *267*).

dibasic. Nevertheless a solution of *sodium hydrogen carbonate* turns red litmus blue, since this salt is also hydrolyzed:

$$HCO_3' + H_2O \rightleftarrows H_2CO_3 + OH'.$$

The ease with which carbonates are decomposed by acids has been pointed out already (§ 96) as a further consequence of weak acid character. A white skin of barium carbonate will form over the end of a glass rod moistened with baryta (or lime water) when the rod is held in the neck of the generator flask and is so brought in contact with the escaping carbon dioxide. What is the equation for this reaction?

To detect small quantities of carbon dioxide air must be excluded in carrying out the experiment, as it contains always about 0.04 per cent of carbon dioxide. In laboratories the percentage is usually higher from the products of combustion of illuminating gas. Generally enough is present to yield a cloudiness in baryta solution, although only after a long time. Consult also § 176 on the preliminary tests with dilute sulphuric acid for the acids contained in solid unknowns.

Quite frequently the solution of a barium *salt*, for instance, barium chloride, is wrongly used in place of baryta to carry out this test. The consequence is that a cloudiness is never observed. This will be clear from the consideration of the equation:

$$Ba^{\cdot\cdot} + CO_2 + H_2O \rightleftarrows \underline{BaCO_3} + 2H^{\cdot}.$$

Hydrogen ion is set free by this reaction, and since barium carbonate is decomposed even by acetic acid, a visible skin of carbonate would never form under these conditions.

138. Reactions of tartrate ion. The test solution is made by dissolving $\frac{1}{20}$ formula weight of tartaric acid in a little water and diluting to a liter after neutralization with sodium hydroxide. With barium ion it yields white *barium tartrate* which is soluble in acids, but is rather difficultly soluble in acetic acid. Ammonium salts also dissolve barium tartrate to a noticeable extent.

Barium tartrate persistently forms supersaturated solutions. The supersaturation can be readily removed by violent shaking.

Lead acetate forms white *lead tartrate,* but little soluble in acetic acid, easily soluble in nitric acid. The precipitate dissolves without difficulty on addition of ammonium tartrate to the liquid. If not enough lead acetate is added to precipitate all the tartrate ion up to the limits of the solubility of lead tartrate, addition of ammonia will cause the precipitate to vanish.

The dissolving of lead tartrate in ammonium tartrate involves the same reaction which was mentioned in the case of lead sulphate (§ 14). By way of supplementing what has already been stated it is well to add that this power of dissolving is not limited to lead sulphate and lead tartrate. In fact the various difficultly soluble lead salts exhibit the same property according to their solubility in water if no reactions take place other than the one referred to. The difference in the behavior of lead tartrate and lead fluosilicate towards ammonium tartrate (§ 111) indicates that the fluosilicate is particularly insoluble in water.

Silver nitrate precipitates white silver tartrate, soluble in acids, in ammonia and in an excess of tartrate ion. Boiling silver tartrate with water causes a blackening from the separation of silver.

The silver separates in the form of a mirror on the glass walls of the vessel on slowly reducing a solution of silver tartrate. This is best effected by heating a solution which contains just enough ammonia to keep the silver tartrate dissolved, to 70° C. on a water-bath.

The solution does not yield a precipitate immediately on adding *potassium ion,* but it gives a precipitate when hydrogen ion is present. According to § 103 acetic acid is best used for this purpose.

By heating tartrates with concentrated sulphuric acid they are decomposed, giving rise to a brown coloration. Carbon monoxide and sulphur dioxide are simultaneously formed. The latter results from a reduction of sulphuric acid. Carbon monoxide burns with a blue flame when ignited.

Heating solid tartrates or tartaric acid in a bulb-tube also effects a decomposition. This is accompanied by the formation of a brown coloration and the evolution of vapors which smell like burnt sugar. Effervescence takes place on pouring acids over the residue left in the bulb-tube owing to an evolution of carbon dioxide. This shows that the tartrates have been changed by heat into the corresponding carbonates.

Tartaric acid is a medium strong dibasic acid. Its normal salts are but little hydrolyzed. The pronounced property possessed by tartrate ion of forming complexes with cathions is regarded as due to its oxygen content (hydroxyl).

139. **The distinction of oxalate- and tartrate ion.** These two ions differ by the fact that barium oxalate is considerably less soluble than barium tartrate in acetic acid. The difference is not great enough, however, to be utilized as a method for the *detection* of *tartrate ion,* especially in mixtures which contain proportionately small amounts of tartrates.

A *sure means* of *distinguishing tartrate ion* from *oxalate ion* is based upon the property possessed by tartrates of forming *complex anions* with many different cathions. This property manifests itself in the solubility of hydroxides of the cathions in question, for example cupric hydroxide, in alkaline solutions of alkali tartrates. Since copper hydroxide yields a strikingly blue-colored solution this reaction affords a delicate test for tartrate ion.

To carry out the test the solution is concentrated as much as possible and is then made distinctly alkaline. A few drops of copper sulphate solution are added and the precipitate of copper hydroxide is filtered after it has stood for a while. The *filtrate* has a *blue color* when the solution contained tartrate ion. The color is made still more striking by adding ammonia to the filtrate.

Cupric ion from the copper sulphate added should be practically completely precipitated by hydroxyl ion in the absence of tartrate ion, so that the copper solution would be decolorized. An alkaline reaction after the addition of $Cu^{\cdot\cdot}$ to the solution shows that the right amount was added.

This reaction affords an *unequivocal test* for *tartrate ion only* when the *solution* contains *no arsenite-* and *ammonium ions.*

Arsenic must be precipitated from the solution by means of hydrogen sulphide. Ammonium ion would interfere with the precipitation of copper by hydroxyl ion on account of the formation of *copper ammonia ion* due to ammonia resulting from $NH_4^{\cdot}$ and OH'. Hence it would cause the copper hydroxide to dissolve in alkali hydroxide, even when tartrate ion is not present. It is removed by *heating* the *alkaline solution before adding copper sulphate.* What is the equation for this reaction?

The test must be carried out in the *cold,* as precipitated copper hydroxide undergoes a change of condition when heated. In this condition it dissolves less readily in tartrate solutions and consequently the test would be less delicate.

The presence of *oxalate ion* along with *tartrate ion* is indicated by the fact that the barium precipitate is then rather difficultly soluble in acetic acid. To make sure that a residue insoluble in acetic acid consists of barium oxalate follow the text of § 125.

GROUP IV.

140. The anions belonging to this group comprise: *chloride-* (Cl'), *bromide-* (Br'), *iodide-* (I'), *cyanide-* (CN'), *ferrocyanide-* ($Fe(CN)_6''''$), *ferricyanide-* ($Fe(CN)_6'''$), *hypochlorite-* (ClO'), *sulphocyanate-* (SCN'), [*iodate-* (IO_3') and *bromate* ion (BrO_3')]. They differ essentially from the anions hitherto dealt with in their reactions with the three typical cathions (§ 110). *Solutions of these anions* (in ordinary concentrations) *give no precipitates with barium ion,* since the solubility products of the corresponding compounds are so large that the supersaturation points of the solutions with respect to these salts are never reached. *The silver precipitates are insoluble in nitric acid.* This is based upon the fact that the hydrogen compounds of these anions are strongly dissociated and the precipitates are difficultly soluble. Hence saturated solutions of the precipitates contain so small an amount of the anion, for example of Cl' from silver chloride, that the reaction:

$$\underline{AgCl} + H^{\cdot} = Ag^{\cdot} + H^{\cdot}Cl',$$

does not occur to a noticeable extent.

Hydrocyanic acid is, however, an exception. As was stated in § 71 it is proportionately little dissociated. An explanation of the inconsistency, which lies in the fact that silver cyanide is nevertheless practically insoluble in nitric acid, will be given later (§ 146).

Similarly hypochlorous acid is a weak acid. The silver precipitate obtained from a solution of a hypochlorite is scarcely soluble in nitric acid. The fact that it is not silver hypochlorite, but silver chloride, explains the inconsistency in this case (compare § 151, (*3*)).

The silver salts of these ions, as well as of those of the two preceding groups, are soluble in ammonia, thiosulphate and potassium cyanide. The reasons for this have already been stated in §§ 11 and 123.

Silver iodide and *silver ferrocyanide* are exceptions. To dissolve the latter a much larger quantity of ammonia, i. e. stronger ammonia, is required than is the case with silver chloride. Silver iodide is practically insoluble in ammonia. These two salts are soluble in thiosulphates and in alkali cyanides, however.

The distinction of the ions belonging to this group will be clear from the following paragraphs.

141. Properties of chloride ion and chlorides. The solution of an alkali chloride containing $\frac{1}{10}$ formula weight per liter is used for

the test solution. *Lead ion* gives a white precipitate of *lead chloride,* which dissolves in hot water (§ 14). *Silver chloride* is soluble in ammonia, alkali cyanide (§ 12) and thiosulphate (§ 123). *Mercurous ion* forms a precipitate of *mercurous chloride* (§ 5) which becomes black with ammonia.

A method often used to distinguish *chlorides* from bromides and iodides is based upon the formation of volatile *chromyl chloride,* CrO_2Cl_2, by heating a solid chloride with potassium dichromate and concentrated sulphuric acid.

To carry out the experiment a dry mixture of the unknown with potassium dichromate is placed in a fractionating flask and a few cc. of concentrated sulphuric acid are poured over the mixture. The side arm dips into a test-tube containing a little ammonia. The end of the delivery tube should not pass below the surface of the ammonia, but should extend down close to the surface. An ordinary tight-fitting cork is used to close the neck of the flask. The dense brown vapors generated on heating the flask escape through the side arm and are dissolved in the ammonia, yielding a yellow solution. This yellowish solution is acidified with sulphuric acid and is tested with hydrogen peroxide for dichromate ion (§ 119). Chromic acid does not distil over in this experiment. When a chloride is present it forms chromyl chloride according to the equation:

$$K_2Cr_2O_7 + 4KCl + 3H_2SO_4 = 2CrO_2Cl_2 + 3K_2SO_4 + 3H_2O \quad (1).$$

The chromyl chloride dissolves in ammonia yielding chromate ion:

$$CrO_2Cl_2 + 4OH' = CrO_4'' + 2Cl' + 2H_2O \quad (2).$$

Hence the appearance of the hydrogen peroxide reaction indicates with certainty that the mixture contains a chloride.

For the behavior of bromides and iodides under similar conditions consult § 143 and also Principles, p. *609*. In case iodides are to be tested for chlorides alkali hydroxide is used in place of ammonia to absorb the vapors distilling over (consult § 144).

Concentrated sulphuric acid decomposes solid chlorides except those which are very little soluble, as lead-, silver- and mercurous chloride, with an evolution of hydrochloric acid gas. The escaping hydrochloric acid is easily recognized by its penetrating odor. A glass rod moistened with silver nitrate becomes covered with a skin

of silver chloride when held in the vapors evolved in this experiment.

In case oxidizing agents, as peroxides or nitric acid, are also present, chloride ion is oxidized to chlorine. This is detected by its green color, unpleasant odor and the property of separating iodine from a solution of potassium iodide. The general equations for the oxidation of chloride ion to chlorine are:

$$Cl' = Cl + ('),$$

or

$$Cl' + (^{\cdot}) = Cl \qquad (3).$$

142. **Differences in the properties of anions and the substances from which they are formed.** The well known fact that chlorine gas dissolves rather abundantly in water makes it possible to distinguish experimentally the properties of chlorine from those of chloride ion. The latter is contained in solutions of chlorides and may be formed from chlorine or may be changed into it. This distinction is not possible in the case of cathions and the substances forming cathions, except for a comparison of the physical properties, because the metals are practically insoluble in water.

A solution of chlorine is found to possess certain properties which chloride ion lacks. Thus, for example, it sets iodine free from a solution of potassium iodide. On the other hand there are certain properties possessed by chlorine solution and chloride ion in common. Thus chlorine solution yields with lead ion a white precipitate, which turns brown, however, when heated. With silver ion it forms a precipitate that has the properties of silver chloride. An exact study of these relations by *Jakowkin* has shown that chlorine water actually contains chloride ion. Its formation results from the following reaction:

$$Cl_2 + H_2O \rightleftarrows H^{\cdot} + Cl' + HClO.$$

The aqueous solution of chlorine gas, when in a state of equilibrium, accordingly contains unchanged chlorine, hydrogen- and chloride ions and also hypochlorous acid. Through an independent change in the concentration of individual substances or ions taking part in the equilibrium the quantities of the others are altered to a definite extent. For example, by adding hydrochloric acid, chlorine and water are formed at the expense of the other components which take part in this reaction, Cl', $H^{\cdot}$ and $HClO$.

The change of color which the lead precipitate undergoes when heated is due to the hypochlorous acid present (§ 151).

This *partial agreement* in the properties of *dissolved chlorine* and of *chloride ion* therefore does not oppose what has been stated already in § 8. It is due to the above-mentioned transformation which chlorine gas undergoes to a certain extent in aqueous solutions.

143. **Bromide ion and bromides.** Like hydrochloric acid the hydrogen compound of bromide ion is a strong monobasic acid. Its alkali salts have a neutral reaction. A solution containing $\frac{1}{10}$ formula weight of the alkali salt per liter is used for the tests.

Lead ion forms white *lead bromide.* By raising the temperature the solubility of lead bromide in water also increases, but not to the same extent as that of lead chloride.

Silver bromide is slightly yellowish. It is insoluble in nitric acid, but dissolves in ammonia, potassium cyanide and sodium thiosulphate. The *solubility product* of *silver bromide* in water is considerably *smaller* than that of *silver chloride.*

When heated with concentrated sulphuric acid and potassium dichromate bromides evolve bromine. (What is the equation?) This dissolves in ammonia to a colorless solution and evolves nitrogen simultaneously. Hypobromite ion is formed on dissolving bromine in ammonia or alkali hydroxide according to the equation:

$$2Br + 2OH' = BrO' + Br' + H_2O \qquad (1).$$

This reacts with ammonia to form nitrogen:

$$3BrO' + 2NH_3 = 3Br' + N_2 + 3H_2O \qquad (2).$$

Chlorine water oxidizes a solution of bromide ion to bromine:

$$Br' + Cl = Br + Cl'. \qquad (3).$$

By shaking a solution containing bromine with *carbon disulphide* (CS_2) or *chloroform* ($CHCl_3$), which do not mix with water and hence form a double layer,[1] the bromine is taken up[2] by these solvents. If an excess of chlorine water is used the color vanishes owing to the formation of lighter colored chlorobromide.

Bromides differ from chlorides furthermore in their behavior towards concentrated sulphuric acid. Dry bromides or their concentrated solutions react with sulphuric acid to form free bromine along with hydrobromic acid.

[1] Carbon disulphide and chloroform are not completely *insoluble* in water. They both dissolve to a slight extent and conversely a little water dissolves in each of them. Hence one layer consists of a solution of one of these substances in water, the other of a solution of water in chloroform or carbon disulphide.

[2] Strictly speaking only the major part of the bromine goes into solution in the chloroform or carbon disulphide. This is evidenced by a *decolorization* of the *aqueous layer* and the formation of a *darker color* in the *other layer.* In all cases where a substance like bromine is brought in contact with two liquids, in each of which it is soluble, distribution takes place according to a definite law (law of distribution). No matter what the absolute amount of the distributing substance is, its ratio in the two layers is always the same [provided it undergoes no chemical changes in either layer].

Bromide ion is much more easily oxidized than chloride ion, i. e. it passes out of the ionic condition more readily. This is shown by the reaction between bromides and chlorine water. Sulphuric acid is reduced to sulphurous acid, which decomposes into sulphur dioxide and water:

$$SO_4'' + 2H^{\cdot} + 2Br' = SO_3'' + H_2O + Br_2,$$

or $$H_2SO_4 + 2H^{\cdot} + 2Br' = SO_2 + 2H_2O + Br_2 \qquad (4).$$

Concentrated sulphuric acid hence acts as an oxidizing agent towards bromide ion, but not towards chloride ion, or only when a stronger oxidizing agent, as alkali dichromate, is also present.

144. Iodide ion and iodides. The tests are carried out with a solution of potassium iodide of the same concentration as the test solutions of bromide or chloride.

Lead acetate gives a yellow precipitate of *lead iodide* (§ 14).

Silver iodide is practically insoluble in nitric acid and ammonia, but is soluble in alkali cyanide and -thiosulphate, and to some extent in alkali halides.

Mercuric ion forms *red mercuric iodide,* soluble in an excess of potassium iodide (§ 33).

A solution of copper sulphate forms a brown precipitate (§ 40) which consists of cuprous iodide and iodine.

From the solution of a *palladous salt* iodide ion precipitates black *palladous iodide,* PdI_2.

Iodides react like bromides with sulphuric acid and dichromates (compare the preceding paragraph).

On testing for chlorides in a mixture which also contains iodides (§ 141), it is better to use alkali hydroxide instead of ammonia for absorbing the vapors distilling over, because iodine and ammonia form compounds which explode readily, especially when dry.

Iodide ion is still more easily oxidized than bromide ion, i. e. it acts as a stronger reducing agent. Thus dichromate- + hydrogen ion and nitrite- + hydrogen ion oxidize iodide ion under certain conditions, but do not oxidize bromide ion. The reaction for the oxidation by nitrite ion follows the equation:

$$NO_2' + I' + 2H^{\cdot} = I + NO + H_2O \qquad (1).$$

This is a very delicate test and may be carried out with low concentrations of hydrogen ion. In making the test a solution of potas-

sium or sodium nitrite is added to the given solution. The mixture is then acidified with a few drops of acetic acid. The resulting iodine is easily detected by means of the blue color it gives with starch solution.[1]

The reaction for the oxidation with dichromate ion may be readily formulated from the general equation (*1*), § 120.

The influence of the nature of the acid upon the reaction may be very strikingly illustrated with this example. A mixture of dichromate ion, potassium iodide and starch solution does not react the same, if acidified with acetic acid, as when dilute sulphuric acid is used. In the latter case the blue coloration appears at once and is distinct, while in the former it makes its appearance only after some minutes and gradually increases in intensity. This difference is due to the *relationship* existing between the *concentrations* of the substances or ions taking part in the reaction and the *velocity* of the transformation, as has already been pointed out in connection with the dissolving of calcium carbonate in acids (§ 96).

Corresponding to its greater capacity for oxidation iodide ion is oxidized by chlorine *and* bromine to free iodine, which separates. It is detected by the deep violet color imparted to carbon disulphide or chloroform on shaking out the solution with either of these substances. When an excess of chlorine water is added the color vanishes owing to oxidation of the iodine to iodic acid:

$$I + 3H_2O + 5Cl = 5H^{\cdot} + 5Cl' + H^{\cdot} + IO_3' \qquad (2).$$

Iodine dissolves much more readily and more abundantly in a solution of potassium iodide or hydriodic acid than in water, owing to the formation of complex triiodide ion, I_3'.

145. The detection of chloride ion in the presence of bromide- and iodide ions. These three ions differ considerably from one another with respect to the solubility of their difficultly soluble salts, for example, the silver salts. However the evaluation of these differences in order to effect a separation of the salts demands much skill. The difficulties relate particularly to the identification of chloride ion, since bromide- and iodide ions are characterized by their reactions with chlorine water.

[1] The starch solution is prepared by stirring 2 g. of soluble starch to a paste with cold water. This is then poured into enough hot water to make 100 cc. solution. After standing for some time the solution becomes clouded and separates a flocculent precipitate.

The reaction is influenced by iodides and hydriodic acid in the sense that the blue color is more distinct when iodide ion is also present. The color vanishes on heating the solution and forms again when it is cooled.

The method described in § 141, which is based upon a distillation with potassium dichromate and sulphuric acid, does not yield as good results as the following method, which utilizes differences in the reducing action of these ions towards permanganate ion.[1] It is easily shown that iodide- and bromide ions are oxidized by permanganate ion + hydrogen ion. It is likewise known that chlorine is formed by a reaction between hydrochloric acid and manganese dioxide and that a reduction of permanganate gives rise to the latter (§ 80). The formation of chlorine then certainly takes place when permanganate is used in place of manganese dioxide, other conditions remaining the same. The following is the general equation for the reaction with permanganate ion in acid solution:

$$MnO_4' + 5Cl'(Br', I') + 8H^{\cdot} = 4H_2O + Mn^{\cdot\cdot} + 5Cl\ (Br, I) \quad (1).$$

The following individual characteristics are exhibited by these three anions. By keeping the concentrations of the halide ions and of permanganate ion unchanged and suitably altering the concentration of hydrogen ion, the oxidation of each of the halides occurs successively. To oxidize iodide ion requires the lowest concentration of hydrogen ion, while chloride ion demands the highest concentration of hydrogen ion. Hence by holding the concentration of hydrogen ion within certain limits bromide- and iodide ion are oxidized and chloride ion is left without undergoing any considerable change of concentration.

Practically the method assumes a very simple form. A potassium permanganate solution containing 3.2 g. $KMnO_4$ and 200 cc. dilute sulphuric acid[2] per liter is used. The solution to be tested is mixed with about its own volume of dilute acetic acid and is heated to boiling in a suitable vessel (Erlenmeyer flask). To this the permanganate solution is added in small portions at a time from a glass tube drawn out to a point until the red color remains permanently, i. e. for several minutes.

[1] This method has been worked out by *Jannasch* and *Küster* for the *quantitative* separation of these ions. For the literature consult Crotogino, Zeitschrift für anorganische Chemie, 24, *227*, 1900.

[2] It is assumed that the dilute sulphuric acid used in this case contains one formula weight of H_2SO_4 per liter. If the solution used is stronger or weaker than this a correspondingly smaller or larger volume must be taken to make up the permanganate solution.

As long as any iodide ion is present the permanganate is decolorized almost immediately. In case no iodide ion is present, or if it has been already oxidized and bromide ion is retained in the solution, the decolorization of the added permanganate takes place more slowly. The rate of the reaction may be increased by adding the permanganate in rather larger portions.

The red solution is then boiled for 30 minutes. In case it becomes decolorized during this operation more permanganate solution is added. Of course the time of heating can be only approximately given since it is assumed that the quantities of bromide- and iodide ion are unknown. In no case should it be assumed that the oxidation of bromide ion is *ended* after 15–30 minutes. This is true for iodide ion if an excess of permanganate is still present. As for bromide ion it must always be determined whether bromine is being evolved or not at the end of this time. This test is very easily made by covering the neck of the flask with a disc of filter paper which has been moistened with a drop of potassium iodide solution. So long as it is escaping bromine will set iodine free from potassium iodide (§ 53). This is detected by its brown color. However this criterion does not indicate with absolute certainty that bromine is no longer escaping. For chlorine will cause a separation of iodine and it may be evolved, especially when the solution contains a great deal of chloride. The oxidation takes place much more slowly than that of bromine under these conditions. Hence the escaping chlorine will also turn the test paper brown at a much slower rate than bromine. In any case the test can lead to a wrong conclusion only through heating the solution too long, so that chloride ion is partly oxidized. This excludes any possible errors that would arise if bromide ion remains in the solution.

Manganese dioxide separates during the heating. This indicates that equation (*1*) does not completely express the actual relations. When the solution is not sufficiently acidified with sulphuric acid permanganate ion does not reduce exclusively to manganous ion, but is partly reduced to manganese dioxide. In this case the concentration of the acid is *purposely* kept low to avoid oxidizing the chloride ion as far as possible. The reaction for the reduction of permanganate with bromide ion to manganese dioxide is expressed by the following equation:

$$MnO_4' + 3Br' + 4H^{\cdot} = \underline{MnO_2} + 2H_2O + 3Br \qquad (2).$$

As a matter of fact the relations are still more complex. Manganese dioxide may also be formed by a reaction between permanganate- and manganous ion:

$$2MnO_4' + 3Mn^{\cdot\cdot} + 4OH' = \underline{5MnO_2} + 2H_2O \qquad (3).$$

The latter reaction takes place not only in alkaline solution, as would be expected, i. e. when hydroxyl ion is present in excess of hydrogen ion, but also in slightly acid solutions. While manganese dioxide will oxidize even chloride ion in strongly acidified solutions, it does not, however, suffice to remove bromide-, and much less chloride ion, from solutions in which acid is lacking. As has already been stated above, it is therefore necessary to have an *excess* of *permanganate* to remove bromide ion.

When it is certain that all bromide ion has been removed up to the limits of its detection by the above test with potassium iodide, a *very dilute* solution of sulphur dioxide[1] is dropped into the flask to remove the excess of permanganate and the precipitated manganese dioxide. It must be added until the contents of the flask are perfectly water-clear. Then the excess of sulphur dioxide is driven out by boiling the solution, and silver nitrate is added. A precipitate indicates chloride ion. To make sure that the precipitate is silver chloride, and not silver bromide, a little ammonia is added. The chloride will go into solution readily.

In case the original solution contained iodide ion it is better to reduce the excess of permanganate ion with alcohol rather than with sulphur dioxide. In acid solutions permanganate oxidizes iodide ion to iodine and also partially to iodate ion:

$$6MnO_4' + 5I' + 18H^{\cdot} = 5IO_3' + 6Mn^{\cdot\cdot} + 9H_2O \qquad (4).$$

Iodate ion is then reduced by sulphurous acid to iodide ion (consult § 121, (*4*), (*5*), (*6*)). Hence by using sulphurous acid to reduce permanganate it is possible that iodide ion would be only *incompletely removed.* It is therefore advisable to reduce the permanganate with alcohol if iodides are originally present. Heating the solution removes any alcohol left after the reduction is completed.

Likewise the behavior of the precipitate towards dilute nitric acid must be tested. If the oxidation of iodide ion has given rise to iodate ion (see equation (*4*) above) the white precipitate which this yields with silver nitrate might be mistaken for silver chloride. This precipitate is *soluble in nitric acid,* however, and is thus easily identified.

[1] The solution of sulphur dioxide must be tested for chloride ion before it is used for this experiment. The test is made by adding silver nitrate and noting if the precipitate is completely soluble in nitric acid. The solution must not be used if it contains chloride ion.

As has already been mentioned an oxidation of chloride ion cannot be avoided by this method. However the *lower* the concentration of chloride ion, the *less easily* it is oxidized. Hence the loss of chlorine through this oxidation becomes smaller and by this method even small amounts of chlorides may be detected with certainty when larger quantities of bromides and iodides are also present. In a solution containing 0.5–1.0 mg. sodium chloride and 100 mg. potassium bromide the chloride ion can be distinctly detected if the bromide ion has been removed as above given.

Cyanide and sulphocyanate ions are also oxidized by permanganate. Hence the method is serviceable to identify chloride ion when cyanide and sulphocyanate ions are present. However complex cyanides (like $K_2Ni(CN)_4$, $K_4Fe(CN)_6$ or $Cu_2Fe(CN)_6$) are not oxidized, or at least not completely, by permanganic acid. Therefore the above test for chloride ion is not directly applicable in this case.

The complex cyanides, even the difficultly soluble ones, can be easily destroyed by boiling the solid salts with a suspension of 3–4 times their weight of pure mercuric oxide, with the addition of 5–10 cc. twice normal sulphuric acid. The decomposition is usually completed in a few minutes. Any slight residue left in case the decomposition is incomplete, is filtered off. The solution contains mercuric cyanide, § 176, the cathions of the metals contained in the salt and the excess of mercuric ion. It may be tested with silver nitrate for Cl' (Br', I'). In this form the test is not specially delicate, however, since silver chloride is much more soluble in the above solution than in water. It is therefore better to precipitate the mercuric ion with hydrogen sulphide. By this means copper and other metallic ions which may also be present are precipitated. The filtrate is now boiled and a current of carbon dioxide, washed free from HCl, is passed through the boiling filtrate for 15–20 minutes, until the escaping steam no longer blackens lead test-paper. Ferrous ion, which will be present when complex cyanides of iron are treated in this way, must be oxidized with permanganate. Then the solution is tested with silver nitrate for Cl', Br' and I'. The test is fairly delicate since it detects less than one milligram of chlorine in the presence of more than 100 times as much cyanide. It must be borne in mind, however, that a very slight reaction (cloudiness) may be observed even when no chloride is present, probably from the formation of a trace of sulphocyanate. (Further details concerning this reaction will be published in the Journal of the American Chemical Society.)

If it is desired to test for chlorides in the presence of bromides or iodides *and* complex cyanides, the procedure is a little more complicated. After boiling the filtrate is saturated with hydrogen sulphide. In order to neutralize the acid which is formed during the precipitation of the sulphides, sodium acetate is added. The filtrate is made ammoniacal in case ions are present which are not precipitated in acid solution ($Ni^{\cdot\cdot}$) and is then saturated with hydrogen sulphide. The sulphides are filtered, acetic acid is added to the

filtrate and it is heated for a short time to expel the excess of hydrogen sulphide. Then it is tested for Cl′ with potassium permanganate as above described.

[Another very serviceable method[1] of separating Cl′, Br′ and I′ differs from the permanganate method by using potassium persulphate, $K_2S_2O_8$, as the oxidizing agent. It works out practically the same as the permanganate method. By regulating the concentration of hydrogen ion in the solution the halogens may be liberated successively, iodide ion requiring the least and chloride ion the most hydrogen ion. *Acetic acid* with a *little sodium acetate* furnishes the proper concentration of hydrogen ion to set free *iodine*, but *not* bromine. Bromine requires a *strongly dissociated acid*, as sulphuric acid. *No chlorine* will be set free unless more than one formula weight of sulphuric acid per liter is present.

To carry out the separation about 10 cc. of the solution are taken and an excess of sodium hydrogen carbonate is added to precipitate the heavy metals. ($NaHCO_3$ is used instead of Na_2CO_3 because the latter is scarcely ever free from chloride.) An excess of acetic acid is added to the alkaline filtrate and then water enough to make 50 cc. of solution. Into this solution about 0.5 g. potassium persulphate crystals are shaken and the mixture is boiled. Iodide ion is hereby oxidized to iodine, which is detected by the carbon disulphide test. In case iodide ion is found to be present the solution is boiled several minutes until all free iodine is driven off and the solution has become colorless. Then a few more crystals of persulphate are again added and the boiling is repeated to make sure that all iodide ion has been oxidized. During the boiling the volume of the solution must always be kept about 50 cc. by adding water from time to time.

Bromide ion is next oxidized to free bromine by adding a few cc. of dilute sulphuric acid[2] and some more persulphate. A portion

[1] This is given through the courtesy of Mr. H. H. Willard, who has worked out the method and uses it at the University of Michigan.

[2] Difficulties are sometimes experienced at this point. They are avoided by making a preliminary test for bromide ion with a small portion of the solution freed from iodide ion. To this a larger volume of dilute sulphuric acid is added (several cc.). In case free bromine is observed, care must be taken that all bromide ion is oxidized in the rest of the solution by increasing the concentration of sulphuric acid if necessary. Under certain conditions this oxidation proceeds slowly when only a little H_2SO_4 is present.

of the mixture is heated until it begins to boil. Any bromine liberated will color the solution yellow to red and is detected by the carbon disulphide test, which also shows if all iodide ion has been oxidized. If bromide ion is present it is expelled in the same manner as iodide ion from the entire solution, the addition of persulphate and the heating being repeated to make sure that all is removed. Always the volume of the solution must be maintained at about 50 cc. to avoid oxidation of chloride ion, which might occur if the concentration of hydrogen ion is allowed to increase. When bromine is all removed silver nitrate is added to the remaining *cold* solution, precipitating chloride ion as silver chloride. Only a few drops of silver nitrate should be added, otherwise silver sulphate would precipitate. This might be taken for silver chloride. From a hot solution silver chloride is rapidly oxidized by potassium persulphate to soluble silver chlorate. (M. Dittrich and H. Bollenbach, Ber. d. d. chem. Gesellschaft, **38**, *747*, **1905**.)

A chlorate in the original mixture would oxidize iodide ion to iodate ion by the above treatment. As this is not decomposed by persulphate the iodide ion would not be detected. In this case silver nitrate is added to the original solution and the precipitated silver halides are heated with dilute nitric acid. Silver chlorate is readily soluble in water and hence is not precipitated. To bring the precipitated silver halides into solution they are heated with some zinc in a little water to which a drop of sulphuric acid is added. Silver reduces to the metallic condition and zinc halides are formed. These are soluble in water. The persulphate separation may then be carried out as above given with the solution of the zinc salts.]

146. Cyanide ion and cyanides. *Cyanide ion* shows a great *similarity* to the *halide ions* in its properties, particularly to *chloride ion.* A noteworthy *difference* exists, however, with respect to the *dissociation* relations of *hydrocyanic acid* and its *salts.* This acid is analogous with the halogen hydrides, but is a *weak* monobasic acid. Cyanides are less dissociated (§ 49) than the other salts with which they may be compared as regards composition. The fact already mentioned in § 71, that soluble cyanides have an alkaline reaction, agrees with these relations.

Cyanide ion has also a very pronounced tendency to form complex

ions, as has been often mentioned. These exhibit the reactions of cyanide ion to only a very slight extent.

The tests are made with a solution of *potassium cyanide* containing $\frac{1}{10}$ formula weight per liter. The solution is neutralized with dilute nitric acid. It forms *no precipitate* with *barium ion*. *Lead* and *silver* ions yield *white precipitates* of *lead* and *silver cyanides*. The latter dissolves in ammonia, thiosulphate and in an excess of potassium cyanide.

A *non-neutralized* solution is much *more active* in precipitating the typical cathions. Furthermore the property of dissolving silver cyanide and other difficultly soluble salts is distinctly increased by adding potassium or sodium hydroxide drop by drop to the neutralized solution containing an excess of potassium cyanide, which is in contact with the silver precipitates. The fact that a *non-neutralized* solution of potassium cyanide has a stronger action in precipitating lead ion, and also gives a distinct precipitate with barium ion, together with the above phenomenon, have a common cause. The concentration of cyanide ion in the neutralized solution is much lower than in an *alkaline* solution containing the same amount of cyanide. This follows from considerations like those holding in the case of carbonate ion (§ 137), on the basis of a reaction between cyanide ion and water:

$$CN' + H_2O \rightleftarrows HCN + OH'.$$

The fact that silver cyanide is scarcely soluble in cold dilute nitric acid, although hydrocyanic acid is a weak acid, leads to the conclusion that the concentration of cyanide ion is very low in a solution saturated with silver cyanide. Silver cyanide dissolves to a noticeable extent in *hot* water and *hot* dilute nitric acid and reprecipitates on cooling the filtered solution.

The following is a particularly delicate test for cyanides. A solution containing cyanide ion is rendered slightly alkaline with alkali hydroxide and is heated a short time with a few drops of a solution of ferrous ion. It is then acidified with hydrochloric acid and a few drops of ferric chloride solution are added. A blue *precipitate* forms, *difficultly soluble* in acids. When only a little cyanide is contained in the solution a *blue coloration* results.

Complex ferrocyanide ion is first formed according to equations (*4*) and (*5*), § **68**. This yields Prussian blue with ferric ion (consult the next paragraph).

Another method of testing for cyanides utilizes the deep *red color* resulting from the reaction between *ferric-* and *sulphocyanate* ions. Sulphocyanate ion, SCN′, differs from cyanide ion merely by its sulphur content. The latter ion takes up sulphur when it is brought

in contact with yellow ammonium sulphide. Hence the mixture to be tested for cyanide ion is heated in a test-tube with dilute hydrochloric or sulphuric acid according to the method described in § 125. The delivery tube dips in yellow ammonium sulphide. The following reaction takes place:

$$CN' + S_n'' = SCN' + S_{(n-1)}''.$$

The contents of the test-tube in which hydrocyanic acid was absorbed are evaporated to dryness on the water-bath. To destroy the excess of ammonium sulphide the residue is acidified with hydrochloric acid and is then tested with ferric chloride for sulphocyanate ion.

Fused alkali cyanides are rather strong reducing agents. They take up oxygen and go over into cyanates:

$$KCN + O = KCNO.$$

For this reason they are frequently used in blowpipe experiments (compare § 48).

Many cyanides are decomposed with effervescence by dilute acids (occasionally even by acetic acid). They are distinguished from carbonates, which have the same property of effervescing, by the fact that the escaping hydrocyanic acid renders silver nitrate cloudy.

When heated to redness cyanides are variously decomposed, now yielding cyanogen, now giving rise to carbide and nitrogen, or metal, carbon and nitrogen.

147. Ferrocyanide ion and ferrocyanides. The test solution contains $\frac{1}{40}$ formula weight of *potassium ferrocyanide*, $K_4Fe(CN)_6 \cdot 3H_2O$, per liter. Ferrocyanide ion, $Fe(CN)_6''''$, is light yellow. With lead ion it yields *white lead ferrocyanide*, but little soluble in dilute nitric acid, easily soluble in the concentrated acid.

Silver ferrocyanide is white. Its saturated solution contains very little silver ion and hence the precipitate is only sparingly soluble in ammonia (compare § 13).

Ferrous ion yields a white precipitate of ferrous ferrocyanide which quickly turns blue:

$$Fe_2Fe(CN)_6 \text{ or } FeK_2Fe(CN)_6.$$

The blue color is caused by an oxidation with atmospheric oxygen. The ferrous salt goes over into ferric ferrocyanide, which may also be prepared directly by adding *ferric ion* to the solution of ferrocyanide ion. The precipitate is called *Prussian blue* and has the composition $Fe_4[Fe(CN)_6]_3$. It is practically insoluble in dilute acids and is decomposed by alkali hydroxides and carbonates (compare § 184, c, on the dissolving of cyanides).

Soluble ferrocyanides are slowly decomposed by dilute acids, giving rise to hydrocyanic acid. Concentrated sulphuric acid decomposes them and forms the corresponding sulphates and carbon monoxide. These reactions are formulated as follows:

$$2K_4Fe(CN)_6 + 3H_2SO_4 = \underline{FeK_2Fe(CN)_6} + 3K_2SO_4 + 6HCN,$$

and $$K_4Fe(CN)_6 + 6H_2SO_4 + 6H_2O = 2K_2SO_4 + FeSO_4 + 3(NH_4)_2SO_4 + 6CO.$$

When heated in the bulb tube ferrocyanides yield simpler cyanides, iron carbide and nitrogen.

According to its composition the hydrogen compound of ferrocyanide ion is a tetrabasic acid. The acid is rather unstable, however, as is evidenced by the behavior of its salts towards other acids. It separates as a crystalline powder on mixing a concentrated solution of potassium ferrocyanide with concentrated hydrochloric acid.

148. Ferricyanide ion and ferricyanides. A solution of *potassium ferricyanide*, $K_3Fe(CN)_6$, containing $\frac{1}{30}$ formula weight per liter, is used for the tests. Ferricyanide ion has a yellow color. *Silver nitrate* yields an orange precipitate of *silver ferricyanide*. The precipitate dissolves in ammonia, but is insoluble in nitric acid.

Ferrous ion precipitates *ferrous ferricyanide*, $Fe_3[Fe(CN)_6]_2$, Turnbull's blue, resembling Prussian blue. *Ferric ion* gives a *brown colored* solution. Cupric ion forms *green cupric ferricyanide*, $Cu_3[Fe(CN)_6]_2$, while with ferrocyanide ion *brown cupric ferrocyanide*, $Cu_2Fe(CN)_6$, results (compare § 40 concerning the composition).

Ferricyanides react like ferrocyanides with dilute and concentrated sulphuric acid, and in the bulb-tube they also give reactions resembling those of ferrocyanides.

Ferricyanic acid is tribasic. It is obtained in brownish green needles by adding fuming hydrochloric acid to a cold saturated solution of potassium ferricyanide.

149. The relations between ferrocyanide- and ferricyanide ion. These two ions have the same chemical composition, but differ in the number of electric charges. They have the same relation to one another that ferrous ion bears to ferric ion. Ferricyanide ion is an oxidation product of ferrocyanide ion. This relation is formulated by the general equations:

$$Fe(CN)_6'''' + (^{\cdot}) = Fe(CN)_6''' \qquad (1a),$$

and

$$Fe(CN)_6'''' = Fe(CN)_6''' + (') \qquad (1b).$$

Thus in the presence of ions which can give up one positive charge or can take up one negative charge, ferrocyanide ion goes over into ferricyanide ion.

Most of the reactions by which ferrocyanide ion is oxidized illustrate the latter class; for example, by oxidizing with chlorine, which forms chloride ion, or with bromine. Reactions in which *ferricyanide ion oxidizes* certain substances are made use of to identify it in the presence of ferrocyanide ion.

Thus *iodide ion* is oxidized to free *iodine* in an acid solution (*2*), *manganous hydroxide* to *manganic hydroxide* in an alkaline solution (*3*):

$$Fe(CN)_6''' + I' = I + Fe(CN)_6'''' \qquad (2),$$[1]

and $Fe(CN)_6''' + \underline{Mn(OH)_2} + OH' =$

$$Fe(CN)_6'''' + \underline{Mn(OH)_3} \qquad (3).$$

In carrying out these experiments potassium iodide and hydrochloric acid, or manganous salt and alkali hydroxide are added to the solution that is to be tested for ferricyanide ion. In the first case a brown color is produced after a few minutes. Addition of

[1] According to experiments of F. G. Donnan and R. le Rossignol, Proc. Chem. Soc. London, the reaction between alkali ferricyanides and alkali iodides in neutral solution is formulated as follows:

$$2Fe(CN)_6''' + 3I' = 2Fe(CN)_6'''' + I_3'.$$

However many phenomena are not satisfactorily explained by this formulation.

starch solution gives the characteristic blue color caused by free iodine. In the second case light colored manganous hydroxide separates at first and goes over rapidly into brown manganic hydroxide. A parallel experiment without ferricyanide ion will show that the oxidation of manganous hydroxide by atmospheric oxygen (§ 80) takes place much more slowly.

In dilute neutral solutions, that is, with low concentrations of the ions taking part, the first reaction proceeds slowly and is incomplete. It takes place *in a much shorter time* and *to greater extent,* however, when some *dilute hydrochloric acid* is added. Acetic acid is considerably less active (consult §§ 69 and 96). Equation (*2*) does not state this, since it does not include hydrogen ion. Under these conditions the reaction is probably the following:

$$Fe(CN)_6''' + H^{\cdot} + I' = HFe(CN)_6''' + I \qquad (2b).$$

This explanation is based upon the assumption that the ion $HFe(CN)_6'''$ is more readily formed than the ion $Fe(CN)_6'''$. It would follow from this that ferrocyanide ion is a less active reducing agent in acid solutions than in neutral or alkaline.

Another method of detecting ferricyanide ion in a solution which also contains ferrocyanide ion is based upon the difference in solubility of the respective lead salts of these anions. *Lead ferrocyanide* is much *less soluble* than *lead ferricyanide.* Accordingly lead acetate is added drop by drop to the solution until the precipitate ceases to form. The precipitate is then filtered. The filtrate is heated and lead ion is precipitated from it with sodium sulphate. The filtrate from this latter precipitation will not contain more than traces of ferrocyanide ion, but may contain considerable quantities of ferricyanide ion. A freshly prepared acidified solution of pure ferrous salt, that is, a solution as far as possible free from ferric salt, is added to this filtrate. Ferricyanide ion is precipitated as blue ferrous ferricyanide. If the solution contains only very small amounts of ferricyanide ion it assumes a deep blue color.

150. The detection of cyanides in the presence of ferro- and ferricyanides. It is specially important to detect cyanides in a mixture also containing ferro- and ferricyanides, because the complex ion cyanides are less poisonous than simple cyanides and certain other complex cyanides, for instance potassium nickel cyanide. Simple cyanides are more quickly decomposed by dilute acids than the complex ions of iron and cyanogen. Hence dilute hydrochloric acid will leave the complex iron cyanides practically unchanged, in

case they are insoluble in water. The soluble complex salts are partly transformed into ferro- and ferricyanic acids respectively, to an extent corresponding to the quantity of hydrochloric acid added.

On adding a salt that is decomposed by acids, for example, a difficultly soluble carbonate, it is decomposed by the ferro- and ferricyanic acids and by the excess of hydrochloric acid. Hydrocyanic acid resulting from simple cyanides or the less stable complex cyanides, does not decompose carbonic acid, however. Ferro- and ferricyanic acids are transformed into non-volatile salts and hydrocyanic acid is distilled off by the method described in §§ 146 and 125, and is then detected as ammonium sulphocyanate.

To *carry out* the *test* from half a gram to a gram of the given mixture is placed in a test-tube and *dilute* hydrochloric acid is added. Then calcium carbonate is added in small portions at a time until the evolution of carbon dioxide ceases. The test-tube is next fitted with a stopper and delivery tube and is heated by dipping it in a larger vessel containing hot water for a few minutes (see Fig. *9*, § 125).

A violent evolution of carbon dioxide will sometimes carry over noticeable quantities of hydrocyanic acid. Hence it may be advisable to attach the delivery tube immediately after the carbonate is added. The escaping carbon dioxide is then also passed into ammonium sulphide. The contents of the ammonium sulphide tube are treated according to § 146.

151. Hypochlorous acid and hypochlorite ion. The test solution must be made up immediately before use, since solutions which have stood for some time contain other ions in consequence of the reactions mentioned later [see equation (*3*)]. It is made by shaking up yellow mercuric oxide or finely triturated red oxide of mercury (Principles, p. *210*, footnote) with chlorine water. It may also be prepared by passing chlorine into water containing mercuric oxide in suspension, until the reaction ceases to be visible. The chlorine is most simply generated from manganese dioxide and concentrated hydrochloric acid. The gas is passed through two wash bottles (Fig. 8, § 122) to take out hydrochloric acid vapors which are carried over with it.

The solution should smell strongly of chlorine. It is left in contact with mercuric oxide, with frequent shaking, until the most

striking characteristics of chlorine (odor and color) are no longer observed. To remove impurities the solution is then decanted from any unchanged mercuric oxide and the mercuric oxychloride, Hg_2OCl_2, which is formed by the reaction:

$$2HgO + 2Cl_2 + H_2O = Hg_2OCl_2 + 2HClO \qquad (1).$$

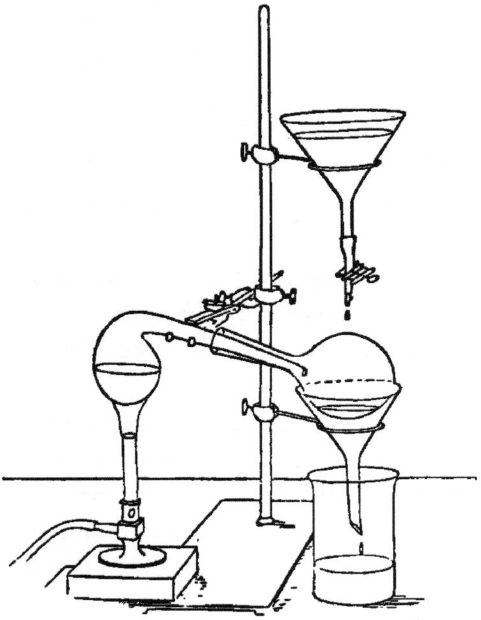

FIG. 10.

It is finally distilled from an apparatus of the form shown in Fig. *10.*

An attempt to neutralize the solution will show at once that this cannot be carried out. The litmus dye is quickly bleached in consequence of the great instability of hypochlorous acid. It decomposes spontaneously with an evolution of oxygen.

Lead ion forms a *white precipitate,* which goes into solution again by warming the contents of the test-tube. On standing the solution finally separates brown *lead peroxide:*

$$Pb^{\cdot\cdot} + ClO' + H_2O = \underline{PbO_2} + Cl' + 2H^{\cdot} \qquad (2).$$

Freshly prepared hypochlorous acid solution gives *no* precipitate with silver ion, or at most a *very slight cloudiness* (opalescence). After the solution has been kept for a while, however, it gives a distinct precipitation with silver ion. This arises from the following transformation:

$$3ClO' = 2Cl' + ClO_3' \qquad (3).$$

The reaction proceeds more rapidly and to an increased extent when silver ion is present, since this yields insoluble silver chloride with the chloride ion and so removes it as fast as it forms.

Solutions of hypochlorites also have strong oxidizing properties. For the same total concentration they probably contain more *hypochlorite ion.* They turn litmus paper blue (from hydrolysis), bleach indigo solution and evolve a continuous current of oxygen in the presence of certain substances, for instance, solutions of nickel, cobalt or manganese salts. The transformation which gives rise to chloride- and chlorate ions (*3*) takes place to a greater extent in alkaline solutions than in neutral solutions, since proportionately more hypochlorite ion is contained in alkaline solutions.

The oxidation of iodide ion to iodine by hypochlorite ion *in alkaline solutions* is especially characteristic. This distinguishes hypochlorite from other ions, which oxidize iodide ion only in acid solutions, or more correctly in solutions in which the concentration of hydrogen ion is greater, for example, Cr_2O_7'', NO_2', ClO_3', IO_3'.

Chlorine is evolved by acidifying hypochlorites or hypochlorous acid with hydrochloric acid, in consequence of the reaction:

$$HClO + H^{\cdot} + Cl' = H_2O + Cl_2 \qquad (4).$$

This is a reversal of the change which chlorine undergoes when dissolved in water (§ 142).

152. Reactions of sulphocyanate ion. The detection of this ion is comparatively seldom an analytical problem. It belongs to this group by reason of the practical insolubility of the silver salt in nitric acid. Of its special reactions the behavior towards ferric ion is striking (consult § 69). The red color of *ferric sulphocyanate* disappears upon addition of *mercuric nitrate* (distinction from acetate ion, § 158). *Mercuric ion* also influences the test for *acetate ion,* but in this case the effect is by no means so striking as with ferric sulphocyanate.

Only concentrated solutions of cupric ion precipitate black cupric sulphocyanate (§ 40). Dilute solutions give rise to a green coloration and separate cuprous sulphocyanate on standing. Black cupric sulphocyanate is also slowly transformed into white cuprous sulphocyanate when left in contact with the solution, even without the application of sulphur dioxide.

Concentrated sulphuric acid decomposes sulphocyanates, forming carbon oxysulphide, COS, which burns with a blue flame.

[**153. Iodate ion and iodates.** The tests for iodate ion are carried out with a solution of potassium iodate, KIO_3, or potassium hydrogen iodate, KHI_2O_6, containing $\frac{1}{10}$ formula weight per liter.

When using commercial potassium hydrogen iodate a residue which is practically insoluble may be left.[1] The solution has a strong acid reaction.

With the typical cathions (§ 110) neutral and acid solutions of iodates yield characteristic white precipitates. Silver nitrate and lead acetate form voluminous precipitates, appreciably soluble in dilute nitric acid. The lead salt is more readily soluble. Barium nitrate yields a rather less voluminous precipitate. It dissolves slowly in moderately concentrated nitric acid, more readily when the acid is very concentrated. This might give rise to errors in connection with group I, i. e. a precipitate of barium iodate might be mistaken for the sulphate or fluosilicate if nitric acid is used. The difficulty would not be experienced when hydrochloric acid is used, since barium iodate readily dissolves in HCl.

Silver iodate is curdy, white, readily soluble in dilute ammonia.

To detect iodates they must be reduced. Reducing agents like *sulphur dioxide* give a black precipitate of free iodine with the test solution, and effect a further reduction to the *iodide*. *Ferrous ion* is the best reducing agent for iodates, yielding free iodine with both neutral and acid solutions. Similarly antimonous chloride, $SbCl_3$, forms only free iodine. Since insoluble oxysalts of antimony are also precipitated the solution must be filtered. The detection of iodine is then carried out with the filtrate by means of carbon disulphide.

With potassium iodide *neutral* solutions of iodates yield a *colorless solution*. When acidulated, even with *acetic acid*, the solution becomes reddish brown owing to the separation of iodine:

$$5KI + KIO_3 + 3H_2SO_4 = 3K_2SO_4 + 3I_2 + 3H_2O,$$

or

$$5I' + IO_3' + 6H^{\cdot} = 3I_2 + 3H_2O.$$

This reaction serves to detect iodates in the presence of iodides, in case other anions which oxidize iodide ion to iodine, as NO_2', NO_3', ClO_3', Cr_2O_7'', are not present. If these ions are present the test is carried out as follows. The silver salts are precipitated, filtered and treated with ammonia, whereby all will go into solution except the *iodide* and ferrocyanide. From the ammoniacal filtrate

[1] Commercial potassium hydrogen iodate is frequently used for the purposes of volumetric analysis to obtain a solution of definite acid titer. It often contains the salt $KH_2I_2O_9$, which is still more difficultly soluble (Principles, p. 455).

the silver salts are reprecipitated by acidulating with acetic or sulphuric acid. This precipitate is then heated for a few minutes with sulphurous acid. By this means *silver iodate* is reduced to *silver iodide*, which is readily detected by its insolubility in ammonia. Silver bromate would yield silver bromide under these conditions. This salt is difficulty soluble in ammonia.

154. Bromate ion and bromates. The test solution contains $\frac{1}{10}$ formula weight of potassium bromate, $KBrO_3$, per liter. The solution has a neutral reaction.

It yields *no* precipitate with *barium-* or *lead nitrate*. Lead acetate, however, gives a slight cloudiness with the solution. *Silver nitrate* forms a *white* precipitate, soluble in dilute ammonia.

When *slightly* acidified with sulphuric acid a solution of potassium bromate gives a *red* coloration with *manganous sulphate*. On boiling a *brown black* precipitate of manganese dioxide separates.[1] Under similar conditions iodates and chlorates show *no change of color* and form *no precipitates*. Bromates may also be recognized in the presence of iodates by the fact that silver bromate yields free bromine when acidified with strong *nitric* or *hydrochloric acid*. This dissolves in chloroform with a characteristic yellow color.

Neutral solutions of bromates yield a *colorless* solution with potassium bromide. When acidulated with sulphuric acid the solution becomes reddish brown owing to the separation of bromine:

$$5KBr + KBrO_3 + 3H_2SO_4 = 3K_2SO_4 + 3Br_2 + 3H_2O.]$$

GROUP V.

The anions of this group comprise *nitrate-, nitrite-, acetate-* and *chlorate ions*. They give no precipitates with the typical cathions except in concentrated solutions. Hence to test for these anions the following characteristic reactions must be carried out separately with the original solution.

155. Nitrate ion and nitric acid. The test solution contains $\frac{1}{10}$ formula weight of potassium nitrate, KNO_3, per liter. A definite volume, about 2 cc., of the solution is mixed with an equal volume of saturated ferrous sulphate solution in a test-tube. On pouring concentrated sulphuric acid carefully down the (dry) walls of the test-tube, which is conveniently inclined for this purpose, the acid sinks

[1] This test is also due to Mr. H. H. Willard, University of Michigan.

to the bottom without mixing very much with the solution and at the surface of contact between the solution and acid a brown ring forms. This is due to nitric oxide which results from a reduction of the nitrate by ferrous ion (§ 54, (*2*)). It dissolves in ferrous salt solutions with a dark brown color. The color disappears when the test-tube is warmed. Hence the brown ring test fails when not enough care is taken to keep the sulphuric acid from mixing with the solution.

Bromide- and *iodide ion* disturb this test for nitrates owing to the fact that concentrated sulphuric acid decomposes their salts, setting free bromine and iodine respectively (§§ 143 and 144). Therefore they must be removed from the solution before the test for nitrates is made. This is effected by precipitating them with *lead acetate* and afterwards separating the lead ion remaining in solution by means of dilute sulphuric acid. The removal of bromide or iodide ion is still more complete if *silver acetate* is used instead of the lead salt.

Silver acetate should be kept in stock always. It is made by shaking up sodium acetate solution with silver nitrate, using an excess of sodium acetate. The silver salt is washed with a mixture of alcohol and sodium acetate solution until it no longer gives a test for *nitrate ion.* To carry out this test some of the silver acetate is dissolved in water. The silver ion is then precipitated from this solution by means of potassium chloride. This is necessary because silver ion is reduced by ferrous ion to metallic silver, which gives rise to a brown coloration, thus interfering with the detection of the brown ring. Then the filtrate is tested for nitrate ion with ferrous sulphate and sulphuric acid, being evaporated if necessary.

The silver ion left in the solution after precipitating bromide- or iodide ion with silver acetate, must likewise be removed by potassium chloride. This may be done before filtering the precipitate of bromide and iodide. The filtrate is then tested for nitrate ion.

Since the above operations necessarily increase the volume of the solution considerably, it is advisable to reduce the volume somewhat by evaporating the solution on a water-bath before making the test for nitrate ion. While being heated the solution must have a *neutral* reaction.

Dichromate ion also interferes with the test for nitrate ion owing to the coloration produced by the oxidation of ferrous ion with dichromate- + hydrogen ion. It is accordingly reduced to chromic ion by sulphurous acid. The chromic ion is then precipitated with ammonia.

Brucine is another very delicate reagent for nitrate ion. A dilute solution containing 0.2 per cent brucine in concentrated sulphuric acid is used and several cc. are taken for each test. A few drops of

a solution containing NO_3' impart a red color to the brucine solution. The color vanishes in a short time.[1]

This reaction is important in testing drinking water for nitrates. In this case the reaction of the sulphuric acid used for the test must first be determined with brucine, since sulphuric acid sometimes contains nitric acid.

Indigo solution is decolorized when warmed with nitric acid or with nitrate solutions plus sulphuric acid (compare the behavior of hypochlorites).

Ammonia is formed by boiling a solution containing nitrate ion with zinc dust [or aluminum wire] and sodium hydroxide. The ammonia is then tested according to § 108. This reaction distinguishes nitrate ion from chlorate ion. The equations are:

$$4Zn + 7NaOH + NaNO_3 = 4Zn(ONa)_2 + 2H_2O + NH_3 \quad (1a).$$

or

$$4Zn + 7OH' + NO_3' = 4ZnO_2'' + 2H_2O + NH_3 \quad (1b).$$

In neutral or acid solution a reduction to nitrite takes place:

$$NO_3' + Zn + 2H^{\cdot} = NO_2' + Zn^{\cdot\cdot} + H_2O \quad (2).$$

Nitrate ion is also indicated by the appearance of a blue coloration on adding a solution of potassium iodide with a little starch solution to a solution upon which zinc has acted in the presence of sulphuric acid. It must always be tested whether the coloration does not appear also before the reduction with zinc and sulphuric acid has been carried out. If this is the case nitrite ion is present.

Nitric acid is a monobasic acid of the character of the halogen hydrides.

The fact that nitric acid oxidizes many substances and goes over into lower oxidation forms, NO_2 and NO, i. e. can give up positive charges, finds a possible explanation in the assumption of a dissociation according to the following equations (compare also § 54):

$$HNO_3 = OH' + NO_2^{\cdot} \quad (3),$$

and

$$HNO_3 + H_2O = 3OH' + NO^{\cdot\cdot\cdot} \quad (4).$$

[1] The vanishing of the color is especially important. According to *Lunge*, Zeitschrift für angewandte Chemie, 15, *1*, 1902, it characterizes nitrate ion only, and is not characteristic of nitrite ion, which resembles nitrate ion in many respects (see the next paragraph). *Winkler*, loc. cit. *170*, states, however, that the above is true only when concentrated sulphuric acid is present in a sufficient excess (about four times the volume of the nitrate solution).

These two reactions accompany the dissociation into $H^{\cdot} + NO_3'$. The extent to which they take place varies with the total concentration and the concentrations of the dissociation products.

By heating nitrates with concentrated sulphuric acid brown vapors escape. They consist of NO_2 or N_2O_4 (Principles, p. *324*). Nitric acid vapors decompose at higher temperatures according to the equation:

$$2HNO_3 = H_2O + \underset{(2NO_2)}{N_2O_4} + O \qquad (5).$$

As the temperature is raised the following reaction also takes place to an increasing extent:

$$2HNO_3 = H_2O + 2NO + 3O \qquad (6).$$

Solid nitrates also decompose with evolution of oxygen when heated to redness. On this is based the use of fused nitrates as oxidizing agents. Nitrates of the alkali- and alkaline earth metals decompose according to the equation:

$$KNO_3 = KNO_2 + O \qquad (7).$$

Those of the heavy metals, as lead, copper, etc., yield oxides of the respective metals and mixtures of the oxides of nitrogen.

156. Nitrite ion and nitrous acid. A dilute solution containing $\frac{1}{10}$ formula weight of sodium nitrite, $NaNO_2$, per liter, is used for the tests. Solutions of sodium nitrite have an alkaline reaction in consequence of hydrolysis. To neutralize them acetic acid is added. They give no very pronounced reactions with the typical cathions (§ 110). *Silver ion* gives a precipitate with this solution. The precipitate dissolves in water and in nitric acid. *Lead ion* yields a yellow precipitate.

Nitrite ion resembles nitrate ion very closely. Hence it is hard to identify and separate them. The difficulty lies in the fact that many properties which are utilized to detect nitrate ion in acid solutions depend upon oxidations which are still more distinctly obtained with nitrite ion. Hence the reactions of nitrate ion just described serve as a criterion when nitrite ion has been shown to be absent by the following means (compare § 155, footnote).

The special tests for nitrite ion in a solution containing hydrogen

ion, have already been mentioned: for example, the oxidation of iodide ion to free iodine (§ 144, (*1*)), which is particularly delicate, or the oxidation of cobaltous- to cobaltic ion (§ 83, (*1*)).

The reducing action of nitrite ion towards permanganate ion is of interest in connection with its quantitative estimation, but is not much used for the detection of nitrite ion:

$$2MnO_4' + 6H^{\cdot} + 5NO_2' = 2Mn^{\cdot\cdot} + 5NO_3' + 3H_2O \qquad (1).$$

According to *Lunge* a solution for detecting nitrite ion in drinking water is prepared by adding 2 cc. of naphtylamine solution to a solution of 0.5 g. sulphanilic acid in 150 cc. dilute acetic acid. The naphtylamine solution may be prepared by heating 0.1 g. α-naphtylamine with 20 cc. water and diluting with 150 cc. dilute acetic acid. 2 cc. of this solution are added to 50 cc. of the drinking water. If this latter contains nitrite ion a red coloration[1] is produced in 5–10 minutes. This is an important test.

Nitrous acid is a monobasic acid of about the character of acetic acid. It is unstable, as is evidenced by its powerful oxidizing action, and further by the fact that it undergoes spontaneous decomposition, i. e. with a loss of energy, according to the following equations:

$$3HNO_2 = HNO_3 + 2NO + H_2O \qquad (2a),$$

or

$$3NO_2' + 2H^{\cdot} = NO_3' + H_2O + 2NO \qquad (2b).$$

Solid nitrites evolve vapors of nitrogen trioxide, N_2O_3, with concentrated sulphuric acid. This partly decomposes into NO and NO_2.

157. Separation of nitrite- and nitrate ions. To test for nitrate ion in a solution containing nitrite ion, the latter must be removed without giving rise to nitrate ion. This condition cannot be entirely fulfilled, i. e. the removal of nitrite ion is not even practically complete. Consequently the detection of nitrate ion with ferrous sulphate and sulphuric acid is always accompanied by the limitation that only relatively larger quantities can be determined in the presence of nitrite ion by means of a *qualitative* experiment. For a slight reaction may be caused by nitrate ion formed on removing nitrite ion, or it may be due to nitrates originally present.

A practical method of destroying nitrite ion is the following. The nitrite solution is neutralized and is heated to boiling with a

[1] Text books of organic chemistry deal with the theory of the above reaction. It depends upon the formation of an azo-dye.

solution of an ammonium salt (chloride), until it no longer gives the blue coloration at once, when acidified and tested with iodide ion and starch solution (§ 144). Then the test for nitrate ion may be carried out.

The reaction between a nitrite and an ammonium salt is summarized by the equation:

$$NaNO_2 + NH_4Cl = NaCl + 2H_2O + N_2.$$

Leaving out the ions which undergo no essential changes of concentration this becomes:

$$NO_2' + NH_4^{\cdot} = N_2 + 2H_2O.$$

An exact study of these relations has shown that it is very essential to have the solution slightly acid, not neutral. These conditions exist, however, on boiling the solution of an ammonium salt, for the ammonium hydroxide resulting from hydrolysis (§ 71) breaks down mostly into ammonia and water. Now evaporation causes a comparatively greater decrease in the concentration of the former product. Hence the conditions are favorable for an increased hydrolysis and for an increase in the concentration of hydrogen ion. This is readily shown by boiling an ammonium salt solution for a few minutes. It will then have a more distinct acid reaction than before.

This fact explains the above phenomenon that nitrate ion is formed during the removal of nitrite ion in this way. For *in the presence of hydrogen ion nitrite ion* goes voluntarily over into *nitrate ion* according to equation (*2b*) of the preceding paragraph.

158. Acetate ion and acetic acid. The test solution contains $\frac{1}{5}$ formula weight of sodium acetate, $NaC_2H_3O_2.3H_2O$, per liter. It yields no precipitates with lead and barium ions. A *leafy crystalline* precipitate forms with *silver ion* only when the concentration is high. This indicates that the solubility product of silver acetate is rather large.

Acetate ion gives a dark red coloration with ferric ion. This is due to the formation of slightly dissociated ferric acetate. Ferric acetate is decomposed by water, quickly when heated, separating basic acetate (compare § 69 and also § 152 on the distinction between ferric acetate and ferric sulphocyanate).

The dark red color is sometimes caused by an oxidation of iodide ion to iodine, in case ferric- and iodide ions are present in such concentrations that the transformation:

$$Fe^{\cdots} + I' = Fe^{\cdot\cdot} + I,$$

can proceed in the sense →. Iodine goes into solution with a red color (§ 144) and volatilizes when heated. In case iodide ion is contained in the solution, it is precipitated with silver nitrate before adding ferric chloride.

With dilute acetate solutions the above test is sometimes indistinct. By combining the reaction with the Bunsen cacodyl test, however, (see below) it is always possible to detect even small quantities of acetate ion. The test is then carried out with the precipitate obtained by boiling the solution with ferric ion.

If tartrate ion is present (§ 138) the characteristic reaction with ferric ion does not take place, or only when ferric chloride is added in excess of the quantity used up in the reaction with tartrate ion. In this and in other doubtful cases the dry mixture of salts is heated in a distilling flask with dilute sulphuric acid. The vapors are condensed in a test-tube and the distillate is then neutralized and tested for acetic acid by any of the characteristic tests.

The Bunsen reaction is based upon the evolution of cacodyl oxide, a very unpleasant smelling gas, on heating arsenic trioxide with solid acetates in the bulb-tube:

$$4NaC_2H_3O_2 + As_2O_3 = 2Na_2CO_3 + 2CO_2 + 2(As(CH_3)_2)O.$$

The trustworthiness of this test is influenced by the fact that other acids of the acetic acid type also give this reaction. However the problem of separating acetate ion from the ions of closely related acids is very rarely met with, and therefore may be left out of account.

A further characteristic reaction of acetates consists in the formation of *acetic ethyl ester,* a pleasant smelling liquid of the composition $CH_3CO_2C_2H_5$. It is produced when an acetate is heated with a few cc. of a mixture made up of *equal volumes* of *alcohol* and *concentrated sulphuric acid.* The characteristic odor becomes distinct after a few minutes. The mixture of alcohol and sulphuric acid should be kept in stock. It contains some ethyl sulphuric acid, which reacts with acetic acid as follows:

$$CH_3.CO_2H + C_2H_5HSO_4 = CH_3CO_2C_2H_5 + H_2SO_4.$$

It is advisable to have on hand some of the pure ester for the purpose of comparison in doubtful cases.

Acetic acid is a medium strong monobasic acid. The greater strength of hydrochloric, nitric or sulphuric acid is shown by the fact that these acids displace acetic acid from aqueous solutions of acetates. Thus a mixture of hydrochloric acid and sodium acetate gives rise to sodium chloride and acetic acid, and the transformation proceeds nearly to an end in the one direction, even when the quantity of sodium acetate taken is only equivalent to the hydrochloric acid. All effects due to the concentration of hydrogen ion[1] are produced to a much slighter extent by acetic acid than by a solution of hydrochloric acid of equivalent concentration. Hence the action of *hydrochloric* or *nitric* acid may be readily *weakened* to a considerable extent by adding a solution of *sodium acetate.* This action of sodium acetate has already been pointed out in §§ 38 and **96.** Over it is superposed the action mentioned in § **89,** which increases in effect as the excess of acetate increases.

When heated to redness acetates decompose, giving rise to carbonates or oxides, or even separating the metal. Simultaneously carbon and volatile organic compounds are formed.

159. Chlorate ion and chloric acid. Tests for chlorate ion are based upon its oxidizing action. It is, however, a much weaker oxidizing agent than hypochlorite ion, which voluntarily gives rise to chlorate ion under certain conditions.

This is a striking illustration of the fact that an ion or substance which forms spontaneously from a second ion or substance is to a greater or less extent less active than the second ion or substance, in the case of a reaction between either of them and a third substance or ion. It is due to the *interrelations* of the two ions or substances. The first contains less free energy than the second, i. e. it has less power of doing work. The change of energy from the initial state of either substance to the state after the reaction with the third substance is greater for hypochlorite ion than in the case of chlorate ion (consult Principles, p. *208*).

The test solution contains $\frac{1}{10}$ formula weight of potassium- or sodium chlorate per liter.

The solution undergoes no visible changes with the typical cathions (§ 110). The following are the general equations for the oxidizing action of chlorate ion in the presence of hydrogen ion:

[1][A characteristic test for acetates, based upon this property, is given by S. R. Benedict, Amer. Chem. J., 32, *480*, 1904. See, also, Benedict and Snell, J. Amer. Chem. Soc., 27, *743*, 1905.]

$$ClO_3' + 6H^{\cdot} = Cl' + 3H_2O + 6(^{\cdot}) \qquad (1a),$$

or

$$ClO_3' + 6H^{\cdot} + 6(') = Cl' + 3H_2O \qquad (1b).[1]$$

Thus chlorate- plus hydrogen ion will oxidize ferrous ion to ferric ion, iodide-, bromide- or chloride ion to the respective halogens, sulphite- to sulphate ion.

To test for chlorate ion in the presence of nitrates the solution obtained by the reduction of chlorate ion with sulphite ion or ferrous ion ($FeSO_4$) in acid solution[2] is precipitated by silver ion. In accordance with the preceding a white precipitate of silver chloride is formed, insoluble in nitric acid. For nitrate ion the evolution of ammonia on heating the solution with zinc dust (or aluminum) and alkali hydroxide is characteristic (§ 155).

The addition of chromate ion has a characteristic effect upon the oxidation of iodide ion by chlorates. The reaction proceeds much more quickly when dichromate ion is contained in the solution. This is shown by dividing a mixture of potassium chlorate, potassium iodide, starch solution and sulphuric acid into two parts, and adding a few drops of dilute potassium dichromate solution to one of them. The blue color arising from free iodine is produced much more rapidly in this case, although at the end of the reaction there is no difference in the intensity of the color in the two cases. Reactions of this sort are known as catalytic reactions (§ 164). Ferrous ion has an effect quite similar to that of dichromate ion.

The transformation between silver chlorate solution and hydrochloric acid gives rise to a solution of chloric acid. The solution slowly undergoes a complex change, producing perchloric acid, $HClO_4$, chlorine dioxide and oxygen:

$$3ClO_3' + 2H^{\cdot} = ClO_4' + H_2O + 2ClO_2,$$

$$2ClO_2 = Cl_2 + 2O_2 \qquad (2).$$

The transformations expressed in equation (*2*) also result by acidifying a chlorate solution with nitric or sulphuric acid. With concentrated acids and solid chlorates a very violent reaction, often explosive, takes place. Hence in testing the behavior of solid salts towards concentrated sulphuric acid according to § 177, it is important to use always *small* quantities of the substances in question.

Chlorates are decomposed with evolution of oxygen when heated in the bulb-tube (consult the following paragraph).

[1] [It is worthy of note that formaldehyde does not reduce IO_3' from a hot acid (HNO_3 or H_2SO_4) solution, while BrO_3' and ClO_3' are each reduced under these conditions to the respective halides or free halogens.]

[2] Of course hydrochloric acid must not be used to acidify the solution in this case.

160. Preparation of potassium perchlorate and potassium chloride from potassium chlorate. Potassium chlorate at first melts when heated. By increasing the temperature above that of the melting point an evolution of gas takes place, which increases with rising temperature. The generation of gas may be made still more violent by adding certain solids, i. e. manganese dioxide, ferric oxide, and to a less extent by quartz sand. The action is catalytic in all these cases (§ **164**). Finally the mass begins to solidify and becomes completely solidified before the total amount of oxygen is given off, which would be set free according to the equation:

$$KClO_3 = KCl + 3O \qquad (1).$$

If the solid mass is now allowed to cool and is extracted with a small quantity of water, it is observed that part of it dissolves abundantly and that a residue is left, which goes into solution in a much larger volume of water. The solution of this residue possesses neither the properties of chloride ion nor those of chlorate ion (oxidation of iodide ion to free iodine, § 159). The residue is *potassium perchlorate,* $KClO_4$. Its solutions contain the ions $K^{\cdot}$ and ClO_4'.

Potassium perchlorate fuses when heated to higher temperatures and evolves oxygen, decomposing according to the equation:

$$KClO_4 = KCl + 2O_2.$$

The fact that it is formed in the above case shows that potassium chlorate in the fused condition, in addition to a decomposition into chloride and oxygen according to (*1*), also undergoes the transformation:

$$4KClO_3 = 3KClO_4 + KCl \qquad (2).$$

The solidification of the fused mass of potassium chlorate before the evolution of oxygen has ceased, is explained by the higher melting points of potassium perchlorate (about 600° C.) and potassium chloride (730° C.), as compared with that of potassium chlorate, 350° C.

Preparation of potassium perchlorate from potassium chlorate. About 30–50 g. of the latter salt are weighed out in a thick-walled flask which has been previously weighed. The flask is heated until

the mass fuses and is then further heated till a violent evolution of oxygen begins. In case it is desired to collect the gas a stopper and delivery tube are fitted to the flask. The gas is collected by upward displacement of water.

As soon as the fused mass begins to solidify the heating is discontinued and the flask and its contents are weighed when cool. As a rule the glass becomes soft and changes its form during the experiment.

The decrease of weight affords a means of estimating approximately the amounts of the products obtained. Assume that 'a' grams of $KClO_3$ were taken. If the reaction followed equation (*1*) exclusively, the decrease of weight, $x = a\frac{48}{122.6}$g., since 122.6 g. $KClO_3$ would lose 48 g. oxygen. A smaller decrease of weight indicates the stimultaneous reaction according to (*2*). By assuming that the mixture left in the flask *no longer* contains any *chlorate* (the assumption is only approximately fulfilled), the quantity of perchlorate formed may be estimated. If the actually observed decrease of weight is 'b' grams, then $\frac{b}{x}$.100 per cent of the chlorate has decomposed according to (*1*), and $100(1 - \frac{b}{x})$ per cent has undergone the transformation (*2*), i. e. $\frac{b \times a}{x}$ g. and $(1 - \frac{b}{x})a$ g. respectively. Since 4 formula weights of $KClO_3$ produce 3 formula weights of $KClO_4$, (*2*), it follows that 4×122.6 g. $KClO_3$ would give 3×138.6 g. $KClO_4$. Therefore, if $a \times (1 - \frac{b}{x})$ g. $KClO_3$ have actually been decomposed according to (*2*), the quantity of potassium perchlorate in the mixture left in the flask will be $a \times (1 - \frac{b}{x}) \times \frac{3 \times 138.6}{4 \times 122.6} = a \times 0.85 \times (1 - \frac{b}{x})$ g. The quantity obtained in the manner given below should be compared with the quantity calculated in this way.

The contents of the flask are loosened with a little warm water and are washed into a beaker. The whole is then heated. If all of the salt mixture does not go into solution, water is added in small portions at a time and the heating is continued until complete solution in the smallest possible volume of liquid is effected. On cooling the solution to room temperature a crop of crystals is obtained. These are separated from the mother liquor by filtration and are washed by repeatedly moistening them with water. They are then tested for a content of *chloride* and *chlorate*. To do this a small portion is dissolved in water. Part of this solution is tested with silver ion, another portion with potassium iodide and hydrochloric or sulphuric acid and starch solution (§ 159) for *chlorate ion*. A blue coloration at the end of a few minutes indicates the presence of chlorate ion.

In this case the salt is recrystallized, i. e. heated with just enough water to effect complete solution and then cooled. The crystals which separate are collected.

The mother liquors are evaporated to a small volume and cooled to room temperature. The crystals formed differ in shape from those of potassium perchlorate. They are separated by filtration. On washing them with water it is noted that nearly all goes into solution again. A small residue of perchlorate may be left. This is added to the solution of perchlorate and is recrystallized with the rest.

The solution obtained by washing the crystals of potassium chloride is mixed with the mother liquors. Essentially only potassium chloride is present, and frequently a little unchanged chlorate. This latter is detected with potassium iodide starch solution + hydrogen ion.

To prepare *pure* chloride the solution is evaporated to dryness. The solid salts are then further heated in a drying oven or air-bath[1] and are finally heated to redness in a porcelain crucible until all chlorate is decomposed into chloride and oxygen. The test for chlorate should now fail.

Owing to the cost potassium chloride is not obtained commercially in this way. The method has been described to point out the possibility of separating the products of several reactions which take place simultaneously.

161. The law of successive reactions. Potassium perchlorate is an intermediate product in the transformation of the chlorate into chloride and oxygen. The reaction proceeds spontaneously at higher temperatures, i. e. is accompanied by a decrease of free energy, and takes place in successive stages. From beginning to end the different stages are $4KClO_3$, $3KClO_4 + KCl$, and $4KCl + 6O_2$ (compare the preceding paragraph). This illustrates a general law which may be applied in many analogous cases. The law was formulated by Ostwald and is expressed as follows (Principles, p. *207*). In all reactions the state that is *most stable* under the conditions in question does not *result* at first, but states of lesser stability form, either the *next less stable* or the *least stable* of *all possible states.*

The system $3KClO_4 + KCl$ is less stable than $4KCl + 6O_2$, i. e. it contains more free energy. Hence it passes spontaneously without any change of the external conditions, as temperature, into potassium chloride and oxygen. However, raising the temperature increases the velocity of the transformation.

Another case which is also included under this law is the formation of sulphur dioxide through the combination of sulphur with oxygen. This repre-

[1] An air-bath may be improvised by constructing a small tripod from brass wire. This is placed on the asbestos wire gauze and supports the evaporating dish in which the solution is evaporated.

sents the first stage. The second stage is the further combustion of the dioxide to trioxide by using catalyzers, for example, finely divided platinum (Principles, p. *283*). This reaction does take place voluntarily, but proceeds so slowly that it cannot be used technically without the help of catalytic agents.

Other cases coming under the head of successive reactions are afforded by the intermediate formation of more readily soluble modifications of a substance (§ 86, footnote, and § 115), when the components are brought together.

162. **The separation of several substances by means of fractional solution.** The following remarks concerning this problem, which is very often presented to the chemist, can by no means include all the possibilities that are offered for consideration. They are intended to add to our knowledge of the methods used in the simplest cases, for instance, in the above case where a mixture of two salts (with three ions, viz. one in common, $K^{\cdot}$, and Cl' and ClO_3') is dealt with.

As has been stated already in § 90 the relations are the following in the present example. Since the two substances contain a common ion, $K^{\cdot}$, the solubility of each is reciprocally influenced by the other.[1] The decrease in the solubility of the more difficultly soluble salt is proportionately much greater than that of the more soluble salt. Nevertheless the solubility of this latter is also lessened. As long as the mixture is extracted with small quantities of water or other solvent, which do not suffice to dissolve all of it, that is, so long as both solids are present, the different portions *saturated* with respect to the two salts will have the same composition. This is evident since the solubility of a substance is *fixed* as long as any of it is present in solid form. In this case only an alteration of external conditions, as temperature (§§ 6 and 30), will change the composition of the solution. Similarly, with *two salts* the *composition* of the *solution* is *invariable* under definite external conditions when the *two salts* are present in the *solid state*.

The invariability of the system ceases when one of the solid salts vanishes, usually the more readily soluble. Then more of the remaining salt will go into solution on further treatment with the solvent, in accordance with the slighter decrease of the solubility through the lower content of the other salt with a common ion in the solution (§ 90). Solutions are finally obtained which contain practically none of the more readily soluble salt. This indicates that the residue consists of only one of the two salts, usually the less soluble. The residue may consist of the more readily soluble salt if it was present in very great excess over the other to begin with, or when the solubility products of the two salts are not very far removed from one another.

Evaporation of the solution saturated with respect to both salts will not serve to obtain the more readily soluble salt in the pure state. The same result is always obtained by repeating the operation of dissolving and evaporating, unless account is taken of the unequal velocities with which different salts often go into solution and separate on cooling. A further separation can be effected only with the aid of another expedient which produces a change of the important property, the solubility. A chemical transformation of one of the substances serves this purpose, as was effected in the above case by heating to redness. Another means consists in altering the solvent

[1] The formation of double salts or complex ions is not taken into account.

bv adding another substance to it, on the supposition that the latter affects the two salts to a different extent.

Those fractions which are *unsaturated* with respect to the more readily soluble substance contain proportionately rather more of the difficultly soluble substance than would be present if the solution were saturated with the other salt. By evaporating the solution and then repeating the operation of dissolving with the residue from evaporation, the more difficultly soluble component can be separated.

The success of this method depends upon external conditions, which may be varied at will, as for example, temperature. This can be estimated approximately when the dependence of the solubility of the single salts upon the external conditions in question is known, taking into account also the dissociation relations. The following table illustrates the above point:

100 g. water dissolve[1] at	0° C.	50° C.	100° C.
g. potassium chloride	28.50	42.8	56.6
g. potassium perchlorate	0.70	6.45	19.90
Ratio of the solubilities $\frac{KCl}{KClO_4}$	40.7	6.64	2.84

It is evident from what has just been stated that the success of a separation by fractional solution is considerably greater at lower temperatures in this case.

163. **Separation by means of fractional evaporation.** Similar considerations hold in the case of separating two salts (with a common ion) from a given solution by removal of the solvent. The removal is effected at a definite temperature by evaporating the solution. The temperature need not be 100° C. A bath at a lower temperature will also serve the purpose. Under these conditions the more difficultly soluble salt will separate first in the majority of cases, since its solubility product is reached first, provided supersaturation is excluded. This is readily effected by means of inoculation. The salt separating first is obtained in pure state so long as the solubility product of the other more readily soluble salt is not reached on further decreasing the quantity of solvent. Afterwards a mixture of the two salts is separated. They are present in the ratio of their total concentrations in the solution saturated with respect to both.

To separate the precipitated *pure substance* and the precipitated *mixture* the precipitates are removed from the liquid at definite intervals by decantation or filtration. After washing the first precipitations consist practically of the less soluble salt. Those fractions which separate last consist of a mixture of the two salts, having a definite composition. Between these extremes occur fractions containing some of the second salt, but less than the amount that is present in the mixture of invariable composition. From these latter fractions the first salt may be separated by repeating the operation.

Other means must be employed to separate the salts in the mixture of constant composition. Chemical methods are most successful.

Disturbing influences have not been taken into account in these remarks.

[1] From Landolt and Bornstein's Tables.

To be considered especially in this connection are the formation of double salts, mixed crystals[1] and mixed amorphous precipitates.[2]

More complex cases may be omitted here since the difficulties are generally more simply overcome by chemical methods.

164. Catalyzers. In the case of the dissolving of a carbonate (§ 96) or of the action of dichromate ion as an oxidizing agent (§ 144) in the presence of different acids, phenomena have been pointed out from which certain deductions can be drawn concerning the concentration of hydrogen ion in various acids. These conclusions are derived from the course of the reactions, which varies in each case with the nature of the acid, and also they are based upon the assumption that the *chemical action* of an ion or dissolved substance is *proportional* to its *concentration.*

In other cases, for example the reduction of chlorate ion by iodide ion, the reaction velocity is very much increased by the mere presence of small amounts of *other substances* which apparently take *no part* in the reaction. Thus dichromate ion or ferrous ion accelerates the reaction in the example cited (§ 159). Similar effects are produced by substances which do not come into the equation for the reaction in the following cases; by water in connection with the action of ammonia on mercurous chloride (§ 17); by the spontaneous decomposition of hypochlorites through cobalt, nickel or manganous salts (§ 151); in the decomposition of chlorates, by manganese dioxide (§ 160), and also in numerous other examples (Principles, pp. *104* and *157*; also The Scientific Foundations of Analytical Chemistry, translation by McGowan, p. *71*).

Those substances which alter the velocity of a chemical reaction without taking any part in it chemically, are termed *catalyzers.* The reactions in question would also proceed *voluntarily without the catalyzer* (but at a quite different rate), and *would take place to the same extent, other conditions remaining unchanged.* If, therefore, a catalyzer is added to a *system in equilibrium,* the *system* will *remain* in *equilibrium.* When the system is not in equilibrium the catalyzer alters the velocity with which equilibrium is reached.

Substances which *increase* the reaction velocity, are termed *positive* catalyzers. Those which retard it are *negative* catalyzers. An example illustrating the action of a negative catalyzer is afforded by the means (mentioned in § 121) of keeping a solution of alkali sulphite from oxidizing with atmospheric oxygen through the addition of small amounts of glycerine or alcohol.

The experience which has hitherto been gained in this very important domain, has shown that there are specific and general catalyzers. The former are active only in certain definite reactions, the latter alter the velocity of certain classes of reactions. Thus iron salts catalyze many oxidation and reduction reactions. Platinum black catalyzes those reactions in which hydrogen or oxygen or both take part. Furthermore hydrogen ion catalyzes many reactions. Past experience has not shown that it is chemically directly active in these cases.

165. Relations between the different states of oxidation of an element. An

[1] Compare, Ostwald, Grundriss der allgemeinen Chemie, p. *180*, 3rd edition.

[2] Küster and Thiel, Zeitschrift für anorganische Chemie, 19, *338*, 1899; 26, *201*, 1901; 33, *129*, 1903.

essential difference exists in the relations between the anions and cathions of the different states of oxidation of an element and their composition. The various cathions of a given metal, for example, $Fe^{\cdot\cdot}$ and $Fe^{\cdot\cdot\cdot}$, $Cu^{\cdot}$ and $Cu^{\cdot\cdot}$, $Hg^{\cdot}$ and $Hg^{\cdot\cdot}$, and $Sn^{\cdot\cdot}$ and $Sn^{\cdot\cdot\cdot\cdot}$, have the same composition and *differ* only with respect to the *number* of positive *charges.*

The relations are quite different for a series of anions composed of two or more elements, which contain the same elements, as the following examples show:

hypochlorite ion, ClO', chlorate ion, ClO_3', perchlorate ion, ClO_4';
nitrite ion, NO_2', and nitrate ion, NO_3';
sulphite ion, SO_3'', and sulphate ion, SO_4'';
phosphite ion, PO_3''',[1] and phosphate ion, PO_4'''.

In these cases the difference between the anions of the various states of oxidation do not relate to the *number* of *charges,* but to the *composition.* A higher oxygen content[2] corresponds to a higher state of oxidation.

Apparently this openly opposes the rule set forth in § 53 that a gain of positive charges or a loss of negative accompanies a transformation into a higher state of oxidation or valence.

The contradiction is readily obviated, however, by taking into account that water plays a role in the transformation from one state to the other. Thus water takes part in the transformation of sulphite ion into sulphate ion by giving up oxygen. Hence hydrogen must either escape as a gas or must form hydrogen ion. The former is not the case, while the latter is, provided a substance is present which can give up positive charges, *i. e.,* acts as an oxidizing agent.

The general equation for this reaction is:

$$SO_3'' + H_2O + 2(^{\cdot}) \rightleftarrows SO_4'' + 2H^{\cdot} \qquad (1).$$

$SO_3'' + H_2O$ corresponds to a lower state of oxidation of sulphur, as compared with $SO_4'' + 2H^{\cdot}$. The latter is an oxidizing agent towards an ion or substance which can take up the two positive charges of the hydrogen ions, or can give up negative charges equivalent to the vanishing of $2H^{\cdot}$. It has already been stated (§ 143) that sulphate- plus hydrogen ion is an oxidizing agent towards bromide- or iodide ions. The reaction then taking place is the reversal of equation (*1*).

Analogous equations may be set up for the relations between other oxygen anions; for example:

$$ClO' + 2H_2O + 4(^{\cdot}) \rightleftarrows ClO_3' + 4H^{\cdot} \qquad (2a),$$

and

$$ClO' + 2H_2O \rightleftarrows ClO_3' + 4H^{\cdot} + 4(') \qquad (2b);$$

$$ClO_3' + H_2O + 2(^{\cdot}) \rightleftarrows ClO_4' + 2H^{\cdot} \qquad (3a),$$

and

$$ClO_3' + H_2O \rightleftarrows ClO_4' + 2H^{\cdot} + 2(') \qquad (3b).$$

[1] Phosphite ion is an anion of phosphorous acid and its salts (phosphites).

[2] Sulpharsenite-, AsS_3''', and sulpharsenate ion, AsS_4''', are analogous with arsenite- and arsenate ions respectively. With these the difference relates to the sulphur content.

It will be well to complete this survey with the relations to the lowest oxidation form in the series of the chlorine anions, chloride ion and hypochlorite ion:

$$Cl' + H_2O + 2(^{\cdot}) \rightleftarrows ClO' + 2H^{\cdot} \qquad (4a),$$

and

$$Cl' + H_2O \rightleftarrows ClO' + 2H^{\cdot} + 2(') \qquad (4b).$$

Bromine + hydroxyl ion acts as an oxidizing agent of the second kind for the oxidation of chloride ion to hypochlorite ion. If hydroxyl ion is lacking the reverse transformation takes place, other conditions remaining the same.

The transformation of chloride ion into chlorine according to § 141, (*3*) is also an oxidation. In order that chloride ion goes over into chlorine, another ion must give up positive charges or take up negative, *i. e.*, must act as an oxidizing agent. The different oxidation forms of chlorine may be arranged in the following series according to the number of positive charges which an oxidizing agent must give up to transform chloride ion into one of the higher oxidation forms:

$$Cl', \ Cl, \ ClO', \ ClO_3', \ ClO_4'.$$

By applying the above equation it becomes evident that the numbers are 1, 2, 6, 8.

In these and in similar cases the substance which gives up positive charges or takes up negative is not defined.

Besides the cases just considered some are also known where the anions corresponding to different degrees of valence, have the same composition and differ only by the number of charges. The anion corresponding to the lower state of oxidation has the greater number of negative charges.

Cases of this kind are afforded by the pairs of anions:

$$Fe(CN)_6'''' \text{ and } Fe(CN)_6''' \text{ and } MnO_4' \text{ and } MnO_4''.$$

166. Relations between the states of oxidation and the oxidizing and reducing action of a substance. It has already been pointed out that the lower states of oxidation are sometimes stronger oxidizing agents than the higher oxidation forms. For example, nitrite ion acts as a stronger oxidizing agent than nitrate ion (§ 156), hypochlorite ion (§ 151) stronger than chlorate ion (§ 159).

It follows from this that the *states of oxidation defined by the number of charges* in the case of cathions or by the *content of the ions in oxygen or sulphur* in the case of anions (§ 165) *do not follow the order in which the ions act as oxidizing or reducing agents.*

Similarly the phenomena observed with metallic compounds show that the cathion with the largest number of positive charges (or a compound corresponding to it) is not necessarily less stable than a cathion of the same metal with fewer changes. (The same holds true for a compound corresponding to it.) The former does not of *necessity* pass voluntarily into the latter in the presence of a third substance which can take up positive charges. On the contrary the reverse changes *may* take place in different cases. Thus, for example, mercurous halides (§ 17) are transformed voluntarily into mercuric halides and mercury. Furthermore stannite ion decomposes in alkaline solutions into stannate ion and metallic tin (§ 47):

$$2SnO_2'' + H_2O = SnO_3'' + 2(OH') + \underline{Sn}.$$

Luther has shown that there exists a numerical relation between the ions or compounds corresponding to the different states of oxidation. That state of oxidation which is the *stronger oxidizing agent* also has a *greater tendency* to *act* as a *reducing* agent.

The following considerations will help us understand these relations. The above equation reads in words as follows: the middle state, stannite ion, passes voluntarily into the higher state, stannate ion (oxidation) and the lowest state, metallic tin (reduction). This relation between stannite- and stannate ion, in which stannite ion acts as a reducing agent, and between stannite ion and tin, in which it acts as an oxidizing agent, is not limited to cases where tin and stannate ion are formed simultaneously from stannite ion. It holds also for other reactions in which stannite ion takes part. Thus it may act as an oxidizing agent in cases where metallic tin and some oxidation product other than stannate ion are formed. The same is also true of other reducing actions of stannite ion, in which stannate ion and some product of reduction other than tin are formed. From this it follows that *stannite ion* not only has a *stronger oxidizing action* than the highest state of oxidation, *stannate ion,* in cases of oxidation, but also that it is a *more efficient reducing agent* than the lowest state of oxidation, *metallic tin.*

These considerations may be applied *qualitatively* to all cases in which the middle state of oxidation voluntarily decomposes into lower and higher forms, irrespective of the influence which the concentration of the substances taking part has upon the transformation of the various states of oxidation. This influence of concentration, however, has a very important bearing upon an exact *numerical* treatment of the cases in question. The qualitative considerations thus afford a prolific source of aids to our understanding with respect to many phenomena which apparently are not at all related to one another.

A case where the middle state of oxidation is voluntarily formed from the higher and lower states, is afforded by iron in its relations to ferrous and ferric ions. Under certain conditions of concentration the reaction

$$2Fe^{\cdot\cdot\cdot} + Fe = 3Fe^{\cdot\cdot},$$

takes place spontaneously. In comparison with ferrous ion, ferric ion is the stronger oxidizing agent in cases where ferric ion reacts with a third substance to form ferrous ion, and where ferrous ion passes into metallic iron. Metallic iron is also a stronger reducing agent than ferrous iron in cases where either of them is oxidized by a reaction with a third substance. For the demonstration of these relations consult the papers of *Luther,* Zeitschrift für physikalische Chemie, **34,** *488,* 1900 and **36,** *385,* 1901.

167. Valence. Hitherto *valence* has been used in a somewhat variable sense. It is referred to the valence of ions on the one hand, and to the valence of elements in their compounds on the other.

The valence of an ion is expressed by the number of positive or negative charges. Thus the ions Cl', Br', NO_3', MnO_4', $K^{\cdot}$, $Na^{\cdot}$, $NH_4^{\cdot}$, are monovalent, SO_4'', CO_3'', MnO_4'', $Mg^{\cdot\cdot}$, $Ca^{\cdot\cdot}$, $Fe^{\cdot\cdot}$, are divalent, PO_4''', $Al^{\cdot\cdot\cdot}$, $Fe^{\cdot\cdot\cdot}$, are trivalent, $Fe(CN)_6''''$ and $Sn^{\cdot\cdot\cdot\cdot}$ are tetravalent. The theory of chemical compounds aims to establish simple relations between the greatly varied compounds of all elements. It explains the limitations that exist with respect to the ratios in which, say, two elements can enter into combinations according to their combining weights to form new substances, by assigning a definite

valence to each element. For compounds which are composed of two elements the following relation exists between the valencies of these elements. The product of the valence with the number of combining weights of the one element is equal to the product of the valence with the number of combining weights of the other. H, Cl, Na, K are monovalent elements, O, S, Mg are divalent, N, P are trivalent, C is tetravalent. The relations appear to be less clear in the case of a compound composed of more than two elements. The difficulty is overcome, however, by assigning a definite valence to certain groups of elements, or radicals, which frequently take the place of one element. Thus the radicals OH, NH_4, NO_3, NO_2, CN are monovalent, SO_4, CO_3 are divalent, etc. Accordingly the above relation will also hold in these cases, *i. e.*, the products of the valence with the number of combining weights of the radicals or elements in each compound are the same. In the above sense valence has the significance of numbers without dimensions. It should always be borne in mind that valence must not be confused with *affinity.*

Application of the above-mentioned rule to the compounds HCl, H_2O, NH_3, NaCl, $MgCl_2$, MgO, CO_2, CH_4, NH_4Cl, $(NH_4)_2SO_4$, etc., shows that the ratios in which the different elements or radicals enter into combinations according to their combining weights, are in fact in accord with their *valence.*

The attempt to extend this relation to *all compounds* has led to contradictions between the assumption and the actual relations. Accordingly the assumption has been broadened in the sense that the valence of an element is not restricted to one value. Thus nitrogen is assumed to be di-, tri-, tetra-, pentavalent, chlorine is mono-, tri-, penta- and heptavalent, sulphur is di-, tetra- and hexavalent.

The student is referred to text-books of general chemistry for information concerning the numerical values of valence, so far as these have not already been given. A further connection exists between the *valence* of the *elements* which form the components of complex ions, and the *valence* of the *ion.* It will serve to clear up the relations met with in these cases.

According to this rule the valence of the complex ion is determined by the difference between the products of the valence with the number of combining weights of the single elements making up the ion. It follows that the valence of one of the elements which constitutes a component of a complex ion, can be determined from the *formula* of the ion, if the valence of the other element or radical is known, and the number of charges are also known. This problem is frequently encountered in practice.

Very often the complex ion contains oxygen. To determine the unknown valence in this case, the ionic valence is *subtracted* from or *added* to the product of the valence with the number of combining weights of oxygen. It is subtracted when the complex ion is an anion, added when it is a cathion. This difference or sum is divided by the number of combining weights of the other element that are contained in the formula weight of the complex ion. The value so obtained is the valence sought.[1] The following examples will serve as illustrations. Suppose it is desired to determine the valence of manganese in permanganate ion. The rule gives

[1] The anions of the organic acids mentioned in §§ 116, 138 and 158 constitute exceptions. This is due to the characteristic behavior of carbon. Thiosulphate ion (§123) also affords a special case.

$$\frac{4 \times 2 - 1}{1} = 7 \qquad (1).$$

Manganate ion, MnO_4'', contains hexavalent manganese,

$$\frac{4 \times 2 - 2}{1} = 6 \qquad (2).$$

What are the valencies of chlorine in the ions ClO', ClO_3', ClO_4', and of sulphur in the ions SO_3'', SO_4''?

The ions $SbO^{\cdot}$ (§ 46), $BiO^{\cdot}$ (§ 37) and uranyl ion, $UO_2^{\cdot\cdot}$ (§ 211) are examples of complex oxygen cathions. The rule gives 3 as the valence of antimony and bismuth, 6 for uranium.

$$\frac{1 \times 2 + 1}{1} = 3 \qquad (3).$$

and

$$\frac{2 \times 2 + 2}{1} = 6 \qquad (4).$$

A close consideration of the relations described above for complex ions, and in particular for those containing oxygen, leads to some interesting deductions. By transposing equations (*1*) and (*3*), the ionic valence is obtained as a value derived from the products of the valencies of the elements with their combining weight factors. Thus $-7 + 8 = 1$ and $-3 + 2 = -1$.

By changing the sign on each side of the equations the valence of permanganate ion is expressed as a negative value, that of antimonyl ion as a positive value. This permits another formulation of the above rule. The ionic valence is the algebraic *sum* of the products of the valence of the single elements with their combining weight factors. By this the valence of oxygen is expressed as a negative value.

The relations are subjected to no change when the complex ion contains sulphur in place of oxygen.[1] The rule also applies to cases which are apparently more complicated. Thus it gives 3 as the valence of iron in the complex ion, $Fe(CN)_6'''$ (§ 148):

$$\frac{6 \times 1 - 3}{1} = 3$$

In accordance with the composition of hydrocyanic acid cyanogen is assumed to be monovalent. By transposition the ionic valence becomes:

$$3 - 6 = -3.$$

It is expressed as an algebraic sum when cyanogen is assigned a negative valence (-1).

These considerations apply also to cobaltinitrite ion (§ 83). From the composition of nitrous acid the nitro-group is monovalent and the summation rule gives it a negative value.

Nitrogen in ammonium ion is trivalent and negative according to the rule

$$-3 + 4 = 1.$$

[1] This is not generally true without certain restrictions, but holds in all cases dealt with in this text.

Hydrogen is monovalent and positive, since it forms monovalent cathions.

It is worthy of note that the elements or radicals which have a negative valence according to the rule, also form negative ions. It is not permissible, however, to reverse the rule, for this would lead to contradictions. While these are not incontrovertible, they would, however, lead us beyond the bounds of the present text.

PART III.

THE COMPLETE ANALYSIS OF A GIVEN UNKNOWN.

Preliminary or Dry Tests.

168. General. The methods and fundamental theory underlying the identification of the cathions and anions contained in a solution have been treated in the two preceding parts. Hitherto the problem of preparing solutions of the given substances that are required for this has been left an open question. This latter phase of the problem now presents itself for our consideration.

In case the complete analysis of an unknown is in question, a so-called preliminary or dry test is first carried out with a small portion of it before testing for cathions and anions. This serves the purpose of indicating the combinations of the components that are present, and will also lead directly to the identification of certain components. The phenomena accompanying the reactions that are brought about with this end in view are more complex and therefore less definitive than are the tests for individual cathions. Hence the phenomena must be considered as a whole in estimating the results of the preliminary tests. To facilitate this, phenomena characterizing certain compounds or components have been tabulated in the following paragraphs and serve as a guide. Dry tests comprise the study of the unknown under the following conditions:

1. When heated in a closed tube (bulb tube).
2. When heated on charcoal or when reduced with sodium.
3. In the phosphor salt and borax beads.
4. In the flame of the bunsen burner.
5. With acids.
6. With water and acids as solvents.

For the sake of completing the results other properties, as odor and color, which are directly evident, are also taken into account in addition to the above reactions.

In case a uniform liquid is presented for analysis a portion of it is evaporated to dryness on the water-bath. With the residue so obtained the dry tests *1–5* are then carried out.

169. **Coarse grained solids must be pulverized** before they can be analyzed. Great care is requisite to prepare for solution very hard objects, as minerals or rocks.

As a rule these are composed of substances of different hardness. Insufficient trituration reduces them to particles of irregular size. To avoid errors the finer particles are mechanically separated from the coarser and the *latter must be again triturated* until the *whole* mass has attained a definite average degree of fineness.

A sieve is used to separate the finer from the coarser particles. It may be readily improvised by stretching some linen of suitable fineness over the neck of a broad soft-mouthed bottle. The powder to be sieved is placed on the linen. A piece of sheet rubber is then stretched over the top of the bottle. By *gently* beating the rubber with a suitable beater (not too hard: a glass rod covered with rubber tubing serves admirably) the separation is readily effected. Any coarser particles left on the linen are then pulverized again and are sieved as before. [When a small jar mill is available, of course the pulverized product need not be sieved.]

170. Experiments with the closed tube have already been illustrated in § 35. Certain phenomena are rendered more distinct by holding the tube in an inclined position. The following changes are most frequently observed:

(A) *Water is driven off:* indicates salts containing water of crystallization; the phenomenon is occasionally accompanied by a change of color or a crepitation (alkali chlorides) due to a shattering of the crystals. Often the salts melt (alkalies), also without giving off water.

(B) *A sublimate forms.* It has the characteristic property of being easily driven from one place to another by heat.[1]

[1] This phenomenon will be understood by applying Watt's principle. Temperature differences existing in a closed space will drive a vaporizable substance from the warmer to the coldest parts. This is due to differences in the vapor pressure of the vapor-forming substance at different temperatures. Higher pressures correspond to higher temperatures. Different pressures, however, cannot remain *co-existent* in the same space, but they become equalized. The vapor seeks a position of lowest pressure (that is, the coldest portion of the space), where it condenses. This continues until the pressure corresponding to the lowest temperature rules throughout the space, when all solid is vanished from the hotter parts. The condition that the space is closed, is not fulfilled in the bulb-tube experiments. The length of the tube, however, retards diffusion between the vapors in the tube and the surrounding atmosphere, thus affording an approximation to the above condition.

A *white* sublimate indicates:

Ammonium salts (in this case the vapors smell of ammonia), *arsenic* and *antimony trioxides, mercury salts.* A sublimate arising from mercury salts can be distinguished from As_2O_3 or Sb_2O_3 by heating it with sodium carbonate. The following reaction takes place:

$$HgCl_2 + Na_2CO_3 = 2NaCl + HgO + CO_2.$$

If the temperature is not too high a red sublimate of mercuric oxide is formed.

A *yellow sublimate* is produced by *arsenic trisulphide* and *mercury compounds.* A sublimate of mercuric iodide has the property of turning red[1] when touched with a solid, the red color spreading out through the mass from the point of contact. *Sulphur,* which deposits in reddish brown drops, especially when quickly sublimed, is formed by the decomposition of polysulphides or other sulphur compounds.

A *gray sublimate, varying to black,* results from *mercury compounds* (finely divided mercury); *arsenic compounds* (in the presence of reducing substances, a mirror of arsenic); *iodides* when oxidizing substances are present; *iodine,* detected by color and odor.

(C) A *change of color* is sometimes due to the transformation of salts of heavy metals (nitrates) into oxides, or in other cases is a purely physical phenomenon.

The power of absorbing light possessed by colored substances, in case the absorption takes place in that part of the spectrum with shorter wave lengths, is altered by a rise of temperature in the sense that a substance absorbs waves of greater length at higher temperatures. Accordingly a yellow substance often becomes red when heated (chromates), a white substance will become yellow (zinc oxide). Compare Principles, pp. *605* and *621.*

Organic compounds frequently become brown or separate carbon (D).

(D) *Volatile decomposition products are evolved.*

Oxygen from peroxides, mercuric oxide, nitrates, chlorates, etc.

[1] This is due to the fact that two modifications of mercuric iodide may be formed, a yellow and a red sort. Below 126° C. the former is unstable as compared with the latter (compare Principles, p. *664*). The intermediate formation of the yellow modification illustrates Ostwald's law of successive reactions.

Carbon dioxide from carbonates and oxalates—forms a cloud on a glass rod moistened with baryta solution.

Carbon monoxide from oxalates and other organic compounds —burns with a blue flame.

Acid vapors which redden moistened blue litmus paper may be:

Sulphur trioxide from the decomposition of sulphates of heavy metals.

Sulphur dioxide from sulphur compounds (sulphides) when air has access.

Nitrogen peroxide from nitrates.

Bromine from bromides in the presence of oxidizing agents. Bromine is distinguished from nitrogen peroxide by coloring starch paper yellow.

Violet vapors of *iodine* (compare *B*).

Cyanogen (caution) from cyanides. It burns with a peach blossom red flame.

Odor of *garlic* (caution) indicates arsenic compounds (§ 44).

Ammonia from ammonium salts (B) or cyanides.

Empyreumatic vapors (C) from organic compounds.

171. Charcoal tests are made with a mixture containing about one part by weight of the solid unknown and two parts by weight of anhydrous sodium carbonate. The mixture is heated in the reducing flame[1] according to § 39. Oxides or easily decomposed carbonates give the reactions without sodium carbonate (compare § 39, footnote).

Compounds of nearly all metals are reduced by heating them in the reducing flame. Easily volatile metals, as lead, bismuth, arsenic, zinc or cadmium, volatilize. Metallic beads of the three last metals are obtained with difficulty. As a rule the beads formed on charcoal are small and hence readily escape detection. In order to facilitate their detection the entire fused mass is cut away from the hollow on the charcoal and is triturated in a mortar. The light particles of carbon, which do not wet easily, are floated off with water, leaving the metallic grains behind.

The hot vapors of certain metals ignite in the air, forming oxides which deposit on the colder portions of the charcoal. The color of

[1] Potassium cyanide is sometimes used as an active reducing agent (compare § 48). Caution! Hood!

these sublimates also gives certain indications of the nature of the metallic components. Arsenic condenses, in part without igniting. It is detected by the garlic odor which is characteristic of the particular modification that results.

The behavior of the metallic grains under the pressure of a pestle is also characteristic. *Ductile* grains are *flattened* without breaking up, while *brittle* grains *fly to pieces.*

(A) The *most important characteristics* of the reducible *metallic components* are thus the colors of the *sublimate* and the reduced *metallic bead,* and also the behavior of the bead under pressure.

Metallic beads without sublimate may be:

Gold, yellow, ductile;

Silver, white, ductile;

Tin, white, ductile, fusible;

Copper, red masses;

Platinum, cobalt, nickel and *iron* form gray masses. The three latter are magnetic.

Metallic beads with a sublimate may be:

Antimony, white, brittle, white *sublimate;*

Bismuth, brittle, brown yellow *sublimate;*

Lead, white, ductile, yellow *sublimate.*

A *sublimate without bead* indicates compounds of the following metals:

Arsenic, white (garlic odor);

Cadmium, brown;

Zinc, white, yellow when hot (§ 170, *C*).

(B) *Green*[1] or *yellow fusions* indicate *manganates, chromates* and *sulphur compounds.*

It is seldom that sulphur compounds can be identified with certainty from the color of the fused mass. To test for them a portion is cut away from the charcoal and broken up. It is then placed on a bright silver coin and is moistened with water. A *black stain* of *silver sulphide*[2] forms when the *original unknown contains sulphur compounds.*

[1] *Green infusible* masses indicate chromium salts.

[2] This is called the "*hepar reaction*" owing to the fact that a mixture of alkali sulphides (liver of sulphur, hepar sulfuris) also gives it.

Sodium sulphide is formed by fusing substances containing sulphur, for instance sulphates, with sodium carbonate in the *reducing flame*. The following equations express the reactions for calcium sulphate·

$$CaSO_4 + Na_2CO_3 = Na_2SO_4 + CaO + CO_2 \quad (1),$$

$$Na_2SO_4 + 4C = Na_2S + 4CO \quad (2).$$

Sulphides and hydrogen sulphide form a black stain of silver sulphide on metallic silver, owing to the reaction :

$$2Ag^{\cdot} + S' = \underline{Ag_2S} \quad (3).$$

By taking into account the effect of atmospheric oxygen the reaction is formulated as follows:

$$2Ag^{\cdot} + S'' + O + H_2O = \underline{Ag_2S} + 2OH' \quad (4).$$

Equation (*3*) indicates that silver ion takes part in the reaction. The surfaces of silver coins are partially oxidized. Silver oxide is slightly soluble in water and hence forms silver ion with the water used to moisten the fusion.

The test for sulphur compounds is *indefinite* if iodides are also present, unless the silver coin is *bright*. A black stain forms under such conditions, but is *not* necessarily silver sulphide. Attempts to explain the reaction between iodide ion and silver have proven that the black stain is produced only when the surface of the silver has not been cleaned before the experiment. On polished silver surfaces an obscure white skin of iodide can be detected after the reaction with iodide ion. This is due to an oxidation of the silver surface in contact with water. The silver ion thus formed then reacts with iodide ion to form silver iodide. It is probable that the black stain produced by iodides on coins that are not freshly polished is due to a reduction of silver iodide by organic impurities (grease), which the coins gather during circulation.

Errors may also be caused by using illuminating gas as a source of heat. It frequently contains enough hydrogen sulphide to give a reaction with silver. In cases of doubt it is therefore advisable to repeat the reduction, using a stearin candle or alcohol flame in place of illuminating gas.

(C) *White unfused masses* indicate compounds of the alkali earth metals, magnesium or aluminum. When heated in the oxidizing flame and moistened with cobalt nitrate solution, these compounds give rise to gray, flesh-colored or blue masses.[1] Arsenates, borates, phosphates and silicates, however, also form blue masses under similar conditions, but these are partially fusible.

172. Reduction by the Hempel method[2] is effected with metallic sodium, magnesium or aluminum. The reduction takes place more

[1] It is not certain whether the colors are due to definite compounds or to *solid* solutions of oxides of the metals in question. Ostwald, Grundriss der allgemeinen Chemie, 3d edition, p. *231*, defines solid solutions to be those homogeneous mixtures of solid substances, the components of which are not in simple stöchiometric proportions.

[2] Zeitschrift für anorganische Chemie, **16**, *24*, 1898.

readily and more rapidly by this method than with the blowpipe, but the formation of characteristic sublimates of oxide is lacking.

To carry out a reduction with metallic sodium a freshly cut cube of sodium of about 1–1.5 mm. edge is spread out on a piece of filter paper of approximately four square centimeters area by means of a knife blade, previously dipped in petroleum, or better with the handle of a horn spatula. The sodium has about the thickness of filter paper after it is thus spread out. On this thin layer of metal is placed some of the *dry* (see below) material to be tested. The metal and paper are then folded once across and rolled up to a cylinder so that the sodium is covered with a double layer of paper. Any projecting edges of the paper are cut off and the cylinder is closely wound with thin iron wire, such as is used by florists.

The cylinder is next ignited on one side in the luminous flame of a bunsen burner. The entire mass burns rapidly, and if a *sufficiently thin* layer of sodium has been used, none of it will be lost by spirting. When the combustion is ended the spiral is held in the inner flame to avoid oxidation, and is then gradually lowered into the tube of the burner where it is cooled. The iron wire is next unwound and the remaining contents of the cylinder are brought into an agate mortar and moistened with water. The liquid will have a distinct alkaline reaction owing to the formation of sodium hydroxide.

In case sulphur compounds are contained in the mixture so tested, hydrogen sulphide is evolved on adding hydrochloric acid to the liquid. This is identified by its odor or with lead acetate paper. The Hempel method is accordingly particularly adapted to the detection of sulphur compounds.

Compounds of *reducible metals* yield *metallic grains.* These are picked out and are then identified by their chemical and physical behavior (for example, by dissolving them in nitric acid and noting the reactions of this solution: consult also the preceding paragraph).

Borates and silicates reduce to boron and silicon respectively. These are not dissolved by acids. To test for them the residue insoluble in acid is filtered, washed and heated on platinum foil in the air. Carbon from the filter paper burns to carbon dioxide, boron is partially transformed into the oxide, B_2O_3, while silicon forms silicon dioxide. The oxide of boron is fusible.

Magnesium or aluminum is used in place of sodium, especially in the case of compounds difficultly soluble in water. This is done in order to combine with the reaction tests for those metallic com-

ponents, the salts of which are volatile and emit light of characteristic colors (K, Na, Li). Powdered magnesium or aluminum is mixed with the unknown on filter paper. The mixture is then rolled up and treated as above described. To extract the product of the reaction dilute hydrochloric acid is used. A drop of the solution is taken up on a platinum loop and evaporated by approaching it cautiously to the bunsen flame. The residue is then heated to redness (compare § 106). Hence these experiments with magnesium or aluminum are very important, particularly to determine whether a silicate contains alkali metals or not (consult § 192). For a further application of the method see § 194.

It must be specially emphasized that the *unknown* is to be *dried before carrying out the reduction with sodium.* This is best effected by heating it on platinum foil. *Salts which contain water of crystallization occasionally react so violently with sodium that ignition takes place on rolling the cylinder.*

It is advisable to keep the metallic sodium used for this method in a glass tube closed air tight, rather than under petroleum.

173. **Phosphor salt and borax bead tests.** It has been described in § 65 how these tests are to be carried out. Hence a compilation of the behavior of different metallic components already mentioned will suffice here.

Color.	Phosphor Salt Beads.		Borax Beads.	
	Oxidizing Flame.	Reducing Flame.	Oxidizing Flame.	Reducing Flame.
Colorless.	Most colorless salts. SiO_2[1] skeleton.	SiO_2, Cu,[4] Mn.	SiO_2.	SiO_2, Mn, Cu.
Gray.	——	Ag, Pb, Bi, Sb, Cd, Zn, Ni.	——	Ag, Pb, Bi, Sb, Cd, Zn, Ni.
Red.	Fe,[2] Ni.	Fe, Cu.[5]	Fe, Ni.	Cu.
Yellow.	Fe, Ag,[2] Ni.	Fe.	Fe, Ag, Ni.	——
Green.	Cr, Cu.[3]	——	Cr, Cu.	Cr, Fe.
Blue.	Co, Cu.[3]	Co.	Co, Cu.	Co.
Violet.	Mn.	——	Ni containing Mn and Co.	——

[1] Silicon dioxide is very little soluble in alkali metaphosphate, rather more soluble in alkali metaborate. Undissolved silicic acid obscures the transparency of the bead (silica skeleton).

[2] The colors of the iron and silver beads become paler on cooling.

[3] The copper bead is green when hot, blue when cold. However these phenomena vary considerably with the concentration of the salt.

[4] This takes place on reducing cupric to cuprous salts.

[5] A leaf of tin foil aids in the reduction of copper. It is attached to the hot bead without coming in contact with the platinum. Tin forms an alloy with platinum that is fusible in the bunsen flame.

174. Flame reactions. Consult § 106 for carrying out these tests.

Yellow light is emitted by volatile *sodium* compounds;
Reddish yellow ... *calcium* compounds;
Red .... *strontium* and *lithium* compounds;
Blue violet .... *potassium* compounds;
Green .... *barium* compounds and *boric acid;*
Pale blue .... *lead, arsenic* and *antimony* compounds.

Small pocket spectroscopes are very serviceable for the exact inspection of flame colors. The lines characteristic of single components are given in the spectral table at the end of the book.

175. Preliminary tests for acids. These tests aim particularly to point out the nature of the non-metallic components that are present. They comprise a study of the behavior of the unknown (dry mixture or solution) towards dilute and concentrated sulphuric acid and towards a mixture of alcohol and sulphuric acid.

In connection with the results of preliminary tests it is to be noted that *difficultly soluble solid salts* weaken the phenomena described in the following paragraphs to a considerable extent. Hence it does not necessarily follow that a given component is *absent* if some particular reaction cannot be obtained. For this reason *special tests* for anions described in the second part of the book are requisite for a definitive distinction.

176. Behavior towards dilute sulphuric acid. A few cc. dilute sulphuric acid are poured over the unknown. There results:

An evolution of a colorless gas (effervescence), which forms a cloud with baryta solution;	CO_2 (from carbonates);	Odorless (§ 137);
	HCN (from cyanides);	Odor of bitter almonds (§ 146);
	SO_2 (from sulphites or from thiosulphates with simultaneous separation sulphur);	Penetrating odor; the behavior towards iodic acid starch paper is characteristic (§ 121).
Brown vapors;	Mixtures of NO, NO_2, N_2O_4;	From nitrites.
Hydrogen sulphide;	Turns lead paper brown;	From sulphides.
Acetic acid;	Detected by its odor;	From acetates.
Hypochlorous acid;	Turns potassium iodide starch paper blue;	From hypochlorites (§ 151).
Oxygen;	Supports combustion;	From peroxides of the alkalies and alkali earths.

Frequently cyanides or sulphites are mistaken for carbonates, owing to the fact that they also evolve gases which render baryta solution turbid, when they are treated with dilute sulphuric acid. Cyanides and sulphites, however, are readily detected by the above-mentioned properties. In case they are present a special test must be made for carbon dioxide.

To distinguish carbonates from sulphites use is made of the difference in the strengths of the corresponding acids. A medium strong acid, as acetic acid, decomposes carbonates with effervescence, but it effects the decomposition of sulphites much more slowly. However it acts only slowly on difficultly soluble carbonates. These are conveniently decomposed by heating them with a freshly prepared solution of alkali hydroxide, free from carbonate.[1] It must be tested for CO_2 previous to use. (What is the reaction involved?)

In case a residue is left the solution is filtered and the filtrate acidified with acetic acid. A distinct evolution of gas, accompanied by effervescence, takes place if the unknown contains a carbonate.

Carbonates and cyanides are more frequently mistaken for one another. Most cyanides have the property of forming mercuric cyanide and the corresponding hydroxide when they are heated with mercuric oxide and water (compare § 183). This serves to distinguish them from carbonates. The reaction between the complex nickel cyanide and mercuric oxide and water is expressed by the equation:

$$K_2Ni(CN)_4 + \underline{2HgO} + 2H_2O = \underline{Ni(OH)_2} + 2KOH + 2Hg(CN)_2.$$

By this operation the anion of soluble carbonates goes into solution. Insoluble carbonates are more or less completely decomposed, forming insoluble hydroxides of the metals in question. Addition of some alkali hydroxide (free from carbonic acid: consult the last footnote) ensures a more complete decomposition of these latter without any loss of carbon dioxide: i. e. the transformation of the carbonate into hydroxide *with simultaneous formation of carbonate ion* (equations?). This scarcely affects the above-mentioned transformation of the cyanide into mercuric cyanide, which is but little decomposed by acids.

The *criterion* for the *presence* of a *carbonate* when the *unknown also contains a cyanide,* consists in a *distinctly visible evolution* of *gas* upon acidification of the solution with *acetic acid.* The solution in question is obtained by heating some of the unknown with mercuric oxide and water to which a little alkali hydroxide is added, until the transformation of mercuric oxide ceases to be visible. The residue is filtered.

It must be borne in mind that absorption of carbon dioxide from the air can give rise to errors. Hence air must not have free access. This is effected by placing a watch glass over the flask in which the heating is carried out. For the same reason the contents of the flask are filtered without delay as soon as they are somewhat cooled.

Only those cyanides which do not react with mercuric oxide, as above mentioned, are excluded from this method of testing. Since these cyanides also react very slowly with *dilute* sulphuric acid, they are never mistaken for carbonates and hence it is not necessary to use this method of detecting carbonates in such cases.

It is not possible to make these tests extremely delicate without *complicating* the *manipulation.* They are, however, delicate enough to exclude gross errors.

[1] For the above purpose pure potassium hydroxide is dissolved in water. The solution is tested for carbonate ion by acidifying a portion of it with acetic acid. If the solution contains CO_3'' a more or less striking evolution of gas takes place.

177. **Behavior towards concentrated sulphuric acid.** In accordance with its stronger action concentrated sulphuric acid gives more pronounced reactions than those just recorded for dilute sulphuric acid in the preceding paragraph.

To illustrate the distinctive action of concentrated sulphuric acid as compared with the dilute acid, a few drops of the latter are added to a small portion of the unknown drop by drop until a reaction is no longer perceptible. Then a few cc. concentrated sulphuric acid are added. As a rule a considerable evolution of heat takes place and therefore it is not necessary to heat the mixture. There takes place:

An evolution of *colorless vapors with a penetrating odor:*

Hydrochloric acid gas from chlorides: This yields a white precipitate of silver chloride on a glass rod moistened with silver nitrate solution, when the latter is brought in contact with the escaping vapors. The precipitate does not disappear on dipping the rod in dilute nitric acid;

Hydrofluoric acid gas from fluorides and fluosilicates: It forms a white precipitate of silicic acid (§ 117) on a glass rod moistened with water and brought in contact with the escaping vapors;

Sulphur dioxide from sulphites or thiosulphates: It turns iodic acid starch paper blue[1] (consult also § 121, footnote);

Acetic acid from acetates;

Hydrogen sulphide from sulphides [it blackens lead acetate paper];

[1] Quite frequently the formation of colorless vapors with a penetrating odor, on heating the unknown with concentrated sulphuric acid, is *wrongly* considered a test for chlorides. Also the reaction with iodic acid starch paper, characteristic of sulphur dioxide, is often *wrongly* considered a test for sulphites. In these cases the penetrating odor is due to sulphur trioxide, formed on heating concentrated sulphuric acid. Some component contained in the unknown may reduce it to sulphur dioxide. [Furthermore HBr and HI also reduce concentrated sulphuric acid to sulphur dioxide, but in these cases the gas mixture is colored by free bromine and iodine respectively.]

The unequivocal reaction for *chlorides* is the white precipitate formed on a glass rod moistened with silver nitrate solution, which is held in the escaping vapors. Bromides and iodides yield bromine and iodine vapors [together with sulphur dioxide. With iodides there may also be formed sulphur and hydrogen sulphide]. *Sulphites* give *sulphur dioxide,* even with *dilute sulphuric acid* [and thiosulphates yield sulphur in addition].

Yellow green chlorine from chlorides when oxidizing agents are also present: It turns potassium iodide starch paper blue;

Yellow chlorine dioxide, ClO_2, *from chlorates:* It explodes when heated (caution!);

Yellow [brown] vapors of bromine, together with *hydrobromic acid, from bromides:* They color starch paper yellow. [Sulphur dioxide is also formed from a reduction of H_2SO_4 by HBr];

Violet vapors of iodine, together with *hydriodic acid, from iodides:* [Sulphur dioxide, sulphur or hydrogen sulphide may also be formed from reduction of H_2SO_4 by HI according to the following equations:

$$H_2SO_4 + 2HI = 2H_2O + SO_2 + I_2,$$

$$H_2SO_4 + 6HI = 4H_2O + 3I_2,$$

$$H_2SO_4 + 8HI = 4H_2O + H_2S + 4I_2.]$$

Brown vapors: N_2O_4 or NO_2 *from nitrates;*
CrO_2Cl_2 (§ 141) *from chlorides* when chromates are also present;

Oxygen from chromates, permanganates (caution!) and *peroxides;*

Carbon dioxide from oxalates;

Carbon monoxide from oxalates, tartrates and cyanides. It burns with a blue flame when ignited.

To determine whether acetates are present or not (compare § 158), a *test* is made with a mixture of *alcohol* and *concentrated sulphuric acid,* in equal parts.

Borates are tested in another portion of the unknown according to § 135. For this a larger volume of alcohol, 10 cc., and only a few drops of concentrated sulphuric acid are used (compare the test for acetates).

178. Behavior of the unknown towards solvents. The following experiments are designed to give certain indications of the method that should be followed in bringing the unknown into solution, in order to carry out the analysis for cathions and anions.

A small portion is first heated with water with frequent stirring. *If all goes into solution the succeeding tests are unnecessary.* When only a part dissolves care must be taken that enough water is used.

The residue is then filtered and a few drops of the filtrate are evaporated on a platinum foil. To avoid loss by spirting the foil is not heated over a free flame but is placed on an asbestos gauze, and the height of the flame is suitably regulated.

In case the unknown contains material quantities of substances that are soluble in water, a distinctly visible residue is left on the foil.

To make sure that the residue left undissolved does not result from the use of too small an amount of water, the solution is decanted and the residue is repeatedly treated with water until it no longer visibly decreases in quantity.

It is then treated in the same way with dilute nitric or hydrochloric acid. If necessary a few drops of the respective concentrated acids may be added to the dilute reagent.

The methods to be used with unknowns which are not soluble in water or dilute acids will be taken up in the following paragraphs.

To Bring Solids into Solution.

179. **General.** The problem of dissolving solids is simple when the results of the preliminary tests just described show that the unknown is completely soluble in water or dilute acids. Water is used as solvent alone only when the unknown is soluble in a volume that is readily handled, i. e. when 1 g., the quantity of a solid usually taken for an analysis, dissolves in 50–100 cc. water.

If the preliminary tests have shown that the unknown is partly soluble in water, partly in an acid, it is always advisable to first extract about 1 g.[1] with water, and then dissolve the residue in the acid. The two solutions are analyzed separately. The extra time and labor involved in this procedure are compensated by the more complete information obtained concerning the nature of the unknown, as will be seen in the following examples.

The analysis of a solution prepared by dissolving cupric chloride, manganous carbonate and sodium carbonate in a dilute acid shows only that the solution contains the cathions $Cu^{\cdot\cdot}$, $Mn^{\cdot\cdot}$ and $Na^{\cdot}$, and the anions Cl' and CO_3''. It does not give any information concerning the combinations of the respective cathions and anions. On the other hand, by using first water and then acid to bring the mixture into solution, the analysis will yield more information.

[1] This quantity will be found sufficient in most cases.

On heating this mixture with water, $CuCl_2 + Na_2CO_3$ react to form $CuCO_3 + 2NaCl$. In the majority of cases, however, the two salts will not be present in exactly stöchiometrically equivalent amounts. Accordingly, when it is found that the aqueous solution contains *cupric ion* and *does not react alkaline,* the conclusion may be drawn that the mixture contains a soluble cupric salt, $CuCl_2$.

When the aqueous solution has a *blue color,* however, it does *not necessarily follow* that it contains a cupric salt which is soluble in water, for cupric carbonate is quite abundantly soluble in sodium carbonate solution. This solution is unstable and it decolorizes when boiled, separating black cupric oxide, CuO.

The conclusion that the mixture contains no soluble manganous salt, i. e. no $MnCl_2$, follows when manganous ion is not found in the aqueous solution. Since it contains $Cu^{..}$ and does not react alkaline, there is not enough CO_3'' present to hold all the $Cu^{..}$ and $Mn^{..}$ in combination.

Conversely when the aqueous solution contains no $Cu^{..}$ or $Mn^{..}$ and reacts alkaline, it follows that the mixture contains more than enough carbonate to react with all the copper and manganous salts, i. e. it has an excess of sodium carbonate. Then *qualitative* experiments alone will not serve to distinguish whether the mixture is composed of $CuCO_3$, $MnCO_3$, NaCl, Na_2CO_3, or $CuCl_2$, $MnCO_3$, Na_2CO_3, or $CuCO_3$, $MnCl_2$, Na_2CO_3, or $CuCl_2$, $MnCl_2$, Na_2CO_3, without the aid of physical methods of separating the salts, or by a special investigation.

Only the aqueous solution need be tested for the ions of the *alkali metals.* This lessens the labor involved in analyzing the acid solution. Most alkali salts are readily soluble in water. Hence it may be assumed that practically all of the alkali salts are contained in the aqueous solution.

In the acid solution tests for the alkali ions may be omitted, even when the unknown is practically insoluble in water. Exceptions are met with when tartrate and pyroantimonate ions are simultaneously present. Their respective potassium and sodium salts are difficultly soluble. Furthermore silicates and certain cyanides constitute special cases which will be considered later (§§ 189 and 184).

The following example also illustrates clearly the advantage of analyzing the different solutions. Assume that the given mixture contains HgO, PbO_2 and MnO_2. Analysis of the hydrochloric acid

solution would indicate the ions $Hg^{\cdot\cdot}$, $Pb^{\cdot\cdot}$ and $Mn^{\cdot\cdot}$. Moreover, since chlorine gas is evolved it would follow that an oxidizing agent for chloride ion is present, i. e. lead or manganese dioxide (or peroxide).

By treating the given mixture with dilute nitric acid before dissolving in hydrochloric acid,[1] it may be easily determined whether the unknown contains lead- or manganese peroxide, or a mixture of both, and also which oxides are present.

Peroxides dissolve only sparingly or slowly when they are heated with dilute nitric acid. Oxides on the other hand dissolve more readily. The above questions can be answered, at least with considerable approximation, from the results of the analysis of the solution prepared with dilute nitric acid, and also from the analysis of the residue left after the above treatment. This is washed and dissolved in concentrated hydrochloric acid.

Since qualitative tests are less definite than quantitative, the question is answered only approximately by the above means. This is a consequence of the less exact definition of the nature of the object under analysis, concentration of acid, method of treatment, in short of the general experimental conditions.

All of the conditions in favor of the one procedure or the other cannot be considered here. It must be left to the chemical tact of each individual to judge when the more tedious procedure of carrying out two or more analyses will yield more complete information concerning the unknown than the shorter method of using only one solvent.

180. Continuation. The most important of the compounds which are scarcely soluble in water, but dissolve in dilute acids, comprise *oxides* (except tin dioxide and the oxides of iron, aluminum and chromium), *peroxides* (these are only sparingly soluble in nitric acid), *oxalates, phosphates, tartrates, cyanides* (consult also § 183) and *sulphides.*

[1] A preference for the use of a mixture of concentrated hydrochloric and nitric acids (aqua regia) to dissolve substances which are insoluble in water is quite general. It is thought that the analysis is more quickly effected by this means. In many cases, however, the conjectured rapidity is doubtful, since the excess of acid used interferes with the even course of the analysis. Furthermore the method often gives rise to undesigned complications which must be overcome, occasionally at the cost of still more time. Reference will be made later to the cases in which *aqua regia* may be *used* to *advantage.*

Chlorides that are difficultly soluble in water, as *silver chloride, mercurous chloride, lead chloride,* the corresponding *bromides* and *iodides,* also *mercuric iodide, sulphates* of *lead, barium, strontium* and *calcium* (§ 182), the above-mentioned *oxides* of *iron, chromium* and *aluminum* and certain *sulphides, cyanides* and *silicates,* are *scarcely soluble* in *dilute acids.*

The following paragraphs treat of the chemical processes intentionally introduced to bring these compounds into solution—the methods of *solution.* The treatment of alloys will be considered in a special paragraph. These are sometimes not attacked at all by dilute acids, or at least only incompletely.

The general method of preparing the solution used in testing for anions (§ 110) will now be considered.

181. To prepare the solution for the tests for anions. The procedure is very simple in case the unknown is soluble in water and the tests for cathions have shown that those of the *light metals* alone are present, e. g. $K^{\cdot}$, $Na^{\cdot}$, $Ba^{\cdot\cdot}$, $Sr^{\cdot\cdot}$, $Ca^{\cdot\cdot}$, $Mg^{\cdot\cdot}$. If the *solution reacts alkaline* in this case, two portions of it are neutralized, *one* with *nitric acid,* the *other* with *acetic* or *sulphuric acid.* The first of these is tested for all anions except NO_3' according to § 110. With the other solution a special test is made for NO_3' according to § 155. *When the solution has an acid reaction it is first neutralized with alkali hydroxide.*

In case the unknown is soluble in water and the aqueous solution contains other cathions besides those just mentioned above, or when it dissolves only in dilute acids, a solution must be prepared which does not contain the cathions in question. This is effected by heating some of the given mixture with a solution of *sodium carbonate* for a few minutes. The cathions which are to be removed from the solution form difficultly soluble (and basic) *carbonates* or *hydroxides* with carbonate ion or with the hydroxyl ion present in a solution of sodium carbonate. In the filtrate from this precipitation are contained the *excess* of *carbonate* and *sodium ions* and the *anions in question* (formulae?). For the above reasons two portions of the solution are separately neutralized (compare § 110).

It is advisable to add a slight excess of acid to the portion neutralized with nitric acid. The solution is then heated to expel dissolved carbon dioxide. If this precaution is not taken mistakes may arise, owing to the reactions of carbonate ion (§ 137). Afterwards the excess of acid is neutralized with ammonia.

Treatment of the unknown with sodium carbonate will not always effect the removal of *all cathions* from the solution to the desired extent. Thus, for example, the ions of arsenic, tin or antimony are not precipitated if tartrate ion is present. Furthermore nickel carbonate does not precipitate when the solution contains complex nickel cyanide, $K_2Ni(CN)_4$. All of these cases may be foreseen, however, from the combination of the results of the preliminary tests with those from the tests for cathions. The particular components in question must be removed by means of a suitable reagent. Most frequently hydrogen sulphide is added to the acidified solution, or ammonium sulphide is used. Then hydrogen sulphide is expelled by heating the filtrate, and ammonium sulphide is removed according to § 85. Sulphate and sulphite ions will be found in the filtrate in consequence of a partial oxidation of hydrogen- or ammonium sulphide. Hence these ions must be tested for beforehand. The reaction with barium ion is so delicate a test for sulphate ion, however, that it is not appreciably affected by any cathions contained in the solution. Therefore the test for SO_4'' may be carried out before removing the disturbing cathions. Sulphites are indicated by the preliminary tests.

A large *excess* of *sodium carbonate* should not be used, because copper and zinc carbonates dissolve in sodium carbonate solution. By boiling the solution for a long time, or by neutralizing the excess of carbonate, these carbonates reprecipitate and can then be filtered out.

The methods of dissolving compounds which are practically insoluble in acids, in order to obtain solutions to be tested for anions, are considered in the following paragraphs. It should be pointed out here that the solutions containing anions are to be neutralized before the tests are carried out.

In case a given mixture dissolves partly in water, partly in dilute acids, or if it must be brought into solution by special means, the remarks relating to the tests for cathions in § 179 are to be followed.

182. To bring difficultly soluble halides into solution. *Chlorides* are transformed into corresponding *hydroxides* or *oxides* by heating them with alkali hydroxide. The residue is filtered. The tests for *anions* are carried out with the *filtrate* after neutralization. The *residue* is washed, dissolved in dilute *nitric acid* and tests for *cathions* are made with this solution.

The same method will also serve for *bromides* and *iodides*.[1] Difficulties are sometimes experienced, however. These are partly due

[1] [Another method is to treat the solid halides with zinc and a very little acid. This forms zinc halides after some time (an hour or two). Reducible compounds of the halogens, for instance silver iodate, also yield halides under these conditions.]

to the fact that the hydroxide of lead is soluble in an excess of alkali hydroxide[1] (§ 14). *Mercuric iodide* on the other hand forms complex anions (§ 33) with iodide ion which results from the reaction:

$$HgI_2 + 2OH' = \underline{HgO} + H_2O + 2I',$$

and accordingly goes into solution.

This circumstance acts as a disturbing factor in a double sense. Mercury must be precipitated by means of hydrogen sulphide from the *solution to be tested for anions.* Furthermore the nitric acid solution of the residue left after treatment with alkali hydroxide does not contain *all* the cathions which were originally present as insoluble halides.

Another disadvantage arising from the use of this method consists in the fact that very difficultly soluble compounds, as silver iodide, are transformed very slowly, and sometimes only partially, into oxide, unless the alkali hydroxide is frequently renewed. The residue filtered after treatment with alkali hydroxide will then be only partly soluble in dilute nitric acid.

For the above reasons it is better to use another method, which, with few exceptions, leads more quickly to the desired end. It is based upon the formation of complex anions between thiosulphate ion and the cathions in question ($Hg^{\cdot}$, $Hg^{\cdot\cdot}$, $Ag^{\cdot}$, $Pb^{\cdot\cdot}$, § 123). The majority of these insoluble halides will dissolve quite abundantly in a solution of sodium thiosulphate.

The given mixture is treated with a solution of sodium thiosulphate containing one formula weight of $Na_2S_2O_3.5H_2O$ to a liter of water. In case solution takes place only slowly, the mixture is gently warmed with the thiosulphate.

Silver halides dissolve most readily. Their complex compounds are also quite stable, even when heated. Of the lead halides the bromide in particular is little soluble.

The complex lead thiosulphate ion decomposes more quickly when heated than the complex silver thiosulphate ion. Mercuric thiosulphate is most unstable. Addition of thiosulphate solution to mercuric chloride or iodide gives rise to a blackening of the surface of the salt, which apparently is due to the formation of a compound containing sulphur. The filtrate, however, also contains mercuric thiosulphate ion, for hydrogen sulphide precipitates mercuric sulphide from the solution. In the present case therefore this method is preferable to that mentioned above with hydroxide.

From the thiosulphate solution, which is separated by filtration from any residue not soluble in thiosulphate, hydrogen sulphide pre-

[1] This also relates to the dissolving of lead chloride. In this case, however, the special method of bringing about the solution with alkali hydroxide may be avoided in consequence of the fact that lead chloride is much more soluble than the bromide or iodide, particularly in hot water.

cipitates sulphides of the metals in question. The precipitate is filtered and hydrogen sulphide is expelled from the filtrate by boiling. It is then heated with a little dilute nitric acid to destroy thiosulphate ion. Finally tests are made for the ions of the halogens.

A *large* excess of nitric acid should be avoided because a solution which contains simultaneously chloride ion, nitrate ion and hydrogen ion, evolves chlorine when heated. Under similar conditions bromine and iodine are also set free, in case the solution contains Br' or I' in place of Cl'. In this way the ions which are to be identified, are removed when an excess of acid is present.

The precipitate obtained with hydrogen sulphide is treated with dilute nitric acid according to § **21**. This decomposes sulphides of lead and silver. The solution so obtained is then precipitated with dilute hydrochloric acid for the bases of group I.

A residue left after treatment with nitric acid is tested for mercury by heating it with a mixture of hydrochloric and nitric acids. From the solution the excess of acid is driven off by evaporation on a water-bath. Then the solution is tested with stannous chloride for mercuric ion (§ **21**, (5) and (6)).

Mercurous compounds, which are indicated by the black color of the residue insoluble in thiosulphate solution, are brought into solution by heating the residue with a mixture of concentrated hydrochloric and nitric acids (3 : 1). This solution is then tested for $Hg^{\cdot\cdot}$ by the method just given.

183. To bring lead-, barium-, calcium- and strontium sulphate into solution these compounds are transformed into corresponding carbonates. The latter differ from the sulphates in that they are *soluble* in acids.

When boiled with a solution of alkali carbonate the sulphates are transposed according to the equation:

$$\underline{MSO_4} + CO_3'' \rightleftarrows SO_4'' + \underline{MCO_3}. \quad (M = Ba, Ca, Sr, Pb).$$

Calcium sulphate is quite *readily* transformed in this way by using a sufficiently concentrated solution of alkali carbonate. *Lead* and *strontium sulphate* are transposed with *much greater difficulty,* while barium sulphate goes over into the carbonate only under special conditions.

The reason for this behavior is due to the reversibility ← of the reaction. This may be shown readily by heating solid barium carbonate for some time with sodium sulphate solution. When acidified the filtrate evolves carbon dioxide. After the residue left on the filter has been *washed* it is observed to be not completely soluble in hydrochloric or nitric acid. From this it is

evident that some barium sulphate has been formed, i. e. that the reaction has proceeded in the sense ← opposed to the dissolving of barium sulphate. This reaction is more pronounced as the solubility of the resulting carbonate becomes *greater* and the solubility of the sulphate in question becomes *smaller*. Reverse conditions hold for the reaction →.

The discoverers of the law of mass action, Guldberg and Waage,[1] have made an exact study of this case. They have found that the above reaction, particularly between barium sulphate and potassium carbonate, proceeds until the concentrations of carbonate and sulphate ions in the solution have the ratio 4 : 1.

Hence to transform a given amount of barium sulphate into the carbonate, at least enough alkali carbonate must be used to give the ratio 4 : 1 between the concentrations of CO_3'' and SO_4'' after the reaction, i. e. after a definite quantity of sulphate ion has been formed and a corresponding amount of carbonate ion has vanished. An estimation of Küster[2] gives the amount of alkali carbonate needed to transform a given quantity of strontium sulphate. In a solution in equilibrium with solid strontium sulphate and carbonate the ratio between the concentrations of the ions CO_3'' and SO_4'' is 1 : 60.

On account of the practical difficulties which arise in connection with the *dissolving* of *barium sulphate* under these conditions (which are not insurmountable, however), the following method is usually adopted.

The sulphate or unknown is mixed with from four to five times its weight of a mixture of potassium and sodium carbonate.[3] At first this is gently heated in a platinum crucible[4]—as long as water is given off. Then it is strongly heated 15–20 minutes until the mass melts. The fusion can be readily removed by placing the hot crucible in cold water, or by gently heating it with some water for a short time. It is then triturated in a mortar and is heated a *few*

[1] Guldberg and Waage, Etudes sur les affinités chimiques (Christiana, 1867); Journ. f. prakt. Chemie, (*2*), **19**, *69*, 1879. The deduction of the law will also be found in Ostwald, Lehrbuch der allgemeinen Chemie, 2d edition, volume *II*, (*2*), p. *305*; Nernst, Theoretische Chemie, 2d edition, pp. *498*, *595*; Van't Hoff, Vorlesungen, Part I, p. *99*.

[2] Verhandlungen der Versammlung deutscher Naturförscher und Aerzte, **1899**, *II*, *128*.

[3] This mixture is used because it melts more readily than either of the carbonates alone, i. e. at a lower temperature. The phenomenon that one substance lowers the melting point of another is very general. It is regulated by definite laws, which, however, cannot be taken up here. Consult Ostwald, Grundriss der allgemeinen Chemie, 3d edition, p. *333*. This is analogous with the phenomenon that salt solutions freeze below 0° C., the freezing point of pure water.

[4] When lead sulphate is present, and in all other cases where compounds which reduce readily to a metal are heated, platinum vessels cannot be used. They would soon become unserviceable owing to the formation of *easily fusible alloys* between platinum and the metal in question.

minutes with water. The alkali sulphate formed and the excess of alkali carbonate go into solution, and are separated from the difficultly soluble carbonate by filtration. A sodium carbonate solution is used to wash the residue on the filter. This prevents transformation of the carbonate into sulphate[1] through a reaction with alkali sulphate, which is contained in the solution wetting the precipitate. When the sodium carbonate solution which runs through the filter is acidified with hydrochloric acid and yields no precipitate of barium sulphate upon addition of barium ion, the washing may be considered satisfactory. This indicates that alkali sulphate, which is retained by the carbonate, has been practically completely removed.

A *solution of the carbonate* in hydrochloric or nitric acid is used in the tests for *cathions.*

It should also be mentioned that an attempt to dissolve the fusion in dilute acid before extracting it with water is wrong. A simple reflection will show that this would lead to the original conditions. For the cathion resulting from the decomposition of the carbonate (formed by the fusion) with acids will yield the corresponding sulphate with sulphate ion.

184. To bring cyanides into solution. *The solubility in water and the stability towards dilute acids* furnish the distinctive characteristics of the method for treating cyanogen compounds. Consult § 178 concerning the determination of the solubility in water. The stability towards dilute acids varies in the following manner. Less stable soluble cyanides give rise to a precipitate within a short time after acidification. Stable cyanides, for instance those of the character of potassium ferro- and ferricyanide, react with dilute acids in this sense (see below) incomparably more slowly.

(a) Of the cyanides readily soluble in water, those which are only

[1] The same reason also holds for the use of a 4–5 fold excess of alkali carbonate in making the fusion. W. Meyerhoffer, Zeitschrift für physikalische Chemie, 38, *314*, 1901, has shown that to transform a definite amount of barium sulphate into barium carbonate a quantity of alkali carbonate *equivalent* to the barium sulphate taken probably suffices; thus, for example, 0.53 g. $KNaCO_3$ for 1 g. $BaSO_4$. The excess of alkali carbonate is designed to prevent the (partial) transformation of barium carbonate into the corresponding (barium) sulphate on extracting the fusion with water. Otherwise the transformation would take place according to the equation:

$$BaCO_3 + KNaSO_4 = BaSO_4 + KNaCO_3.$$

It has already been frequently stated that reactions which can proceed in two directions are limited in the one sense → by the presence of a large amount of one of the resulting products, in this case $KNaCO_3$.

slightly stable towards dilute acids and which therefore form a precipitate[1] within a short time after acidification, are brought into solution by means of dilute hydrochloric or nitric acid. These cyanides belong to the following types:

$$KM_I(CN)_2,\ KM_{II}(CN)_3 \text{ or } K_2M_{II}(CN)_4.$$

The solutions of these salts contain chiefly the anions:

$$M_I(CN)_2',\ M_{II}(CN)_3' \text{ and } M_{II}(CN)_4''.$$[2]

Only very small amounts of the ions $M_I^{\cdot}$, $M_{II}^{\cdot\cdot}$ and CN' are formed through secondary dissociation of the complex ions (compare §§ 12 and 81). Accordingly the solutions exhibit none of the properties that are characteristic of simple ions, or at any rate they show them to only a very limited extent.

The fact that the cyanides $M_I(CN)$ or $M_{II}(CN)_2$ precipitate on acidifying the solutions, indicates that the solutions also contain the ions or substances from which the complex ions are formed. In other words the formation of a complex metallic cyanogen ion → is limited by the reaction proceeding in the opposite direction:

$$\underline{M_I(CN)} + CN' \rightleftarrows M_I(CN)_2'.$$

For the addition of hydrogen ion leads to the formation of hydrocyanic acid (§ 146) in consequence of the reaction $H^{\cdot} + CN' = HCN$. This, however, produces a *decrease* in the concentration of the cyanide ion, with which $M_I(CN)$ and $M_I(CN)_2'$ were in equilibrium. Therefore it gives rise to the conditions for the further progress of the reaction, that is for further decomposition of the complex cyanide.

The separated cyanide goes into solution again on increasing the amount of acid added, or after the acid has reacted for some time. This takes place as soon as hydrocyanic acid begins to escape. The latter is formed from the reaction:

$$\underline{M_I(CN)} + H^{\cdot} = M_I^{\cdot} + HCN.$$

[1] This is quite frequently regarded as a first group precipitate in testing for cathions. It may be distinguished from the latter, however, by its property of vanishing when heated with dilute acids. A further difference consists in the fact that *chloride ion* is necessary for the *precipitation* of *first group ions* (§ 5), while *hydrogen ion* determines the precipitation of *cyanides*. Accordingly *hydrochloric, nitric* or *sulphuric acids* will *decompose cyanides*, but only *hydrochloric acid* precipitates first group components.

[2] Consult § 167 for the relation between the valence of the anions and that of the metal M.

The above equations show clearly the results obtained in bringing cyanides into solution. The *complex ion* $M_I(CN)_2'$ is transposed into the cathion $M_I^{\cdot}$. Similar considerations hold for the reactions of the ions $M_{II}(CN)_3'$ and $M_{II}(CN)_4''$. How are these formulated?

(b) Cyanogen compounds of the type of *ferro-* and *ferricyanides*, which are *soluble* in water, are decomposed by concentrated sulphuric acid, and not by hydrochloric acid. In comparison with sulphuric acid concentrated hydrochloric acid is less active, because a larger percentage of it escapes during the heating, without acting on the cyanide.

The cyanide is stirred up with concentrated sulphuric acid to a thin paste in a platinum crucible. The crucible is then heated over a bunsen burner until the evolution of white fumes of sulphur trioxide ceases. To bring into solution the basic sulphates that are thus formed, the residue is cooled and moistened with concentrated hydrochloric acid. After the acid has reacted for some time, water is added and the mixture is heated, if necessary, until solution is effected. As a rule, however, this labor is unnecessary since the identification of the simple cathions and, in particular, of the complex anions according to §§ 147 and 148 presents no difficulties.

It is necessary to decompose soluble cyanides of this nature (see above) only in those cases where it is uncertain whether corresponding complex (cyanogen) anions of other metals than iron, for example, of cobalt or manganese, are present or not.

(c) *Difficultly soluble* cyanogen compounds are also sometimes decomposed in the above-mentioned way (b) to show if they contain *potassium* and *sodium*. However, since they undergo a complete decomposition (compare § 147), it is still necessary to prove which metallic components form complex anions with cyanogen, and which have the function of a cathion. Thus, for example, the decomposition of cobaltous ferricyanide, $Co_3(Fe(CN)_6)_2$, and of ferrous cobalticyanide, $Fe_3(Co(CN)_6)_2$, with concentrated sulphuric acid gives the same analytical results, although the two salts are essentially different in their properties.

The distinction can be made by a method similar to the one used for analyzing difficultly soluble sulphates or chlorides, namely by heating the compounds in question with a solution of alkali carbonate or -hydroxide. The respective carbonate and hydroxide of the

metal which is acting as a cathion is obtained by this means. The complex anions, however, are brought into solution, as is shown by the equations:

$$Fe_3(Co(CN)_6)_2 + 6OH' = \underline{3Fe(OH)_2}^1 + 2Co(CN)_6''',$$

and $$Co_3(Fe(CN)_6)_2 + 6OH' = \underline{3Co(OH)_2} + 2Fe(CN)_6'''.$$

The hydroxide is filtered, washed and dissolved in dilute nitric acid. In case all does not dissolve, the transformation with hydroxyl ion has not been complete. In this case the treatment with alkali hydroxide must be repeated.

The aqueous solution containing the anion and the excess of alkali carbonate or hydroxide, is neutralized with dilute nitric acid. It is then tested with the typical cathions according to § 110. In consequence of the solubility of aluminum-, chromium- and zinc hydroxides in alkaline liquids, a precipitation occasionally takes place on neutralizing hydroxyl ion. The precipitate is then filtered, washed, dissolved in acids and tested for cathions.

Many alkali cyanides are difficultly soluble in water. Hence besides the treatment with alkali carbonate or -hydroxide, always a special treatment with concentrated sulphuric acid or mercuric oxide (d) is made to prepare a solution which can be tested for ions of the alkali metals.

(d) Cyanides of the character just mentioned (c) can be prepared for the tests for *potassium* or *sodium* in still other ways than by heating them with concentrated sulphuric acid. By means of the reaction which takes place on heating cyanides with *mercuric oxide*[2] a solution is obtained, even with the majority of difficultly soluble cyanides (consult § 176). In case the insoluble compound contained potassium or sodium, this solution contains the corresponding ions. It is filtered from the excess of mercuric oxide and the insoluble hydroxides, and is tested in the usual way. A direct test for potassium and sodium without previously removing the cathions that are precipitated by the group reagents is impracticable.

The remaining metallic components (except $Hg^{\cdot\cdot}$) are tested for

[1] To avoid complicating the relations still more, the possible transformations, like those of § 149, (3), are not taken into account. The relations which have just been discussed are not essentially influenced by this.

[2] In this case the addition of alkali hydroxide (§ 176) is unnecessary.

in the solid residue left after filtration. For this purpose it is washed and dissolved in dilute nitric acid. The solution is then tested for cathions by the methods already described. However it need not be tested for cathions of the alkali metals, since these are contained in the solution filtered from the above residue.

The *solution* may also contain ZnO_2''. For zinc hydroxide is formed in the reaction with mercuric oxide and redissolves when the solution contains sufficient hydroxyl ion (compare § 81).

To effect the solution by means of mercuric oxide can serve only in place of the decomposition of difficultly soluble cyanides by concentrated sulphuric acid. The method is more simple, however, and can be carried out more rapidly. To distinguish the nature of the cathions and the *complex anions* (consult (c)), a special treatment with alkali carbonate or -hydroxide must always be made.

185. To bring fluorides into solution. Although many fluorides dissolve in dilute or concentrated hydrochloric- or nitric acid, as has already been mentioned in § 117, this property is not utilized in analysis, because impurities would be introduced into the solution in consequence of the action of hydrofluoric acid upon glass. *Fluorides must be decomposed by heating them with concentrated sulphuric acid in platinum vessels,* at first on the water-bath and finally over a free flame. Platinum withstands the action of hydrofluoric acid.

Not infrequently *far too much* concentrated sulphuric acid is used for this purpose. To evaporate this acid requires a long time. It is therefore more suitable, especially with less difficultly soluble fluorides, to heat the powdered fluoride with dilute sulphuric acid, using a slightly larger volume of the acid than corresponds to the quantity of fluoride taken, and adding a few drops of concentrated sulphuric acid. In calculating the number of grams of sulphuric acid required, it may be arbitrarily assumed that the unknown consists of pure *calcium fluoride,* CaF_2. One formula weight of this, 78.1 g., requires one formula weight of sulphuric acid, 98.1 g.

In case it is desired to measure the acid by volume, and not by weight, the weight must be calculated to volume. This is a very simple calculation if it is known how many g. H_2SO_4 are contained in a definite volume of the acid. When the concentration of the acid is unknown, its specific gravity is determined with an aräometer.

The content of acid corresponding to the density found is then taken from specific gravity tables for sulphuric acid, which are always kept in chemical laboratories.

The excess of sulphuric acid is driven off by heating the platinum vessel over asbestos in the hood. The residue is then treated with dilute acid. Sulphates which do not go into solution by this means are treated according to § 183.

186. The solution of oxides. *Calcined oxides* of *aluminum, chromium* and *iron* dissolve so slowly when they are heated with concentrated acids that the method is not adapted to bring them into solution. The reaction velocity can be increased by raising the temperature. Since this is limited by the volatility of the acids used, (acid-) or *hydrosulphates* are applied. At high temperatures they decompose according to the equation:

$$2KHSO_4 = K_2SO_4 + H_2SO_4.$$

This is readily recognized by the escaping fumes of sulphur trioxide. Hence to render the oxides soluble they are mixed with from six to ten times their weight of potassium hydrosulphate, and the mixture is heated from 10–20 minutes in a covered crucible until it melts. In the case of ferric oxide the following reaction takes place under these conditions:

$$Fe_2O_3 + 6KHSO_4 = Fe_2(SO_4)_3 + 3K_2SO_4 + 3H_2O.$$

The fused mass is dissolved in water when cold. If the temperature is raised too high the sulphates lose sulphur trioxide (Principles, p. *569*), and are transformed into *basic-* (=hydroxy-) or oxv-sulphates, which are less soluble in water. To bring these into solution some hydrochloric acid must be added.

In case the dry test has shown that the mixture of oxides contains chromic oxide (green phosphor salt beads in the oxidizing and reducing flames), the method given in § 57 for separating chromic- and ferric hydroxides may be combined directly with the above fusion. This is done by fusing the substance with several times its weight of potassium nitrate, or potassium chlorate, and sodium carbonate. By this means a chromate is formed. It goes into solution on extracting the fused mass with water, while aluminum- and ferric oxides remain undissolved. These are dissolved in acids, or are rendered soluble by a fusion with potassium hydrosulphate.

The fusion with potassium hydrosulphate may be carried out in a platinum vessel, but for the oxidation of chromic oxide, however, a platinum crucible may be taken *only when* all of the hydrosulphate which has not been used up in the transformation of the oxides, has been decomposed. This is the case as soon as white fumes of sulphur trioxide cease to be evolved on prolonged heating.

Tin dioxide and *antimony trioxide* are rendered soluble by heating them with from four to six times their weight of a mixture containing equal parts of anhydrous sodium carbonate and sulphur. The mixture is placed in a covered porcelain crucible and is heated by means of a bunsen burner with the flame only three to four centimeters high. The heating is continued until sulphur no longer burns from beneath the crucible cover. Sodium salts of sulphostannic and sulphoantimonic acids are formed by this means. They are readily soluble in water. For the analysis of the aqueous solution consult § **25**.

In case any other oxides have been left undissolved in consequence of an insufficient treatment of the mixture (which may contain oxides soluble in acids) with acids, these are also transposed into sulphides by this means. The sulphides are left undissolved on extracting the mass with water. Consult the following paragraph for the methods employed to bring them into solution.

187. The decomposition of sulphides. Only sulphides of the alkali metals and ammonium are soluble in water. Most insoluble sulphides are decomposed by heating them with hydrochloric acid or nitric acid (consult §§ **21, 26, 75**). As a rule, however, native sulphides are more slowly transposed by this means than those freshly precipitated.

A substance or mixture which evolves hydrogen sulphide in the preliminary tests with acids, or which is otherwise indicated as a sulphide in the dry way, is heated with nitric acid of about 1.2 specific gravity until no longer any products of the reduction of nitric acid (§ **155**) escape.

The strongly acid solution is next diluted somewhat, and is filtered from any insoluble residue, which may contain antimonic and stannic acids, arsenic and mercuric sulphides or lead sulphate.[1] The filtrate is tested for the cathions of groups II, III, IV.

[1] Lead sulphate forms through oxidation of the sulphide with nitric acid.

Bismuth is indicated if a cloud forms on diluting the solution in order to render the precipitation with hydrogen sulphide more complete (see §§ 19, 31). For it is not impossible that the filtered residue contains basic bismuth nitrate as a result of diluting the nitric acid solution before filtering.

The residue left after treatment with nitric acid is first *washed* and then gently *heated with ammonium sulphide*. By this means arsenic trisulphide, antimonic and stannic sulphides, antimonic and stannic acids are dissolved. Lead sulphate and basic bismuth nitrate are transformed into the corresponding sulphides.

Crystalline stannic sulphide (mosaic gold) is sometimes much more stable towards solvents. By fusing it with a mixture of sodium carbonate and sulphur (consult § 186) it is transformed into sulphostannate.

Since mercuric sulphide would volatilize under these conditions, a mixture which is taken to be mosaic gold and is correspondingly tested, must be heated with aqua regia (see below) in order to bring *mercuric sulphide* into solution before it is fused with sodium carbonate and sulphur.

A *residue* left after filtering the ammonium sulphide solution is first *washed* and then heated with *dilute* nitric acid in order to dissolve lead and bismuth sulphide. (This step is omitted if the preliminary tests exclude these metals.) The solution is added to the first nitric acid solution, which is to be tested for cathions.

Mercuric sulphide is brought into solution by heating it with aqua regia (§ 21) and is identified according to the methods there given.

Sometimes a residue will still remain undissolved. In this case the nature of the residue must be determined by suitable tests (phosphor salt bead for silicates, hepar test for sulphates.) It is then brought into solution by the proper methods when a sufficient quantity is present to warrant the tests.

188. Dissolving metals and alloys. A convenient preliminary test to determine the nature of the metals consists in heating a small quantity of the object with water to the boiling point. Since the glass of a test-tube is readily cracked when it is heated directly over the flame, owing to local superheating through a more rapid conduction by the metal, it is best to dip the test-tube into a larger vessel filled with boiling water.

A number of alloys which contain cadmium, lead, bismuth and tin melt from 65–95° C. Hence if the alloy melts in the above test it is quite certain that it consists of the above-mentioned metals.

Amalgams of the alkali- and alkali earth metals evolve hydrogen with water at ordinary temperatures, more rapidly when heated. Simultaneously drops of mercury gather at the bottom of the test-tube. How are these reactions for the decomposition of amalgams of alkali metals by water formulated?

The remaining preliminary tests are restricted to the tests on charcoal. Alloys are broken up by means of pliers or some other suitable instrument (file) for the method of dissolving described below.

Most metals go into solution when heated with nitric acid of specific gravity 1.2. Tin and antimony are changed into insoluble tin dioxide and antimonic acid respectively: gold and platinum remain unchanged. As soon as the action of the acid ceases, reduction products in the form of brown vapors are no longer evolved. The solution, together with any undissolved residue, is next evaporated on the water-bath to dryness. The residue is treated with water to which a little nitric acid has been added. The undissolved portion is now filtered, washed and warmed with ammonium sulphide. By this means tin dioxide and antimonic acid are brought into solution.

A residue remaining after treatment with ammonium sulphide may also contain bismuth sulphide which is formed by the action of ammonium sulphide upon the residue left undissolved by acidulated water after treatment with nitric acid. To bring it into solution it is first washed thoroughly and is then heated with dilute nitric acid. This solution is added to the first nitric acid solution. Gold and platinum are dissolved by heating with aqua regia.

Any residue left undissolved after the action of the above reagents is usually composed of silica, SiO_2. To dissolve it consult the following paragraphs.

189. Silicates. The importance of silicate analysis in geology warrants a more comprehensive treatment than can be given here. Accordingly the following statements are designed to furnish merely the principles most essential for further training in this specialized domain.

Natural silicates are all practically insoluble in water except those of the alkali metals. The solution processes which constitute a part of the geological reactions consist in reactions with water which proceed under existing conditions in the sense that the silicates are hydrolyzed, that is, decomposed into silicic acid and base. It may

be assumed as a first approximation that differences in the stability of individual silicates towards water stand to their solubility in water in the following relation. Water acts more rapidly and more completely the greater the solubility of the silicate in water.[1]

The most customary methods of dissolving silicates consist in the use of substances which promote hydrolysis, or in transforming difficultly soluble into more easily soluble silicates. Apart from the effect of carbon dioxide, which plays an important role in nature (Principles, p. *421*), the action of water consists in the formation of the corresponding base and silicic acid, as is expressed by the equation:

$$Na_2SiO_3 + 2H_2O = 2NaOH + H_2SiO_3.^{2}$$

According to the laws governing chemical equilibrium this transformation will proceed further if the concentration of the resulting products is decreased. Hence the simplest means of increasing hydrolysis is to neutralize the resulting hydroxyl ion with hydrogen ion (hydrochloric acid). This also keeps the resulting silicic acid from remaining in colloidal solution.

190. Decomposition of silicates with hydrochloric acid is practically carried out as follows. The silicate is first ground and *sieved* (§ 169). Concentrated hydrochloric acid is then added and the mixture is evaporated to dryness on the water-bath. During the evaporation it is frequently stirred with an instrument that is not attacked by hydrochloric acid (glass rod or porcelain spatula), in order to keep the separated silicic acid from protecting the unchanged silicate against further action of the acid.

The soluble portion of the evaporation residue is taken up with dilute hydrochloric acid. This is effected by moistening the residue with a few drops of *concentrated* hydrochloric acid and diluting with water after 15–20 minutes, when basic chlorides which may have formed during the evaporation are transformed into soluble chlorides. These now go into solution, while silicic acid remains behind.

Silicic acid will go into solution partially and this disturbs the tests for cathions in the third group (§ 51). Hence in cases where it is desired to separate the silicic acid as completely as possible, the

[1] References to this are contained in Braun's Chemical Mineralogy, Leipzig, 1896, pp. *398*. . . .

[2] This is an *ideal* formulation and the sense of the reaction and the nature of the reaction products, rather than their composition, are indicated.

filtrate is again evaporated with hydrochloric acid and is then treated as above described.

To show that only silicic acid remains in the filtered residue it is heated with hydrofluoric acid in a platinum crucible under the hood. Silicic acid volatilizes as silicon tetrafluoride under these conditions:

$$SiO_2 + 4HF = SiF_4 + 2H_2O.$$

For this purpose the precipitated silicic acid is thoroughly washed and dried on the filter in a suitable box. The whole is then placed in a platinum crucible supported in an oblique position in order to give free access to air required for incinerating the filter, and the crucible is heated to redness. When cold the residue is slightly moistened and is mixed with a few drops of concentrated sulphuric acid and about 5 cc. hydrofluoric acid. The contents of the crucible are then evaporated to dryness on the air- or water-bath. It may be necessary to heat the crucible directly over the burner to drive off the sulphuric acid.

When the silicate is only partly decomposed by hydrochloric acid, a residue consisting of corresponding sulphates is left after treatment with hydrofluoric and sulphuric acids. This residue can be dissolved by heating it with hydrochloric acid. The solution is then added to the other hydrochloric acid solution, which is tested for cathions. In case the amount is rather large it is brought into solution according to § **60**.

Not infrequently silicates contain *titanic acid*. This is left behind after treatment with hydrofluoric acid. For the methods employed to distinguish it consult § **222**.

191. Fusion of silicates with alkali carbonate. The majority of silicates which are usually the object of analysis, belong to the less soluble varieties and therefore are only partly decomposed by hydrochloric acid.

The question whether a silicate is decomposed sufficiently rapidly and completely by hydrochloric acid is settled by boiling a small portion of the sieved material for a few minutes with concentrated hydrochloric acid. The liquid is then filtered and to the filtrate are added ammonia and sodium phosphate. In case the amount of precipitate formed is considerable, a larger portion of the silicate is treated according to the above method (§ 190); Kolbeck, Plattner's Blowpipe Practice, p. *89*, footnote.

Silicates which are only incompletely decomposed by hydrochloric acid are transposed into corresponding alkali silicates by fusion with four to six times their weight of alkali carbonate (or a mixture of potassium- and sodium carbonate). Schematically this transformation is expressed as follows:

$$Al_2(SiO_3)_3 + 3Na_2CO_3 = 3Na_2SiO_3 + Al_2O_3 + 3CO_2.$$

On heating the mixture the temperature must be raised only gradually, since the contents of the crucible may be spirted out if carbon dioxide is evolved too rapidly.

Pyro- or polysilicates (Principles, p. *420*) of the general formula $nMO.mSiO_2$ (where M is the symbol of a divalent metal) evolve more carbon dioxide. The corresponding acids, so-called polysilicic acids, give ordinary metasilicic acid, H_2SiO_3, by subtracting $(m - n)H_2O$.

Finally the liquid contents of the crucible are heated 15–20 minutes over a blast lamp. It is then chilled, i. e. dipped into cold water while still hot. The fused mass now easily separates from the crucible walls. It is broken up in a mortar and the excess of carbonate is decomposed by addition of dilute hydrochloric acid. Silicic acid separates in this liquid as a gelatinous mass. A few drops of concentrated hydrochloric acid are added to the liquid, which is then evaporated to dryness on a water-bath. By this means the alkali silicates formed are completely decomposed and silicon dioxide separates.[1] For the separation of the chlorides from silicic acid consult § 190.

192. Evaporation of silicates with hydrofluoric acid. It is evident that the method just described in the preceding paragraph can be applied only when the silicate in question is not an alkali silicate, or when the tests for alkali metals can be left out.

According to Bunsen to determine if a given silicate contains alkali metals, a small portion is heated on a platinum loop in the fusion zone with gypsum that is free from potassium and sodium. There are formed calcium silicate and the corresponding alkali sulphates, which impart characteristic colors to the flame. To clear up any uncertainties that may exist, a comparative test is made without add-

[1] Concerning the behavior of alkali silicates towards acids consult Principles, p. *420*.

ing gypsum to the mineral, or a mixture of a silicate which contains potassium, sodium or lithium, is heated with gypsum in the fusion zone of the flame. The Hempel method of reduction with metallic magnesium (§ 172) is *better adapted* to settle this question, however. For the detailed description of the operations consult § 172.

In case the dry tests have indicated an alkali silicate, solution is brought about with hydrofluoric acid, which reacts with silicates to form gaseous silicon tetrafluoride and fluorides of the metals contained in the silicate. Schematically the reaction may be represented by the equation:

$$Na_2SiO_3 + 6HF = 2NaF + SiF_4 + 3H_2O.$$

According to the method to be described the fluorides are transformed directly into sulphates.

The silicate is placed in a platinum crucible or evaporating dish and is moistened with a mixture made up of equal volumes of concentrated sulphuric acid and water. Then from 5–10 cc. hydrofluoric acid are added in small portions. The mixture is heated on a water-bath and is stirred from time to time with a platinum wire until hydrofluoric acid and silicon tetrafluoride are no longer evolved. It is best to repeat the evaporation with hydrofluoric acid. Then by heating the platinum vessel on an asbestos plate the excess of sulphuric acid over the quantity required to transpose the fluorides into sulphates is driven off. Finally the residue of sulphates is dissolved in water, or is brought into solution by the method described in § 183.

193. **The methods employed when the unknown contains organic compounds.** The following explanations relate to cases in which complex compounds containing one of the organic ions dealt with in §§ 116, 138, 158 occur as components. As has already been stated, for example in § 70, troubles are occasionally met with in such cases. The reactions of the cathions which form the second constituent of the complex ions are weakened to such an extent that variations in the above described procedure of testing for cathions become necessary.

The simplest means of destroying the complex compounds consists in heating the substances in question to redness with free access of air. By this means the organic constituents of the complexes are

destroyed. The metallic components are for the most part practically non-volatile and are left behind except for slight losses, which, with proper care, are negligible. The residue is dissolved in water, to which acid may be added if necessary, and the solution is then tested by the customary method.

Since the disturbing effect of the above-mentioned anions consists in the formation of complex ions, and this influences the different cathions to a variable degree, it is occasionally of interest to determine which cathions exhibit "abnormal reactions." This problem is settled in the following way, but only with the limited certainty which in general accompanies qualitative methods. The original solution in question is first tested by the ordinary methods for cathions.

It must be borne in mind that occasionally the third group cathions are precipitated by ammonium sulphide, that is, in the fourth group, under these conditions. Therefore the solution of the ammonium sulphide precipitate must be tested for cathions of the preceding groups with the proper group reagents.

However, the fact that the precipitate obtained with ammonium sulphide contains hydroxides or sulphides of cathions from other groups, affords evidence that *these cathions* have abnormal reactions, i. e. form complexes with the organic components.

After applying the group reagents there remain in the solution (apart from $Mg^{\cdot\cdot}$, $K^{\cdot}$, $Na^{\cdot}$) only those cathions which form complexes. The nature of these cathions can then be determined by another complete analysis of the solution of the residue left after destroying the excess of ammonium compounds (consult § 101). Otherwise this solution is used only for the tests for $K^{\cdot}$ and $Na^{\cdot}$. The alkali ions are determined according to § 100 in the solution from which the cathions that formed complexes, have been precipitated.

Apart from the above-mentioned special aim, to *heat* the *substance to redness* only after the *analysis carried out in the usual manner has indicated disturbances,* is better than to heat it to redness directly (before carrying out the analysis for cathions), in case the dry tests have shown that compounds of the anions in question are present. For in many cases it is not necessary to destroy the organic compounds.

194. **Analysis of organic compounds** for the most commonly occurring elements follows directly the problems explained in the preceding paragraph. The elements in question are *carbon, hydrogen, sulphur,* the *halogens* and *nitrogen.* To detect *carbon* in a compound a mixture of the substance in question with freshly heated copper oxide is heated in an appropriate tube (bulb-tube). The mixture is covered with a layer of copper oxide and the vapors evolved on heating are passed into baryta solution. A precipitation of barium carbonate is caused by carbon dioxide resulting from the oxidation of carbon.

The separation of water on the cooler portions of the tube indicates that the compound contains hydrogen.

On the one hand the methods of detecting sulphur in organic compounds consist in a *reduction* to a compound which yields *sulphide ion* when dissolved. On the other hand they consist in an oxidation to sulphate ion, which can be unequivocally identified by means of barium ion. Sodium is used for the reduction. A small quantity of the compound is heated in a bulb-tube with a piece of sodium the size of a pea. The sodium is freshly cut from a larger piece and is dried between filter paper. The tube is dipped in cold water while still hot and cracks. In this way its contents are readily removed and can be brought into solution. Any sodium that has not been decomposed will react very violently with water, hence caution is necessary. The filtrate contains sodium sulphide in case sulphur was present in the compound. Upon addition of acetic acid and lead acetate a black precipitate of lead sulphide forms. The Hempel method of reduction described in § 172 is especially adapted to the detection of small quantities of sulphur.

Oxidation to sulphate ion may be effected in various ways, viz. by heating with concentrated nitric acid, hydrochloric acid and potassium chlorate, hydrochloric acid and bromine, or by fusion with sodium carbonate and potassium nitrate.

The detection of *chlorine, bromine* and *iodine* is effected in the following manner. A portion of the substance in question is heated in a bulb-tube with a piece of sodium and the contents of the tube are brought into solution as above described. However the solution must be acidified with nitric acid before silver nitrate is added. The Hempel method of reduction is also recommended in this case if only small quantities of the original substance or mixture are presented for analysis.

According to the method of *Carius* the substance is heated with fuming nitric acid and silver nitrate, which forms the silver halide directly.

The proof that a compound contains *nitrogen* is based upon the detection of *ammonia* or *cyanide ion.* Certain compounds form ammonia when heated with potassium hydroxide or soda lime (a mixture of sodium- and calcium hydroxides) in a test-tube. Ammonia is identified by the means mentioned in § **108.**

However many compounds which contain nitrogen do not yield ammonia in the above process. Hence it is more certain to transform nitrogen into cyanide by fusing the compound with freshly cut potassium or sodium, dried between filter paper. The fusion is brought into solution in the manner described above for sulphur. The solution is then tested for cyanide ion by means of the Prussian blue test (§ **146**). In case the compound contains but very little nitrogen, the precipitate forms only after some time, and at first a blue green coloration is alone produced.

Certain compounds which are easily decomposed demand more specialized treatment. The details are to be found in text books of organic chemistry.

PART IV.

THE RARE ELEMENTS.

195. The properties of the ions of the rare elements that are most important for analysis. These elements are arranged in groups according to the behavior of their ions towards the ordinary group reagents. The groups, however, bear no special resemblance to corresponding groups of the commoner metals taken up in *Part I.* They are arranged in this particular way[1] to facilitate the explanation of certain irregularities met with in the ordinary course of analysis, which indicate that compounds of rare elements are present. Further any similarities as well as differences in the behavior of corresponding compounds are more easily impressed upon the memory.

The following table gives a general retrospect of the most striking color reactions of the rare elements in the phosphor salt bead and in the non-luminous bunsen flame.

The phosphor salt bead is colored:

	In the oxidizing flame;	In the reducing flame;
Red;	By cerium (hot);	By titanium and tungsten (blood red if iron is present);
Yellow;	By cerium (colorless cold);	——
	By uranium (yellow green cold);	——
Green;	——	By vanadium;
	By molybdenum (hot);	By molybdenum (cold);
Blue;	——	By tungsten (cold);
Violet;	——	By titanium, didymium.

The bunsen flame is colored:

Carmine red;	By lithium salts;
Green;	By thallium salts;

[1] For a system of qualitative analysis, including nearly all the metallic elements, consult *A. A. Noyes,* Technology Quarterly, 16 and 17.

Blue green;	By ignited tellurium;
Sky blue;	By ignited selenium (odor of horse radish);
Blue violet;	By indium salts;
Violet;	By rubidium and caesium salts;

(compare also the spectral table).

GROUP I.

196. Hydrochloric acid precipitates the ions of thallium (TlCl), and molybdenum, tungsten, niobium, tantalum. The acids are precipitated in the case of the last four elements. The precipitates go into solution for the most part in an excess of acid, and hence these four elements also belong in another group. Consequently their properties will be mentioned later. Under certain conditions, however, tungstic acid does not dissolve in acids. This point will be taken up again to indicate its connection with the related elements.

197. Thallium forms mono- and trivalent cathions. The former are colorless, the latter light yellowish. Thallous ion is more stable, hence thallic ion may be formed from it only by the action of very strong oxidizing agents, as chlorine or permanganates in acid solution.

With soluble halides thallous ion, for instance a solution of thallous nitrate, forms precipitates that are sensitive to light. The solubility decreases from the chloride to the iodide. The latter salt is yellow and serves best to identify thallous ion by reason of its great insolubility. Thallous halides are not much more soluble in acids, ammonia or potassium cyanide than in water. They are soluble in thiosulphates, however,—especially the chloride. It is hence probable that a complex thiosulphate ion exists.

Thallous hydroxide, -phosphate and -borate all dissolve readily. The carbonate is somewhat difficultly soluble. Hence alkali carbonates will form a white precipitate with concentrated solutions of thallous ion.

The chromate is yellow, quite difficultly soluble in acids. Chloroplatinic acid forms a bright yellow precipitate of thallous chloroplatinate, Tl_2PtCl_6, very difficultly soluble in cold water.

From neutral solutions hydrogen sulphide precipitates brown black thallous sulphide incompletely. By decreasing the concentration of the resulting hydrogen ion through addition of an acetate, the pre-

cipitation becomes practically complete. In its solubility relations thallous sulphide lies between the sulphides of iron and zinc. Ammonium sulphide precipitates thallous ion practically completely. Corresponding to the weaker cathion character of thallic ion its solutions are hydrolyzed to a greater extent than thallous salts.

With alkali hydroxide thallic ion forms a brown precipitate of hydroxide, which has the formula TlO(OH) when dry. Thallic halides decompose into thallous halides and free halogen on standing. The chloride has the greatest stability.

The vapors of thallous compounds emit a green light, which is very important for the analytical detection of this element. This property serves to identify thallous chloride in a mixture with the chlorides of first group ions (§ 5).

GROUP II.

The precipitate obtained with hydrogen sulphide may contain sulphides of the following elements: *platinum* (§ 198), *palladium* (§ 199), *rhodium* (§ 200), *iridium* (§ 201), *osmium* (§ 202), *ruthenium* (§ 203), *gold* (§ 205), *selenium* (§ 207), *tellurium* (§ 208), *molybdenum* (§ 206), *germanium* (§ 209) and *gallium* (§ 223).

198. Platinum exists in di- and tetravalent forms. Cathion characteristics are only slightly developed (Principles, p. *751*). Accordingly the properties of the cathions have very little analytical importance. Complex anions with the halogen elements, however, play an especially important role. They have the composition $PtCl_4''$ and $PtCl_6''$ respectively, and are termed chloroplatinite and chloroplatinate ion. Platinum is also di- and tetravalent in these ions (consult § 167).[1] The latter is more stable and hence more important. Consequently its properties will alone be mentioned.

With chloroplatinic acid potassium- and ammonium ions yield characteristic precipitates, already mentioned in §§ 103 and 109. The acid may be prepared by dissolving platinum in aqua regia and evaporating the solution to dryness.

Alkali iodide gives rise to a dark red color.

[1] The valence of chlorine is assumed to be known. According to the formula HCl, it is monovalent. From the rule referred to above, it follows that the valence is expressed by a negative value.

The details of this reaction have not yet been worked out. Iodine separates to a limited extent, according to the equation:

$$PtCl_6'' + 2I' = PtCl_4'' + 2Cl' + I_2.$$

On shaking out the solution with chloroform only a pale pink color is obtained.

With silver nitrate the solution forms no silver chloride,—in accordance with the fact that it contains very little chloride ion,—but a bright brown precipitate of the composition Ag_2PtCl_6 is obtained. It is very little soluble in ammonia, but dissolves comparatively easily in alkali cyanide and in thiosulphate.

On passing hydrogen sulphide into a solution of chloroplatinic acid a black precipitate of platinic sulphide, PtS_2, forms after some time (since the solution contains very little platinic ion). The precipitate is insoluble in hydrochloric or nitric acid, but dissolves in aqua regia. In this respect it agrees with mercuric sulphide. The two sulphides are distinguished by the greater volatility of the latter. Platinic sulphide also dissolves in solutions of alkali- and ammonium polysulphides. Any residue left after digestion with polysulphides must nevertheless be tested for platinic sulphide.

Chloroplatinic acid differs from chloroauric acid by its greater stability towards reducing agents (ferrous sulphate). *It does not reduce* to the metallic state in *acid solutions,* but may be reduced *in alkaline solution. Chloroauric acid* reduces in *acid* solution. Other reducing agents, as formic acid or a mixture of glycerine and sodium hydroxide, reduce chloroplatinic acid to metallic platinum; stannous chloride reduces it to chloroplatinous acid. Solutions of chloroplatinous acid have a darker color and oxidize to chloroplatinic acid with chlorine, thereby becoming lighter in color.

Platinum tetrachloride is formed by heating chloroplatinic acid in an atmosphere of chlorine. Its aqueous solutions have the properties of a dibasic acid.[1] This is due to the fact that water forms with $PtCl_4$ the compound $PtCl_4.H_2O$. The anions resulting from the dissociation of this compound are $PtCl_4OH'$ and $PtCl_4O''$ The solutions are not very stable, however, owing to the further action of water, by which chlorine is replaced with hydroxyl and the ions $H^{\cdot}$ and Cl' are formed simultaneously. This hydrolysis is catalytically influenced (§ 164) by illuminating the solution and also by means of platinum black.[2]

[1] Miolati, Zeitschrift für anorganische Chemie, 22, *445*, 1901.

[2] Kohlrausch, Zeitschrift für physikalische Chemie, 33, *257*, 1900.

199. Palladium forms divalent and tetravalent ions. The latter are unstable; hence the properties of the divalent ions alone have analytical importance.

Potassium iodide precipitates palladous ion as black *palladous iodide*, PdI_2, a very delicate test. The precipitate dissolves to a brown solution in an excess of iodide ion. It is also soluble in ammonia, forming an ammonia complex.

Mercuric cyanide precipitates yellow white amorphous palladous cyanide, $Pd(CN)_2$, soluble in ammonia and potassium cyanide. In the latter case a complex ion probably forms.

Hydrogen sulphide precipitates black palladous sulphide, PdS, which is insoluble in ammonium sulphide, but dissolves in hydrochloric or nitric acid.

The reducing agents mentioned in the preceding paragraph reduce palladous ion to metallic palladium. In addition carbon monoxide and alcohol effect a similar reduction.

Amongst the higher oxidation forms should be mentioned a complex ion, $PdCl_6''$, analogous with chloroplatinate ion. It yields difficultly soluble compounds with potassium- and ammonium ions.

200. Rhodium. The following reactions relate to the trivalent ion, which is the most stable ion formed by rhodium under ordinary conditions.

With alkali hydroxide a solution of rhodium chloride yields a black brown precipitate of hydroxide upon addition of some alcohol. Without the addition of alcohol a yellow precipitate of a hydrated hydroxide forms slowly. It is soluble in an excess of alkali hydroxide.

When heated with potassium nitrite the rhodium solution gives an orange yellow precipitate of potassium rhodium nitrite. The sodium salt is much more easily soluble than the potassium salt. Hence if sodium nitrite is used, some potassium chloride (potassium ion) must also be added to the solution.

Hydrogen sulphide precipitates rhodium sulphide from a hot solution. The precipitate is insoluble in ammonium sulphide, but dissolves in nitric acid.

Towards reducing agents rhodium chloride solution behaves like the solution of platinic chloride.

201. Iridium exists in three oxidation forms. The compounds of the lowest, divalent, form are very unstable. Analogous with platinum, the higher tetra- and trivalent forms yield anions $IrCl_6''$ and $IrCl_6'''$.[1]

These latter ions bear the following relations to one another. Reducing agents, as sulphur dioxide, hydrogen sulphide, ferrous ion and stannous ion, reduce chloroiridate ion to chloroiridite ion. Chloroplatinate ion does not have this characteristic of being easily reduced. [Sulphur dioxide, however reduces it to chloroplatinite ion, especially from an acid solution. Peter Klason, Ber. d. d. chem. Gesellschaft, **37,** *1360,* **1904.**]

With potassium and ammonium ions *chloroiridate ion* forms crystalline brownish yellow precipitates, which, however, are darker than the corresponding chloroplatinates. *Chloroiridite ion* forms no difficultly soluble compound with potassium salts.

Hydrogen sulphide reduces iridic compounds to iridous forms. With the latter it then gives a brown precipitate of the sulphide, Ir_2S_3, which is soluble in ammonium sulphide.

When a solution of iridium chloride is heated with an excess of potassium nitrite a white precipitate forms, which is difficultly soluble in acids. This reaction is *especially characteristic of iridium salts.*

202. Osmium forms compounds in which it is di-, tri-, tetra-, hexa- and octavalent. Only the properties of the highest oxygen compound, perosmic acid, and of the compounds in which osmium is tetravalent, have analytical importance.

Osmium tetroxide is formed by the action of strong oxidizing agents upon finely divided metallic osmium, for example, by heating to redness in an oxygen atmosphere, by heating with concentrated nitric acid or aqua regia, by fusing with a mixture of potassium hydroxide and potassium nitrate or -chlorate. Characteristic of perosmic acid, OsO_4, are its great volatility and penetrating odor, resembling that of chlorine. The tetroxide volatilizes even on heating aqueous solutions of osmic acid or from its salt solutions. After acidification with hydrochloric acid the distillate from these solutions

[1] According to the rule mentioned in §§ 53 and 165, the ion $IrCl_6''$ corresponds to the higher valence of iridium. Assuming the valence of chlorine to be equal to unity from the formula HCl, it follows that this corresponds to the tetravalent form (§ 167). The rule then gives a negative value for the valence of chloride (compare also § 197).

yields with hydrogen sulphide a black brown precipitate of the tetrasulphide, OsS_4, insoluble in ammonium sulphide.

The fact that a solution of alkali osmate yields a distillate which contains perosmic acid indicates that the salt is strongly hydrolyzed. It is further evident that perosmic acid has only weak acid properties.

Indigo is decolorized by perosmic acid. With ferrous salt it yields a precipitate of osmium dioxide.

In the potassium salt, K_2OsCl_6, osmium is tetravalent. This salt is formed by heating osmium with potassium chloride in an atmosphere of chlorine. It is readily soluble (distinction from corresponding platinum salts). In the solution are contained both the anion $OsCl_6''$ and the cathion $Os^{\cdot\cdot\cdot\cdot}$. A brown precipitate of osmium tetrahydroxide, $Os(OH)_4$, forms slowly through the action of alkali hydroxide, ammonia or alkali carbonate.

Silver nitrate yields a dark olive green precipitate. Through the action of ammonia it decomposes and leaves a brick red residue.

Hydrogen sulphide precipitates brown osmium disulphide, OsS_2.

Reduction to the divalent form takes place when the solution is heated with tannic acid or with alcohol and hydrochloric acid, the color changing to blue. Sodium formate and metallic zinc reduce tetravalent osmium to the metallic state.

When heated with dilute nitric acid oxidation to perosmic acid takes place. The tetroxide distils over with steam.

203. **Ruthenium** is very closely related to osmium with respect to the diversity of its combining ratios. Ruthenium tetroxide is formed under conditions analogous to those which give rise to osmium tetroxide, but the transformation requires higher temperatures and stronger oxidizing agents. The tetroxide is crystalline, very volatile and has a penetrating odor.

Corresponding to the fact that ruthenium tetroxide is less easily formed, it is also more unstable than osmium tetroxide. The sublimation of the dry tetroxide is accompanied by an explosive decomposition. From its solutions it may be reduced to the lower oxidation forms (distinction from osmium tetroxide, which volatilizes unchanged).

The trivalent compounds are the most stable of the lower oxidation forms. They result from spontaneous decomposition of the higher forms, or through the action of reducing agents, as alcohol.

The solution of ruthenium trichloride is yellow. It becomes

clouded on long standing, or when heated, through formation of hydroxide. Alkali hydroxide yields a brown precipitate of the hydroxide, $Ru(OH)_3$.

A solution of the trichloride or of potassium ruthenium chloride, K_2RuCl_5, yields a yellow solution with potassium nitrite (Gibbs). This assumes a characteristic carmine red color on adding a few drops of ammonium sulphide. Ruthenium compounds are distinguished from compounds of the other platinum metals by the fact that solid potassium ruthenium nitrite is soluble in alcohol. To detect ruthenium ammonium sulphide is added to the alcohol solution. It gives a carmine red color in case potassium ruthenium nitrite is contained in the solution.

Ruthenium chloride also gives a purple red colored solution after standing for some time with potassium sulphocyanate, in case ions of the other platinum metals are not present.

When hydrogen sulphide is passed into the solution of ruthenium chloride a dark precipitate of indefinite composition is formed slowly. It dissolves in nitric acid and is also slightly soluble in ammonium sulphide, yielding a characteristic blue solution.

Zinc reduces the trichloride. At first the solution assumes an azure color, probably from the formation of the dichloride, and finally it is entirely decolorized through reduction to metallic ruthenium.

204. Separation of the platinum metals. The essential features of the following method have been set forth by F. Mylius and R. Dietz,[1] especially for the qualitative analysis of a given mixture with respect to the metals mentioned in §§ 198–203.

(a) The metals are presented for analysis in *compact form* as an *alloy*. When in this condition it is a matter of extreme difficulty to bring them into solution. The alloy is first fused with several times its weight of zinc in an atmosphere of illuminating gas. When cold the fused mass is extracted with warm hydrochloric acid to dissolve zinc and any other metals that are present, which do not belong to the platinum group. The platinum metals are changed by this means into a form in which they may be brought into solution more readily than in the original compact alloy, with the exception of iridium and ruthenium. The mixture left after treatment with hydrochloric acid is next dried and heated to redness on a porcelain

[1] Berichte der deutschen chemischen Gesellschaft, 31, *3187*, 1898.

boat in a current of oxygen. If osmium is present it volatilizes under these conditions as the tetroxide, which is collected in sodium hydroxide (forming a yellow solution) and tested according to § 202. A *non-volatile* residue is then mixed with sodium chloride and the mixture is gently heated in a current of moist chlorine. The product is extracted with a small quantity of water. In case all does not go into solution the above process is repeated. A residue left finally after a second extraction with water may contain iridium and ruthenium. The aqueous solution may contain chlorides of platinum, palladium, rhodium, iridium, ruthenium and also gold chloride. By shaking out the solution with a little ether the latter may be removed. If gold is present the ethereal layer will have a yellow color. In this case the process of shaking out with ether must be repeated twice.

The aqueous solution, after separating the ether layer, is then freed from the ether by gentle warming. Details of the further separation and identification of Pt, Pd, Rh, Ir, Ru are given below.

(b) A *solution* containing platinum metals is offered for analysis. In this case the solution is acidified with dilute nitric acid and heated in a retort until it boils. The vapors are collected in sodium hydroxide, which is tested for osmium as above given. The solution thus freed from osmium is next cooled and shaken out with a little ether to remove gold as above mentioned. Then it is mixed with ammonium acetate and formic acid and boiled in a flask with a return condenser for several hours. An open glass tube about a meter long serves admirably to condense any volatile compounds. By this means a black precipitate of the noble metals is obtained through a reduction with formic acid.

The reduced metals are washed and extracted with warm hydrochloric acid to remove metals not belonging to the platinum group. The residue is then dried, mixed with sodium chloride and heated gently in a current of moist chlorine. If this product does not all dissolve in a little water the process is repeated with the insoluble residue. Then, as was already stated above, any portion that is left undissolved after a second extraction with water, is set aside to be tested further for iridium and ruthenium. The aqueous solution may contain chlorides of platinum, palladium, rhodium, iridium and ruthenium. It must be as nearly saturated as possible, and to this end may be evaporated, if necessary.

Identification of the noble metals contained in the concentrated aqueous solution. Addition of saturated ammonium chloride solution precipitates platinum as yellow, crystalline ammonium chloroplatinate. If the solution contained chlorides of iridium and ruthenium their corresponding salts will also be precipitated here. In this case the precipitate has a darker color, which is characteristic.

The precipitate is filtered and washed, and the filtrate is set aside to be tested for palladium and rhodium. In case the precipitate is dark colored it is redissolved in a little warm water. By addition of hydroxylamine hydrochloride to the solution iridium and ruthenium compounds reduce to lower oxidation forms, which yield no precipitate with ammonium chloride. Then the solution is cooled and platinum reprecipitated as above by ammonium chloride.

To separate iridium and ruthenium the filtrate from the precipitation of platinum is evaporated to dryness. By heating the evaporation residue in a current of hydrogen reduction to the metallic condition takes place. The reduced metals are then fused in a silver crucible with a mixture of potassium hydroxide and potassium nitrate. When cool the fused mass is extracted with water, which dissolves ruthenium but leaves iridium behind. On saturating the solution with chlorine and distilling it from a retort the formation of a colored distillate indicates ruthenium, which is volatilized as the tetroxide. The vapors are passed into alcohol acidified with dilute hydrochloric acid, as long as they escape colored, yielding a yellow brown solution. By making this solution ammoniacal and heating it with thiosulphate a deep red violet color is produced (confirmation of the presence of ruthenium).

The residue of iridium left from fusion with potassium hydroxide and niter is washed, dried and brought into solution again by heating it with sodium chloride in a current of chlorine. The solution yields a precipitate with ammonium chloride.

This same method also serves to separate iridium and ruthenium in the residue left after removing osmium and twice heating in a current of chlorine with sodium chloride.

To identify and separate palladium and rhodium the filtrate from the precipitation of platinum (iridium and ruthenium) is slowly evaporated to dryness with an excess of ammonia. The residue is next taken up with a very small amount of warm dilute ammonia. Any insoluble dark particles, which contain ruthenium for the most

part, are filtered. On cooling the solution rhodium is separated as yellow crystals of the composition $Rh(NH_3)_5Cl_3$ (chlorpurpureo chloride). In the filtrate palladium is detected by the addition of concentrated hydrochloric acid, which precipitates palladosammin chloride, $Pd(NH_3)_2Cl_2$. A reduction takes place by heating this precipitate to redness in a current of hydrogen. The reduction product is then warmed with concentrated nitric acid, whereby it goes into solution. The solution yields a precipitate with mercuric cyanide.

205. Gold is monovalent and trivalent. The compounds of monovalent gold are quite unstable in solutions and undergo spontaneous decomposition into the lower and higher oxidation forms, as is illustrated by the decomposition of the chloride:

$$3AuCl = AuCl_3 + 2Au.$$

Hence only the properties of the trivalent compounds are essential for the identification of gold.

The solution of gold chloride contains the cathion $Au^{\cdot\cdot\cdot}$, and also to a certain extent the anions $AuOCl_3H'$ and $AuOCl_3''$[1] from the acid H_2AuOCl_3, which is formed on dissolving gold chloride in water. (Compare the remarks concerning the corresponding compound of platinum tetrachloride, $PtCl_4.H_2O$ or H_2PtOCl_4 in § 198).

Alkali hydroxide precipitates red brown auric hydroxide, $Au(OH)_3$, from a *concentrated* solution of the chloride. The precipitate dissolves in an excess of alkali hydroxide, forming aurates, and is also soluble in acids. This shows that the hydroxide has *amphoteric* character (§ 42, footnote). Aurates contain the anion AuO_2'.

Ammonia behaves quite differently. It forms a dirty yellow precipitate of the composition $AuN_2H_3.3H_2O$. In the dry state this explodes violently when warmed or when triturated. Hence it is called "fulminating gold."

Auric ion is quite unstable in the presence of oxidizable substances.

[1] It has already been stated that gold is trivalent in the acid H_2AuOCl_3. This also becomes evident by applying the rule given in § 167. The sum of the valencies of the elements which can form elementary anions is five. Of this sum chlorine makes up three parts (compare § 198, footnote) and oxygen two (oxygen can form a divalent anion, O''). The sum of the valencies of the elements which form positive ions (hydrogen), and hence the valence of the complex ion, is two. The difference gives three as the valence of gold.

Thus ferrous ion, stannous ion, zinc, copper, formaldehyde, oxalic acid, sulphurous acid and many other substances are oxidized by auric ion with the separation of metallic gold. Especially characteristic is its behavior towards stannous ion. This is oxidized to stannic ion, and a precipitate forms, the color of which varies from pink to reddish purple according to the concentrations of the solutions used. Consequently the precipitate is called gold purple. According to *Zsigmondy* it is a mixture of colloidal gold and colloidal stannic acid, which is formed from stannic ion by hydrolysis:

$$Sn^{\cdot\cdot\cdot\cdot} + 4H_2O = Sn(OH)_4 + 4H^{\cdot}.$$

The reaction is very delicate, even in extremely dilute solutions, because the coloration is sharply defined. By heating the gold solution and adding it to the stannous chloride the delicacy may be increased.

Another very delicate reaction takes place (*Vanino* and *Seemann*) when an alkaline solution of hydrogen peroxide is added to gold chloride solution. A black precipitate of gold forms and oxygen is evolved. If the solutions are extremely dilute a more or less deep blue coloration results instead of the black precipitate.

In the cold hydrogen sulphide precipitates black aurous sulphide, while with a hot solution it yields a brown precipitate of metallic gold. The reactions are formulated as follows:

$$8AuCl_3 + 9H_2S + 4H_2O = 4Au_2S_2 + 24HCl + H_2SO_4,$$

and $$8AuCl_3 + 3H_2S + 12H_2O = 8Au + 24HCl + 3H_2SO_4.$$

The precipitates dissolve in warm solutions of alkali polysulphides, forming sulphaurates containing the ion AuS_2'. (What are the equations for these reactions?) On acidifying these solutions the sulphide reprecipitates. Auric sulphide is not decomposed by acids, except aqua regia.

The following is another characteristic of gold. It forms chloroauric acid, $HAuCl_4$, analogous with chloroplatinic acid, on dissolving gold or gold sulphide in aqua regia and evaporating the solution. This dissociates into the ions $H^{\cdot}$ and $AuCl_4'$, as is shown by the fact that the solution yields with silver nitrate a tawny precipitate of the composition $AgAuCl_4$. If the precipitate is washed and dilute am-

monia is poured over it on the filter, the color changes to the dirty yellow of fulminating gold, and the filtrate yields a precipitate of silver chloride when acidified.

Potassium cyanide gives a colorless solution with gold chloride. It contains the complex anion $Au(CN)_4'$.

The ease with which gold compounds are reduced is also a characteristic property. Metallic gold is ductile.

206. Molybdenum. Of the various states of oxidation of this element the properties of the oxygen compounds in which molybdenum is hexavalent are of special analytical interest. They are formed by treating natural molybdenum compounds with oxidizing agents, as nitric acid or aqua regia. The trioxide, as well as the acids which are derived from it by taking up water are the most stable of the oxygen compounds. *Molybdenum trioxide* is *soluble* both in *acids* and *alkalies*. In order to bring the calcined trioxide into solution it must be fused with alkali carbonate.

By acidifying the solution of a molybdate the white slightly soluble acid, H_2MoO_4, is obtained. Molybdic acid dissolves in acids and in alkalies. Hence it has amphoteric character. With phosphoric, arsenic and other acids molybdenum trioxide forms complex acids (consult Principles, p. *739*).

Solutions of *alkali molybdates*, containing the ion MoO_4'', give with acid solutions of phosphates (§ 127) or arsenates (§ 133) a *characteristic yellow precipitate, but only when ammonium ion is simultaneously added,* in accordance with the fact that ammonium is a component of the precipitate.

Acid solutions of molybdic acid are reduced by reducing agents as zinc, tin, ferrous ion, stannous ion, mercurous ion. At first a blue coloration results, and finally a brown coloration, characteristic of the trivalent form of molybdenum.

Hydrogen sulphide precipitates brown trisulphide, MoS_3, from a solution of molybdic acid. Here also a blue coloration is observed at first.

The precipitation as sulphide is incomplete. The conditions are most favorable for its precipitation when the filtrate from the fourth group is acidified. This is due to the fact that molybdenum trisulphide dissolves in ammonium sulphide, forming sulphomolybdate ion. The latter also forms when ammonium sulphide is mixed with a molybdate solution. The sulphide reprecipitates, however, on acidifying the solution (in accordance with the relations already taken up in § 24). Acids slowly decompose molybdenum trisulphide.

To transform the trisulphide into molybdic acid it is dried and heated in an inclined crucible with the cover off (roasted). In this

way the trioxide is formed. This latter is then dissolved in ammonia or alkali hydroxide. Molybdic acid is precipitated from this solution by addition of an acid.

Potassium sulphocyanate gives a yellow coloration with an *acid solution* of *molybdic acid.* The corresponding compound dissolves in *ether* to a *red yellow* solution, which changes to a *fuchsine red in contact with air.* By addition of a reducing agent, as zinc, thiosulphate or hydrogen sulphide, the reaction can be made much more distinct. The blue or brown solutions of the lower states of oxidation also exhibit the same phenomenon, as would be expected from the effect of the reducing agents. Addition of ammonia or sodium phosphate will decolorize the solution, while acidification makes the color appear again. This reaction is characteristic of molybdenum compounds.

In the *reducing flame* the *borax bead* is colored *green* to *brown* by *molybdenum compounds.*

207. Selenium, like sulphur, forms two series of oxygen compounds, the hydrates of which correspond to sulphurous and sulphuric acids. The properties of selenous and selenic acid have greatest analytical importance. These acids bear the following relation to one another. Selenic acid is formed from selenous acid only by the action of strong oxidizing agents (chlorine, fusion with potassium nitrate). Accordingly hydrochloric acid transforms selenic into selenous acid with evolution of chlorine. Indigo solution is decolorized by selenate ion in the presence of hydrogen ion.

This latter property is made use of to distinguish selenate ion from sulphate ion. The two ions agree in giving with barium ion a precipitate that is practically insoluble in dilute acids.

In acid solution selenous acid acts as an oxidizing agent towards sulphur dioxide, hydrogen sulphide, stannous ion and metals like zinc and iron. Selenous acid goes over into selenium in oxidizing these substances. When the reaction proceeds quickly it forms a red precipitate. With slower reactions selenium precipitates black. The red selenium is transformed into the black variety at higher temperatures.

Selenous acid oxidizes hydrogen sulphide to sulphur, which forms with selenium a bright yellow precipitate of *variable composition.* The equation for this transformation is formulated as follows:

$$\underbrace{2H^{\cdot} + SeO_3''}_{} + \underbrace{4H^{\cdot} + 2S''}_{} = 3H_2O + 2S + Se.$$

Under ordinary conditions, however, this is not the only reaction, but oxidation of hydrogen sulphide to sulphuric acid also takes place simultaneously. Hence the precipitate has a variable composition. The precipitate dissolves in potassium cyanide and in ammonium sulphide.

In the reducing flame selenium compounds evolve an odor resembling that of decayed horse radish.

The non-luminous bunsen flame is colored sky blue by selenium compounds.

208. **Tellurium** is the analogue of selenium with respect to the composition and many other properties of its compounds.

When ignited, tellurium burns with a bright blue flame to the dioxide, TeO_2, which is rather insoluble in water. The corresponding hydrate, tellurous acid, H_2TeO_3, is formed by oxidizing tellurium with nitric acid. It is also rather insoluble and has amphoteric character.

Sulphur dioxide and other reducing agents reduce tellurous acid to tellurium, which separates as a black precipitate (compare § 207).

When fused with potassium cyanide tellurium forms potassium tellurium cyanide. Tellurium reprecipitates from the solution of this salt, however, when it is left in contact with air (distinction from selenium). Concentrated sulphuric acid dissolves tellurium to a red purple solution. On diluting the solution tellurium precipitates as a black powder. Selenium on the other hand dissolves in concentrated sulphuric acid to a green solution and separates as a red powder on diluting the solution.

Telluric acid is formed from tellurous acid by the action of strong oxidizing agents. It has more pronounced acid character than tellurous acid. Like selenic acid it acts as an oxidizing agent towards hydrochloric acid. Tellurate ion is distinguished from selenate and sulphate ions by the fact that its barium salt is more readily soluble than barium selenate or sulphate.

When freshly formed, telluric acid is quite soluble in water, but it dissolves slowly. The hydrate, $H_2TeO_4.2H_2O$, loses water when heated. The product obtained is much less soluble in acids and alkalies. It does not consist of the anhydrous acid, H_2TeO_4, but contains the anhydride, TeO_3, and the dioxide, TeO_2, in quantities vary-

ing with the conditions under which it is formed, according to Gutbier.[1]

With a solution of telluric and tellurous acids hydrogen sulphide forms a brown precipitate, soluble in ammonium sulphide. Apparently the precipitate has not a constant composition (compare the preceding paragraph).

209. **Germanium.** Owing to its limited distribution in nature the description of the most important analytical properties of the compounds of this element can be quickly disposed of. More detailed information will be found in a paper by Cl. Winkler,[2] the discoverer of the element.

The sulphide of germanium resembles the sulphides of antimony and arsenic in its solubility relations. It may be distinguished from these sulphides by the fact that greater concentrations of hydrogen ion are required to decompose the anion of sulphogermanic acid, GeS_3''

$$GeS_3'' + 2H^{\cdot} = GeS_2 + H_2S.$$

The sulphide is white and dissolves in water. Hence it is precipitated by hydrogen sulphide only when a large excess of hydrochloric or sulphuric acid is present.

The disulphide can be transformed into dioxide by roasting and then boiling down the product with nitric acid. The dioxide is white and dissolves in water, acids and alkalies.

GROUP III.

Hydroxides of the following metals may be contained in the precipitate thrown down with ammonia in the ordinary course of analysis: *indium* (§ 210), *uranium* (§ 211), *beryllium* [or glucinum] (§ 212), *zirconium* (§ 213), *thorium* (§ 214), *yttrium* (§ 215), *cerium* (§ 216), *lanthanum* (§ 217), *praseodymium* and *neodymium* (§ 218), *niobium* [or columbium] (§ 220), *tantalum* (§ 221), *titanium* (§ 222) and *gallium* (§ 223).

210. **Indium** occurs chiefly in zinc which is obtained from certain zinc blendes. It is separated from zinc by the action of dilute hydrochloric or sulphuric acid. Zinc alone will dissolve as long as any of it remains in contact with the acid. In this way the percentage of indium increases in the undissolved portion. This latter is then brought into solution by the action of nitric acid.

Ammonia precipitates indium ion as the hydroxide, $In(OH)_3$. The precipitate is soluble in alkali hydroxide. Very important in connection with the separation of indium from accompanying ions is the precipitation of basic indium sulphite, $In_2(SO_3)_3 + In_2O_3 + 8H_2O$, by boiling a solution containing indium ion with sodium hydrogen sulphite.

Hydrogen sulphide precipitates yellow indium sulphide. With alkali sulphides the precipitate forms sulpho-salts.

[1] Zeitschrift für anorganische Chemie, 29, *24*, 1902.

[2] Journal für praktische Chemie, 34, *177*, 1886.

The blue violet color imparted to the bunsen flame by indium vapors is characteristic. In the spectrum this appears as two lines (compare the spectral table).

211. **Uranium** has a great diversity of combining ratios. Two classes of its compounds need be considered here, *uranous salts* in which it is *tetravalent*, and *uranyl compounds* containing hexavalent uranium. For analytical purposes the latter are more important.

The term " uranyl " compounds is due to the fact that the solutions of these salts contain an oxygen cathion, $UO_2^{\cdot\cdot}$. The term is selected from analogy with the ions bismuthyl, $BiO^{\cdot}$, and antimonyl, $SbO^{\cdot}$.

The corresponding hydroxide, $UO_2(OH)_2$ (and also $U(OH)_6 = UO_2(OH)_2 + 2H_2O$), accordingly acts as an acid, i. e. it forms the ions $UO_4'' + 2H^{\cdot}$. Thus alkali hydroxide and ammonia react with solutions of uranyl salts, for instance uranyl nitrate, $UO_2(NO_3)_2$, to form difficultly soluble uranates (for example of the composition $Na_2U_2O_7.6H_2O$). The relations between uranyl and uranate ions are represented qualitatively by the equation:

$$2UO_2^{\cdot\cdot} + 4OH' \rightleftarrows U_2O_7'' + 2H^{\cdot} + H_2O.$$

It is evident from the remarks already made in § 29 that the ratio between uranyl- ($UO_2^{\cdot\cdot}$) and uranate (U_2O_7'') ions preponderates in favor of the latter in alkaline solutions.

Of the *typical reactions* of *uranyl salts* the following deserve mention. Concentrated solutions of uranyl ion give *yellow precipitates* with solutions of *alkali-* and *ammonium carbonates*. These are *soluble* in an *excess* of *carbonate*. *Sodium phosphate* forms yellow white *uranyl hydrogen phosphate*, UO_2HPO_4. If ammonium and acetate ions are also present, uranyl ammonium phosphate, UO_2-NH_4PO_4, precipitates.

Hydrogen sulphide yields *no precipitate* of uranyl sulphide. *Ammonium sulphide,* however, precipitates brown uranyl sulphide, UO_2S. The precipitate dissolves in dilute acids and also in solutions which contain carbonate ion.

Hydrogen peroxide solution gives a *yellowish white* precipitate with concentrated *uranyl salt* solutions. Dilute solutions of uranyl salts are also precipitated with hydrogen peroxide by adding ammonia to neutralize the hydrogen ion formed by the reaction. The composition of the precipitate is not yet definitely known. It is claimed to be UO_4. The precipitate dissolves in ammonium carbonate and in dilute acids. With hydrogen peroxide the acid solutions give the same reaction as dichromate ion (§ 119).

For the separation of uranyl ion from third and fourth group ions the solubility of its difficultly soluble compounds in ammonium carbonate is of importance.

Uranyl compounds and uranates are yellow. By the action of reducing agents, as zinc, they are transformed into green salts, the solutions of which probably contain uranous ion.

212. Beryllium. Of this element only salts are known, the solutions of which contain divalent colorless ions. The properties of the ions $Be^{\cdot\cdot}$ and BeO_2'' resemble in many respects those of the ions of trivalent aluminum compounds, $Al^{\cdot\cdot\cdot}$ and AlO_3'''. Therefore the distinguishing characteristics of beryllium and aluminum ions must be particularly emphasized.

Alkaline solutions of *beryllium hydroxide* are *less stable* than corresponding solutions of *aluminum hydroxide.* This is shown by the fact that *beryllium hydroxide separates* from its alkaline solutions on long standing or by warming them.

The phenomenon that the hydroxide at first dissolves and then separates again has the following explanation. Beryllium hydroxide exists in two distinct states, each with its own properties (Principles, p. *546*). The soluble modification forms when precipitation takes place quickly, i. e. when the supersaturation with respect to beryllium hydroxide is high. It is, however, unstable as compared with the hydroxide slowly separated from an alkaline solution, and hence passes voluntarily into the latter.

Heat accelerates this alteration, which is also accompanied by a decrease of the total chemical activity of beryllium hydroxide.[1] This is shown by the fact that the reprecipitated hydroxide is less soluble in alkali- and ammonium carbonate solutions than freshly precipitated hydroxide (which has not been brought into solution and then reprecipitated). It has also lost the property of readily absorbing carbon dioxide. The alteration not only takes place in alkaline solutions as above described, but also on drying freshly precipitated hydroxide or by heating it under water, ammonia or ammonium carbonate solutions. The phenomenon of an alteration of aluminum hydroxide is also known, but it is *by no means* so pronounced as is the case with beryllium hydroxide.

A method devised by Havens[2] aims at a practically complete separation of aluminum and beryllium ions. It is based upon the fact that hydrated aluminum chloride is practically insoluble in a mixture of concentrated hydrochloric acid and ether. Beryllium chloride, on the other hand, is soluble in this mixture.

[1] F. Haber and G. van Ordt, Zeitschrift für anorganische Chemie, **38**, *377*, 1904.

[2] Zeitschrift für anorganische Chemie, **16**, *15*, 1898.

213. Zirconium is tetravalent in its analytically important compounds. The soluble salts contain colorless tetravalent zirconium ion. It is precipitated as *hydroxide* by *alkali hydroxide, ammonia* and *ammonium sulphide*.

The hydroxide dissolves in acids. It is insoluble in an excess of alkali hydroxide, or at any rate dissolves to only a slight extent, corresponding to its weak acid properties. The properties of the precipitated hydroxide vary more or less with the conditions under which it is formed, analogous with the behavior of beryllium hydroxide. When precipitated from hot solutions it is much more difficultly soluble in acids.

Alkali- and *ammonium carbonate* precipitate *basic carbonate*, easily soluble in an excess of the precipitant, especially in ammonium carbonate. Reprecipitation occurs on boiling this solution.

Potassium sulphate separates zirconium ion as *potassium zirconium sulphate*. The precipitate is difficultly soluble in a solution of potassium sulphate (by reason of the influence of sulphate ion, § 90). It dissolves to a noticeable extent in acids. Sodium sulphate yields no precipitate with zirconium ion.

Oxalic acid forms a white precipitate of *zirconium oxalate*. It is much more soluble in hydrochloric acid, and particularly in ammonium oxalate, than in water, but is *not* more soluble in *oxalic acid*.

Hence *zirconium oxalate* is not precipitated by strong acids from an *ammonium oxalate* solution (distinction from thorium oxalate).

Hydrogen peroxide gives a *white precipitate* with zirconium ion. When this precipitate is washed and dissolved in dilute sulphuric acid, the solution gives the same reaction with hydrogen peroxide as $Cr_2O_7'' + 2H^{\cdot}$ (compare § 119).

A mixture of hydroxide and sulphur precipitates on boiling a zirconium salt solution with *thiosulphate*. (Yttrium and didymium salts do not give this reaction.)

Turmeric paper, moistened with the hydrochloric acid solution of a zirconium salt, assumes a bright red brown color when dried on a water-bath. Titanic acid interferes with this test, but the disturbing influence may be destroyed by reducing the titanic acid with zinc.

According to Delafontaine[1] zirconium compounds are separated from thorium compounds by fusion with potassium hydrogen flu-

[1] Marc Delafontaine, Chemical News, 75, *230*, 1897.

oride, KHF_2, and extraction of the fused mass with water. Potassium zirconium fluoride, K_2ZrF_6, goes into solution.

214. Thorium. The reactions of thorium ion resemble those of the ions of the allied metals, zirconium and yttrium.

The *hydroxide* is practically insoluble in water and in alkali hydroxide, but dissolves in dilute acids. At higher temperatures it loses water and becomes less soluble.

Thorium carbonate dissolves in alkali carbonate and is still more readily soluble in ammonium carbonate.

Thorium ion yields a difficultly soluble double salt with *potassium sulphate*, but gives no insoluble compound with sodium sulphate.

Thorium oxalate differs from zirconium oxalate in the fact that it is precipitated by acids from its solution in ammonium- or alkali oxalate. This is probably due to a decrease in the concentration of oxalate ion by the action of hydrogen ion, which gives rise to undissociated oxalic acid. The oxalate ion served to keep thorium oxalate in solution as a complex ion. Corresponding to its proportionately lower dissociation acetic acid is less active in this respect than the stronger acids.

The precipitation of thorium ion by boiling[1] a thorium solution one to two minutes with a solution of potassium azoimide, N_3K (Principles, p. *346*), is very important in connection with the separation of thorium from the ions of other rare elements. The reaction resembles that of the hydrolysis of ferric acetate, and is expressed by the equation:

$$Th^{\cdot\cdot\cdot\cdot} + 4N_3' + 4H_2O = \underline{Th(OH)_4} + 4N_3H.$$

Cerium, lanthanum and didymium ions give precipitates with this reagent on longer boiling, while zirconium and yttrium ions are quantitatively precipitated in the cold.[2]

When heated with *thiosulphate,* thorium ion behaves like zirconium ion. It differs from the latter, however, by forming a white precipitate with fluoride ion (§ 213).

Thorium salt solutions yield a white precipitate with *hydrogen peroxide.* A solution of this precipitate in dilute sulphuric acid gives the hydrogen peroxide reaction (§ 119).

[1] Dennis and Cortright, Zeitschrift für anorganische Chemie, 6, *35*, 1894; also 13, *412*, 1897 (Dennis).

[2] Curtius and Darapsky, Journal für praktische Chemie, 61, *412*, 1901.

These relations indicate that here and also in the cases mentioned in §§ 211 and 213, peroxides do not come in question in the sense of peroxides like those of lead or manganese, but that the precipitates are difficultly soluble salts of hydrogen peroxide.

For those who desire more detailed information concerning thorium and the related elements, the following references to the literature will be found very serviceable.

On the separation of thorium and cerium salts consult E. Hintz and H. Weber, Zeitschr. f. analyt. Chem., 36, *676*, 1897; also E. Hintz, Zeitschr. f. analyt. Chem., 37, *94*, 1898, "Investigation of commercial incandescent mantles."

On the precipitation of thorium and zirconium ions as basic salts of organic acids, see Haber, Monatshefte für Chemie, 18, *687*, 1897.

On the preparation of pure thorium compounds see Muthmann and Böhm, Berichte d. d. chem. Gesellschaft, 33, *42*, 1900; also Muthmann and Baur, loc. cit., *2028*.

On the analysis of monazite consult Drossbach, Zeitschr. f. angewandte Chemie, 14, *655*, 1901.

[Consult further R. Böhm, Die Darstellung der seltenen Erden, 2 vols., 1905.

A complete list of the papers that have appeared concerning cerite and yttria earths and thorium is contained in the "Bibliographie der seltenen Erden," by R. J. Meyer, 1905, Zeitschr. f. anorganische Chemie.]

215. Yttrium. The properties of yttrium compounds resemble those of the compounds of thorium and zirconium. Consequently only the distinguishing properties need be mentioned here.

With alkali hydroxide and *tartrate ion* yttrium ion yields a precipitate of *yttrium tartrate*. Thorium and zirconium ions are not precipitated by alkali hydroxide in the presence of tartrate ion.

The double salt of yttrium with potassium sulphate is more readily soluble in a solution of alkali sulphate than in water. In hydrochloric acid the oxalate is scarcely soluble. It is a little soluble in ammonium oxalate. Fluoride ion also yields a precipitate with yttrium ion (compare §§ 213 and 214).

Yttrium compounds are very often accompanied by the compounds of other elements, the separation of which has not yet been satisfactorily carried out. Collectively these are termed *erbium compounds*.

The reactions of the soluble erbium compounds resemble very closely those of yttrium ion. They may be distinguished by the fact that these solutions absorb certain definite light rays. The precipitate of yttrium oxalate is colored light red in case it contains erbium oxalate.

A method proposed by Muthmann and Böhm[1] to separate yttrium ion from the ions of accompanying elements, is based upon the fractional precipitation of the chromates.

[1] Ber. d. d. chem. Gesellschaft, 33, *42*, 1900.

According to Urbain[1] a separation of the yttria earths may be based upon the solubility relations of the ethyl sulphates.

216. Cerium forms a colorless cerous ion and an orange red ceric ion, corresponding to its tri- and tetravalent states respectively. These have the following relation to one another. In neutral or acid solutions cerous ion is more stable than ceric ion, while the reverse is the case in alkaline solutions.

Accordingly *cerous ion* is only a weak *reducing agent* in *neutral* or *acid* solutions. It may be oxidized, however, by strong oxidizing agents, as $MnO_4' + 8H^{\cdot}$. Ceric ion is correspondingly an excellent oxidizing agent (for example towards sulphur dioxide and hydrochloric acid). In alkaline solutions the relations are different, for *cerous hydroxide* goes over into *ceric hydroxide* by the action of *atmospheric oxygen* under these conditions.

Alkali hydroxide, ammonia and *ammonium sulphide* precipitate cerous ion as white hydroxide, easily soluble in acids, practically insoluble in alkali hydroxide.

Alkali- and *ammonium carbonate* precipitate white *carbonate*. The precipitate is scarcely soluble in an excess of ammonium carbonate (distinction from beryllium-, thorium- and zirconium carbonates).

No cerous hydroxide is precipitated by heating a solution of *cerous salt* with *sodium thiosulphate*. This behavior serves to distinguish cerium from thorium salts,[2] and a method of separating these ions is based upon it. The fact that *cerous oxalate* is very much less soluble than thorium oxalate in a solution of *ammonium oxalate*, affords another method of separation.

With *potassium sulphate* cerous ion yields a difficultly soluble salt of the composition $K_3Ce(SO_4)_3$. The salt is practically insoluble in an excess of potassium sulphate.

The behavior of cerous ion towards *hydrogen peroxide* is especially characteristic on account of the great delicacy of this reaction. Neutral solutions (or acid solutions containing acetate ion) give with hydrogen peroxide a reddish yellow coloration or a precipitate having the color of ferric hydroxide. The precipitation is made more complete by the addition of ammonia. In place of hydrogen

[1] Comptes rendus, **126**, *835*, 1898.

[2] E. Hintz and H. Weber, Zeitschr. f. analyt. Chem., **36**, *676*, 1897; also **37**, *94*, 1898 (Hintz). Compare § 212.

peroxide a solution of sodium peroxide in ice water may also be used.[1] The brown precipitate dissolves in dilute acids, and this solution also gives the hydrogen peroxide reaction (compare § 119).

Cerous ion is transformed into ceric ion by boiling the solution with nitric acid and lead peroxide. The color changes to yellow (distinction from the ions of the related elements, lanthanum and didymium).

In the oxidizing flame cerium salts impart a red color to the phosphor salt bead.

217. **Lanthanum** forms a colorless trivalent ion. It differs from cerous ion in being non-oxidizable, but in other respects resembles it closely.

In addition to its stability towards oxidizing agents the solubility relations of the sulphate and the behavior of basic lanthanum acetate towards iodine are characteristic. The solubility of lanthanum sulphate decreases with rising temperature. At 100° C. it is ten times less soluble than at 0° C. (according to *Muthmann* and *Rölig*). Hence a large portion of the dissolved salt will separate on evaporating the solution. 100 parts of water dissolve 3 g. at 0° C., and at 100° C. 0.3 g. $La_2(SO_4)_3$.

Basic lanthanum acetate is obtained by precipitating with ammonia a lanthanum salt solution which contains acetate ion. It gives a reddish blue coloration, resembling that imparted to starch paste by iodine, when small pieces of solid iodine are strewn over the surface. The color may be seen still more distinctly by bringing a drop of an iodine solution in contact with the surface of the lanthanum solution. The reaction is, however, not specially delicate.[2]

The insolubility of ammonium lanthanum nitrate, $La(NO_3)_2 + 2NH_4NO_3 + 4H_2O$ is of considerable importance in connection with the separation of lanthanum ion from the ions of the related metals.[3]

With *hydrogen peroxide* lanthanum ion yields a white amorphous precipitate on addition of ammonia or alkali hydroxide. This has the composition La_2O_5, according to *Melikoff* and *Pissarjewsky*.[4] Its solution in dilute sulphuric acid gives the hydrogen peroxide reaction.

[1] Mengel, Zeitschr. f. anorganische Chemie, 19, *67*, 1898.

[2] H. Behrens, Archiv. Neerland sc. exact et nat. (*2*), 6, *67*, 1902. Chem. Centralbl., 1902, I, *296*. This reaction has also been closely investigated by W. Biltz and I. Behre, Ber. d. d. chem. Gesellschaft, 37, *719*, 1904.

[3] v. Scheele, Ber. d. d. chem. Gesellschaft, 32, *409*, 1899.

[4] Zeitschr. f. anorganische Chemie, 21, *70*, 1899.

Muthmann and *Rölig*[1] base a method of separating lanthanum ion from the ions of praseodymium and neodymium upon the precipitation of the two latter from a neutral solution by means of magnesia.

218. **Praseodymium** and **neodymium** were regarded as a single element, *didymium,* for a long time. They resemble cerium and lanthanum closely. Their soluble salts have characteristic absorption spectra,[2] by means of which they may be identified. Neodymium shows bands in yellow and green, praseodymium in blue and violet. For the separation of the ions of these elements consult the papers of Auer v. Welsbach,[3] Schottländer (see § 219) and v. Scheele (§ 217).

219. **Methods of separating the ions of the cerite metals.**[4] Auer v. Welsbach[5] has devised a method of separating cerium salts from those of lanthanum and didymium.[6] If the oxides obtained from the oxalates are heated in air, the oxide of cerium goes over into the dioxide. The separation is based upon the weak basic properties of cerium dioxide. When a mixture of these oxides is treated with concentrated nitric acid (1 : 1), cerium dioxide forms basic ceric nitrate, scarcely soluble, while lanthanum and didymium oxides go into solution. To prepare the pure cerium salt the basic nitrate is dissolved in concentrated nitric acid, and a concentrated solution of ammonium nitrate is added to this solution. Cerium ammonium nitrate, $(NH_4)_2Ce(NO_3)_6 + xH_2O$ separates in purple red crystals on evaporating the solution.

The filtrate from the cerium precipitation contains chiefly ions of lanthanum and didymium. By adding ammonium oxalate a precipitate of the oxalates is obtained. These are then transformed into the oxides by heating them to redness. One half of the oxides is dissolved in dilute nitric acid and into this solution the other half is stirred. The mixture of oxides reacts with the nitrate solution, producing an increase of the lanthanum ion in the solution and an increase of the basic didymium nitrate in the precipitate. The color

[1] Ber. d. d. chem. Gesellschaft 31, *1721*, 1899.

[2] Brauner, Proc. of the Royal Soc., London, 1897–98, *70*.

[3] Monatshefte f. Chemie, 6, *477*, 1896.

[[4] For general information on the methods of separation consult R. Böhm, Die Darstellung der seltenen Erden.]

[5] Wiener Monatshefte, 5, *508*, 1884.

[6] The term "didymium" always refers to the two components, neodymium and praseodymium (compare § 218).

of the thin paste goes over simultaneously from a deep brown to a dirty pale reddish tint. Repetition of this same process with the liquid and solid phases effects a separation of the compounds of these elements that suffices most purposes.

Schottländer[1] gives valuable details concerning the carrying out of the separation.

O. N. Witt and *W. Theel*[2] have adopted v. Knorres'[3] recommendation to use ammonium persulphate as an oxidizing agent for separating cerous ion from praseodymium. The latter is precipitated with cerous nitrate by the method given by v. Scheele (§ 217). The nitrate solution of these metals is heated with ammonium persulphate, the resulting acid being neutralized with sodium carbonate solution. Ceric ion is formed and precipitates under these conditions as basic sulphate or nitrate.

According to Meyer and Marckwald[4] basic ceric sulphate is dissolved in ten times its weight of concentrated nitric acid (specific gravity 1.4). With this is mixed one and a half times as much concentrated ammonium nitrate solution. Cerium ammonium nitrate crystallizes out, as has been already mentioned. The precipitate is purified, if necessary, by crystallization from concentrated nitric acid.

220. Niobium. The compounds of this element vary considerably in their properties from compounds of the elements phosphorus, arsenic and antimony, which are, nevertheless, related to niobium in certain respects.

Niobium pentoxide is also called niobic acid in accordance with its more pronounced acid characteristics. Its solubility relations are complicated and are not yet satisfactorily explained. The longer the acid is exposed to high temperatures the less reactive it becomes.

The different means of transforming niobic pentoxide into soluble compounds are the following, arranged in decreasing order of effectiveness: fusion with potassium hydroxide, potassium carbonate or potassium hydrosulphate, heating with concentrated sulphuric acid.

Niobic acid separates from acid solutions on diluting and boiling them, and from alkaline solutions upon acidification. When freshly

[1] Ber. d. d. chem. Gesellschaft, 25, *378*, 1897.
[2] Ber. d. d. chem. Gesellschaft, 33, *1315*, 1900.
[3] Zeitschr. f. angewandte Chemie, 10, *719*, 1897.
[4] Ber. d. d. chem. Gesellschaft, 33, *3003*, 1900.

precipitated it dissolves in strong acids and in potassium hydroxide. Sodium hydroxide does not have a solvent action upon it, owing to the fact that the sodium salts of polyniobic acids are difficultly soluble in water. The general formula of these acids is:

$$mH_3NbO_4 - nH_2O \text{ (where } n < \tfrac{3}{2}m).$$

Examples are the acids $H_8Nb_6O_{19}$, $H_6Nb_4O_{13}$, $H_4Nb_2O_7$, etc.

Niobic acid is more readily soluble in hydrofluoric than in hydrochloric acid. The fluoride forms a readily soluble double salt with potassium fluoride of the composition $2KF.NbF_5$, and also $2KF.NbOF_3$. *The corresponding salts of tantalum are much more difficultly soluble.* Hence a separation of the compounds of these two elements can be effected by this means.

Another difference between the properties of niobic acid and those of the closely related tantalic acid (§ 221) consists in the fact that acid solutions of niobic acid yield deep blue and brown colorations upon reduction, and give corresponding precipitates on standing. Solutions of tantalic acid do not change color when reduced with zinc.

Niobic acid dissolves in the phosphor salt bead and imparts in the reducing flame a violet, blue or brown color, according to the quantity present. By adding ferrous sulphate the color is changed to red.

The lower oxidation forms, NbO and NbO_2, have but little analytical importance.

221. Tantalum. The compounds of tantalum are very similar to those of niobium. Hence a reference to the preceding paragraph suffices for the general description, and only the more striking distinctions need be mentioned here.

Tantalic pentoxide is still less soluble than the corresponding niobic pentoxide. Hence calcined or natural tantalic acid is most readily brought into solution by fusion with potassium hydroxide. Sodium hydroxide may also be used for this purpose, but is less satisfactory, since sodium tantalate is difficultly soluble in water.

The volatility of tantalic acid when heated with ammonium fluoride affords another means of removing tantalum pentoxide from a given mixture.

Precipitated tantalic acid dissolves in potassium hydroxide, hydrochloric acid, hydrofluoric acid and concentrated sulphuric acid. It

is only very sparingly soluble in dilute sulphuric acid. Thus tantalates are decomposed by dilute sulphuric acid with separation of sparingly soluble tantalum sulphate and tantalic acid.

Niobic and tantalic acids are precipitated by ammonia, even from a fluoride solution. Hence the third group precipitate must also be tested for these compounds.

A solution of tantalic acid in hydrofluoric acid yields with potassium fluoride a precipitate of the composition $2KF.TaF_5$. A basic salt separates on heating. *The corresponding fluoride of niobium is much more readily soluble.*

An acidified solution of a tantalate gives a yellow precipitate with tannic acid, while a niobate yields an orange red precipitate under similar conditions. Tantalic acid is readily soluble in the phosphor salt bead, but imparts no characteristic color to it.

222. **Titanium.** Compounds of this element are rather widely distributed in nature, but are found usually in small quantities. It occurs particularly as titanium dioxide, TiO_2, corresponding to the tetravalent oxidation form. The lower and higher forms are of lesser importance and hence need only be mentioned.

A knowledge of the solubility *relations* of the dioxide and the corresponding hydrates, and their behavior towards different substances used to bring them into solution, is of special analytical importance. Natural compounds which contain the dioxide, or the calcined dioxide, are transposed by long continued fusion with sodium carbonate or potassium hydrosulphate into compounds which dissolve either in water or in hydrochloric acid. Concentrated sulphuric acid and hydrofluoric acid are also very active as solvents. Since titanium fluoride is volatile, sulphuric acid is added to hydrofluoric acid in the latter case to limit the loss of titanium compounds.

The complicated solubility relations of the hydrates of titanic acid are not very satisfactorily explained as yet. To the effects of acids and bases, caused by the amphoteric character, are added also the characteristics of colloids (§ 77) and the variation in the properties of the hydrates on long standing or by warming (§ 212). The freshly precipitated and washed hydrate (precipitated, for example, from acid solutions by alkali hydroxide, ammonia or ammonium sulphide) is soluble in very dilute acids. Corresponding to its weak acid nature the precipitated *acid* is but little soluble in an excess of alkali hydroxide.

Solutions which contain titanium as a cathion, for instance, those obtained by dissolving freshly precipitated hydroxide in acids, are

rather unstable. On diluting such a solution with water and boiling, a precipitate of titanic acid forms. This separation is made complete by adding sodium acetate, as a consequence of the reaction mentioned in § 38, (*1*). Since the precipitation takes place from a hot solution a less soluble form of the acid results. The wash water must be weakly acidified to avoid dissolving the precipitate during the washing operation.

Solutions of titanium salts are precipitated by *sodium phosphate* and by *sodium thiosulphate* as basic titanium phosphate and titanic acid respectively.

With a solution of *titanic acid* in *hydrofluoric acid, potassium fluoride* forms a salt of the composition K_2TiF_6, analogous with a fluosilicate. The salt is rather difficultly soluble.

In its properties *titanic acid* is quite closely related to *niobic, tantalic* and *silicic acids*. Hence it is important to know the properties which permit these acids to be separated or distinguished.

After fusion with potassium hydrosulphate and extraction of the fused mass with cold water, tantalic acid remains undissolved. The qualitative test for niobic acid takes advantage of the solubility of potassium niobate, which is obtained by fusing niobic acid with potassium hydroxide in a silver crucible. Potassium titanate is much less soluble.

Titanic acid is distinguished from *silicic acid* by the volatility of silicon tetrafluoride, formed on heating with hydrofluoric and sulphuric acids. Titanic acid remains behind (compare the second section of § 222).

Zinc and *tin* reduce acidified solutions of titanium salts, giving rise to violet and blue colorations respectively. These colors are characteristic of trivalent titanium salts. Alkali hydroxide precipitates from the colored solution a corresponding blue hydroxide. In contact with air, however, it is oxidized to titanic acid.

A *particularly delicate test* for *titanium salts* is afforded by adding *hydrogen peroxide* to their slightly acidified solutions. A yellow to orange coloration results, and on further addition of alkali hydroxide, ammonia or ammonium carbonate, a correspondingly colored precipitate forms, which goes into solution again on *increasing* the *amounts* of the last named substances.[1]

[1] Classen, Method of separating titanium and iron salts, Ber. d. d. chem. Gesellschaft, 21, *370*, 1888.

In the reducing flame *solid titanium compounds* impart a yellow color to the phosphor salt bead. The color passes over to violet in the cold. By dissolving some ferrous sulphate in the bead the color becomes brown or red.

223. **Gallium.** This very rare element forms di- and trivalent compounds. The latter are more important.

In its properties gallium ion resembles aluminum ion. The hydroxide dissolves in an excess of alkali hydroxide or ammonia, and also in acids. On account of the latter property a dilute solution of ammonium acetate, acidified with acetic acid, is used to precipitate gallium ion completely. This is effected if the concentration of acetate is not made too high.

Hydrogen sulphide precipitates white gallium sulphide in the presence of acetate ion. It is likewise precipitated by ammonium sulphide. Zinc does not reduce gallium ion in acid solutions. This property is utilized to separate gallium from other cathions which are precipitated by this means. A test for gallium ion which is particularly characteristic and delicate is its behavior towards *alkali ferrocyanides.* With these it gives a precipitate from *neutral* or *acid* solutions.

At very high temperatures the chloride emits a characteristic violet light. To carry out the experiment an electric arc is formed between two carbon electrodes which are dipped in a solution of the chloride.

GROUP IV.

224. **Tungsten.** The most important and the most stable combining form of this element is the one from which the trioxide is derived. Compounds corresponding to the dioxide have very little importance in analysis, since hydrated forms of the trioxide, tungstic acids, are always obtained on working up the natural compounds.

Tungstic acids of the composition H_4WoO_5 (white) and H_2WoO_4 (yellow) have amphoteric characteristics (§ 42). Their acid properties are more pronounced than the basic. Thus they dissolve readily in alkali hydroxide and ammonia. In strong acids they dissolve to a slight extent and in weak acids are practically insoluble. Hence weak acids are good precipitants of tungstic acid from solutions of tungstates.

In addition to the salts of these two acids there are also the so-called metatungstates.[1] They are formed by boiling solutions of tungstates with the

[1] The prefix "meta" is used in chemistry to denote rather variable characteristics. In inorganic chemistry it often serves to distinguish certain compounds from others, generally designated as ortho-compounds, in cases where a *distinctly* perceptible *difference* in properties accompanies *slight differences* in *composition* (usually a smaller content of water).

trioxide. Their solutions are characterized by a greater stability towards acids. They separate tungstic acid only on long continued boiling.

It is to be noted that tungstic acid easily forms colloidal solutions unless a little acid is added to the water with which it is in contact.

The trioxide is a yellow powder formed from the two hydrates at higher temperatures. It has solubility relations similar to those of the hydrates, but is rather less soluble. Fusion with alkali carbonate changes the trioxide into soluble tungstate:

$$WoO_3 + Na_2CO_3 = Na_2WoO_4 + CO_2.$$

Similarly a tungstate is formed by fusing it with potassium hydrosulphate, sulphur trioxide being simultaneously evolved.

On extracting the fused mass with water undecomposed sulphate dissolves first. The tungstate is less soluble in a sulphate solution, but dissolves in water after removal of the sulphate. Under similar treatment titanic acid goès at once into solution.

The solubility of tungstic acids in ammonia distinguishes them from silicic acids.

The following reactions are characteristic for tungstate ions.

Ferrous sulphate separates a brown precipitate of dioxide. The behavior of tungsten dioxide towards acids distinguishes tungstates from molybdates. Tungsten dioxide is *not* turned blue by acids.

Stannous chloride gives a yellow precipitate with the solution of a tungstate. It becomes blue on adding hydrochloric or sulphuric acid and warming. The reaction is a very delicate test for tungsten, in case other acids which give a blue coloration when reduced, are not present.

The same phenomenon is also obtained by using other reducing agents, as zinc or tin + acid. The blue compound corresponds to pentavalent tungsten. It is *transformed* into a brown tetravalent compound on long continued action of a reducing agent.

Tungsten compounds are not precipitated in the fourth group in the ordinary course of analysis, owing to the formation of a sulphotungstate ion, WoS_4. From solutions of sulphotungstates, however, tungsten trisulphide is precipitated by acidification with hydrochloric acid (i. e. in the filtrate from the fourth group). It is probably owing to the low concentrations of tungsten ion ($Wo^{\cdot\cdot\cdot\cdot\cdot\cdot}$) in solutions acidulated with strong acids that tungsten trisulphide is not precipitated by hydrogen sulphide. This behavior accords with the weaker basic character of tungstic acid (see above).

In the reducing flame tungsten compounds color the phosphor salt bead blue. The color becomes blood red upon addition of ferrous salts.

225. Vanadium. The compounds of this element are analogous to those of the pentavalent elements phosphorus and arsenic. As is the case with these latter elements, five different combining forms are also known for vanadium. For the most part only the highest forms, the pentoxide and the different vanadic acids derived from it, need be considered here.

The properties of vanadic pentoxide vary with the different methods of preparing it. Thus the color may be brown to red. It is rather difficultly soluble in water, but dissolves both in acids and alkalies. In acid solution, however, the compounds derived from the pentoxide are less stable. This is shown by an evolution of chlorine on dissolving the pentoxide in hydrochloric acid.

Vanadate solutions may be prepared by dissolving the pentoxide in alkali hydroxide. The dissociation relations of these solutions are quite complicated, in agreement with their behavior. The relations existing between chromate ion and dichromate ion, mentioned in § 98, are repeated in this case, only with greater complications. Accordingly the properties of vanadate solutions depend expressly upon the reaction towards litmus, i. e. upon the concentration of hydrogen- or hydroxyl ion. Metavanadate ion, VO_3',[1] is stable in neutral solution, that is, it exists in comparatively high concentrations in such a solution. Addition of alkali hydroxide transposes it to the colorless pyrovanadate ion, V_2O_7'''', while acids change it into condensed vanadate ions, V_6O_{17}'''' or HV_6O_{17}'''. These transformations are indicated bv the following equations, which bring forth the effect produced upon the transposition of the different ions by adding bases or acids.

$$2VO_3' + 2OH' \rightleftarrows V_2O_7'''' + H_2O \qquad (1),$$

$$6VO_3' + 2H^{\cdot} \rightleftarrows V_6O_{17}'''' + H_2O \qquad (2),$$

and

$$6VO_3' + 3H^{\cdot} \rightleftarrows HV_6O_{17}''' + H_2O \qquad (3).$$

The following are the most important distinguishing tests for vanadium compounds. With *ammonium ion* a soluble vanadate yields difficultly soluble *ammonium vanadate*. As a rule solid ammonium chloride is added to the vanadate solution in carrying out the experiment.

[1] Metavanadate ion has not the simple formula ascribed to it here. From the experiments of Düllberg, Zeitschr. f. physikalische Chemie, 45, *129*, 1903, it is probable that metavanadate solutions contain an ion of the composition VO_3'. For the sake of simplicity, however, these relations are left out of the above equations.

In the presence of hydrogen ion vanadate solutions yield a characteristic red brown color with *hydrogen peroxide.* The color vanishes on adding an excess of peroxide.

Reducing agents, as zinc, hydrogen sulphide, sulphurous acid and oxalic acid, reduce acidified solutions to compounds of the tetra-, tri- and divalent forms.[1] These have respectively green, blue and violet colors.

Hydrogen sulphide does not precipitate a sulphur compound. Addition of ammonium sulphide gives rise to a sulpho-ion, from which the pentasulphide separates when the solution is acidified. Nitric acid oxidizes the pentasulphide to vanadic acid.

Vanadium compounds impart a green color to the phosphor salt bead in the reducing flame.

GROUP VI.

226. **The ions of the sixth group.** The ions which have to be mentioned here agree with the ions of the alkali metals in forming salts which are for the most part easily soluble in water. Accordingly their compounds occur to very slight extent as solids in nature. They are found in limited quantities in certain mineral springs and in salt deposits.

227. **Lithium** forms only a monovalent colorless ion, which resembles the ions of the alkali earth metals in certain properties. Thus it yields a difficultly soluble *carbonate* and *phosphate.*

The carbonate is precipitated when a solution of lithium chloride is mixed with a solution of alkali carbonate. Heating the solution favors the precipitation. The solubility of the carbonate decreases as the temperature is increased. At 100° C. only half as much dissolves as at 10° C.

The solubility of lithium carbonate is influenced by alkali salts (chloride, nitrate, sulphate, acetate, hydroxide) and especially by ammonium salts. When these are present it is much more readily soluble.[2] Hence the insolubility of the carbonate does not afford a satisfactory test for lithium ion, for its precipitation by alkali carbonate is less complete and therefore the reaction is a less delicate test if one of these salts is present.

A far better test is the precipitation as difficultly soluble *lithium phosphate,* Li_3PO_4. To neutralize the hydrogen ion formed on pre-

[1] For the reactions of vanadous compounds consult A. Piccini and L. Marino, Zeitschr. f. anorganische Chemie, 32, *68*, 1902.

[2] G. Geffcken, Zeitschr. f. anorganische Chemie, 43, *197*, 1905.

cipitating with disodium hydrogen phosphate a little ammonia is added until the solution has a slight alkaline reaction. Otherwise the precipitation would be incomplete, as lithium phosphate is soluble in acids.

Another characteristic test for lithium salts is the *carmine red light* emitted by heating them to redness.

The prism mentioned in § 106 is used to identify lithium in the presence of potassium and sodium salts. The light emitted from sodium compounds is absorbed by very thin layers of indigo solution. From potassium salts the light appears sky blue, violet and finally carmine red through very thick layers of the indigo solution. Lithium salts emit carmine red light. It is distinguished from potassium, however, by the fact that the intensity significantly decreases from the thinner to the thicker end of the prism.

The solubility of lithium chloride in a mixture of alcohol and ether is used to *separate lithium salts* from the *alkali* salts. Alkali chlorides are practically insoluble in this mixture. Salts other than the chlorides are transposed by evaporating them to dryness with concentrated hydrochloric acid.

To transform the sulphate into the chloride, however, another method must be used, since sulphuric acid is much less volatile than hydrochloric acid. In consequence sulphate would always be left behind on heating with hydrochloric acid.

The sulphate ion is precipitated with lead acetate. From the filtrate (which contains the ions $Li^{\cdot}$, $C_2H_3O_2'$ and $Pb^{\cdot\cdot}$) lead ion is removed by means of hydrogen sulphide. Then the acetate solution is evaporated with hydrochloric acid.

To separate lithium- from potassium ion the latter is precipitated with chloroplatinic acid. In the filtrate the excess of chloroplatinic acid is reduced by heating the solution with oxalic acid. The latter is then destroyed by evaporating the solution to dryness and heating the evaporation residue to redness. Finally the residue is extracted with very dilute hydrochloric acid. With this solution the lithium test is made.

Magnesium ion may be removed according to § 101. Likewise the barium ion, which is introduced into the solution with baryta, may also be precipitated as carbonate, since lithium carbonate does not precipitate when an excess of carbonate ion is present.

228. Rubidium and caesium. The compounds of these metals are very similar to those of potassium. This is shown, for example, by the fact that their ions are monovalent and colorless and are precipitated by the anions $PtCl_6''$, $HC_4H_4O_6'$ and ClO_4', which yield

difficultly soluble potassium compounds. The following are the quantitative differences in the solubility of these compounds. The solubility products decrease for the chloroplatinates and increase for the hydrogen tartrates in the order of increasing combining weights: K, **Rb**, Cs.

The double salts of rubidium and caesium analogous with potassium aluminum sulphate show a still more pronounced decrease of solubility than the chloroplatinates. Hence by fractional crystallization of a mixture of these alums the individual salts can be obtained approximately pure. The greater solubility of the double sulphates as compared with the solubility of the chloroplatinates, however, limits the application of this method in cases where it is desired to separate caesium ion as *completely* as possible.

The insolubility of a double salt of caesium chloride with antimony trichloride, $SbCl_3.6CsCl$, in concentrated hydrochloric acid[1] affords a separation of caesium- and rubidium ions. The method has the further advantage that it is easy to obtain the caesium chloride from the double salt owing to its decomposition by water. Antimony chloride is hydrolyzed and separates antimonic acid. Anv antimonous ion left in the solution is then precipitated by hydrogen sulphide.

A further distinction is afforded by the behavior of the carbonates towards alcohol. Caesium carbonate dissolves in alcohol, while rubidium carbonate is practically insoluble.

The spectroscope (§ 107) is used in qualitative tests for these elements, since their volatile compounds emit light of very characteristic color. The rubidium flame has several characteristic bands, two in red and two in violet, while the caesium flame has two bands in light blue (compare the spectral table).

[1] The insolubility of caesium antimony chloride is also sometimes used to detect very small quantities of antimony ion: consult Denigès, Comptes rendus, 133, *688*, 1901.

ANALYTICAL TABLES.

TABLE I. TESTS FOR CATHIONS.

GROUP I. Reagent: **Dilute hydrochloric acid** (§ 5).

The precipitate may consist of: AgCl, HgCl, $PbCl_2$ (compare also § 18). Upon extraction with hot water (§ 7):

<table>
<tr>
<td rowspan="2">**$PbCl_2$** goes into solution: The solution yields a white precipitate of **$PbSO_4$** with dilute *sulphuric acid* or an *alkali sulphate* solution (SO_4''). For the reactions of $Pb^{\cdot\cdot}$ consult § 14.</td>
<td colspan="2">**AgCl** and **HgCl** are practically insoluble in hot water. *This residue is treated with ammonia* (§ 7):</td>
</tr>
<tr>
<td>**AgCl** goes into solution. Upon *acidification* this solution yields a white precipitate of **AgCl**. For the reactions of $Ag^{\cdot}$ consult § 11.</td>
<td>In case the residue is turned black with ammonia, it contains **HgCl**. For the reactions of $Hg^{\cdot}$ consult § 17.</td>
</tr>
</table>

GROUP II. Reagent: **Hydrogen sulphide** (§ 19).

The precipitate may consist of (§ 20):

A: (copper sub-group) HgS, PbS, Bi_2S_3, CuS, CdS:

B: (tin sub-group) As_2S_3, (As_2S_5), Sb_2S_3, Sb_2S_5, SnS, SnS_2.

1. Separation of the Sulphides of the Copper Sub-Group, *A* (§ 21).

The precipitate is first washed with hydrogen sulphide solution and is then heated with dilute nitric acid:

HgS remains undissolved.[1] The residue is heated with a mixture of concentrated hydrochloric and nitric acids —about 3 vols. HCl to 1 vol. HNO_3 (aqua regia). HgS goes into solution as $\mathbf{HgCl_2}$. With *stannous chloride* the solution forms a *white* precipitate of **HgCl**, becoming finally *gray* **Hg.** Aqua regia must previously be expelled. For the reactions of $Hg^{\cdot\cdot}$ consult § 33.

PbS, $\mathbf{Bi_2S_3}$**, CuS** and **CdS** go into solution. The solution is tested for $\mathbf{Pb^{\cdot\cdot}}$ with dilute sulphuric acid (§ 22). Nitric acid is expelled and $\mathbf{Pb^{\cdot\cdot}}$ precipitated as $\mathbf{PbSO_4}$, which is filtered. The *filtrate* may contain $\mathbf{Bi^{\cdot\cdot\cdot}}$, $\mathbf{Cu^{\cdot\cdot}}$ and $\mathbf{Cd^{\cdot\cdot}}$. *Ammonia* is added till the *filtrate* is *alkaline.*

$\mathbf{Bi(OH)_3}$ precipitates. The precipitate is filtered, then dissolved in *as little* dilute nitric acid *as possible.* This solution is neutralized by adding ammonia drop by drop until the precipitate at first formed barely disappears on shaking up the solution. Then a solution of potassium or sodium chloride is added, precipitating **BiOCl.** For the reactions of $Bi^{\cdot\cdot\cdot}$ consult § 37.

$\mathbf{Cu^{\cdot\cdot}}$ and $\mathbf{Cd^{\cdot\cdot}}$ remain in solution owing to the formation of *complex ammonia cathions,* § 13. The formation of a deep blue color indicates the presence of $\mathbf{Cu^{\cdot\cdot}}$ in the nitrate solution. For the reactions of *cupric ion* consult § 40.

In case $Cu^{\cdot\cdot}$ is absent, the colorless solution is weakly acidified and then saturated with *hydrogen sulphide* to test for $\mathbf{Cd^{\cdot\cdot}}$ (§ 23). A yellow *precipitate* of **CdS** forms. For the reactions of $Cd^{\cdot\cdot}$ consult § 41.

If the ammoniacal solution contains copper ammonia cathions, *potassium cyanide* is added until it is *decolorized* (§ 23). The colorless solution is then saturated with hydrogen sulphide. Sometimes a *dirty yellow* precipitate forms. This indicates that the separation of $Pb^{\cdot\cdot}$ or $Bi^{\cdot\cdot\cdot}$ as black sulphides has been incomplete. In this case the precipitate is redissolved and the separation is carried out again.

2. To Test for the Sulphides of the Tin Sub-Group, *B* (§ 24).

A portion of the original sulphide precipitate is digested with *ammonium-* or *alkali sulphide* (§ 32). By this treatment sulphides of the tin sub-group go into solution, while the sulphides of the copper sub-group remain undissolved. In case the precipitate does not dissolve completely in ammonium sulphide, the presence of sulphides of sub-group *B* may be detected by *acidifying* the *filtrate,* which yields a *colored flocculent* precipitate in addition to a milky

[1] Sometimes mercuric sulphide is decomposed, in particular by concentrated nitric acid (consult § 33).

separation of sulphur. The acid used for this should never be added from the reagent bottle, but from a separate vessel, and any that is unused should be thrown away. A larger portion of the precipitate is then taken for the *separation* of the sulphides of the two subgroups (§ 25).

3. Separation of the Sulphides of the Tin Sub-Group, *B* (§ 26).

The precipitate, which may contain sulphides of *arsenic, antimony* and *tin,* is boiled with concentrated hydrochloric acid for two or three minutes:

A residue may consist of *sulphur* and **arsenic trisulphide** (consult § 26, footnote *2*). It is heated with concentrated *nitric acid.* To the *solution* so obtained are added a *magnesium* and an *ammonium salt solution* together with *ammonia* sufficient to produce an *alkaline* reaction. In case the residue contained **arsenic trisulphide** a white crystalline precipitate of **$MgNH_4AsO_4$** forms, sometimes only after vigorously shaking the solution. For the reactions of the *ions of arsenic* consult §§ 42–44. For the *detection of very small amounts of arsenic* consult § 45.

The solution may contain ions of **antimony** and **tin.** **Antimony ion** is detected by placing a few drops of the solution on a platinum foil and placing one end of a U-shaped strip of *tin foil* in the liquid, leaving the other end in contact with the platinum [or by means of the silver coin test (§ 26), care being taken to completely remove hydrogen sulphide from the solution before making this last test]. A dark stain forms on the platinum [or silver (slowly in the latter case)] if **antimony ion** is present. For the reactions of *antimony* consult § 46.

Mistakes are occasionally made at this point. They arise from the appreciable solubility of *copper sulphide* in ammonium sulphide. Consult § 32 for the recognition of these sources of error and the means of overcoming the difficulties.

To detect **stannous ion** the excess of hydrochloric acid is removed by evaporating or neutralizing the solution. **Antimony** and **tin** are then precipitated from this solution by means of granulated zinc. The metallic *precipitate* is filtered, washed with water and warmed with dilute hydrochloric acid, to which some concentrated acid may be added if necessary. It is essential that the quantity of metal used for this experiment is *completely* dissolved. Otherwise $Sb^{\cdots}$ and $Sn^{\cdot\cdot}$ might be reprecipitated after they are dissolved away from the zinc and the latter again comes in contact with the solution. If **stannous ion** is present the solution gives with *mercuric chloride* a white precipitate of **mercurous chloride.** [Or the hydrochloric acid solution is boiled with pure iron wire until colorless, and then is tested with mercuric chloride.] For the reactions of the *ions of tin* consult §§ 47 and 48.

GROUP III. Reagent: Ammonia + ammonium chloride (§ 51).

The *filtrate* from the second group is heated to expel *hydrogen sulphide.* It is then heated with a few cc. concentrated *nitric acid* to oxidize ferrous- to ferric ion. If the solution has not been treated with hydrogen sulphide the oxidation with nitric acid is carried out

only in case the original solution contains ferrous ion (consult § 51). To the solution are then added 5–10 cc. *ammonium chloride solution* and also *ammonia* drop by drop until an alkaline reaction is produced. The resulting precipitate is filtered after being heated for a few minutes in contact with the solution. As little ammonia as possible should be used (consult § 51, footnote). The precipitate may contain:

A. *Hydroxides* of *aluminum, iron, chromium, manganese, cobalt* and *nickel*;[1]

B. *Aluminum phosphate, iron phosphate* and *magnesium ammonium phosphate,* and *fluorides, oxalates* and *phosphates* of the *alkali earth metals,* which are brought into solution by acids;

C. *Silicic acid;*

D. *Sulphates* of the *alkali earth metals.*

A. Separation of the Hydroxides (§ 55).

The precipitate is washed and dissolved in dilute hydrochloric acid. Excess of alkali hydroxide is then added, until the liquid exerts upon the skin the characteristic action of alkalies. After a few minutes the undissolved portion is separated from the solution by filtration.

[1] As a rule hydroxides of manganese, cobalt and nickel are not precipitated by ammonia, but they may be precipitated under certain conditions. These are not yet definitely established.

Hydroxides of *iron* and *manganese* remain undissolved. To detect **manganese hydroxide** a portion of the precipitate is fused on a platinum foil with *sodium carbonate* and *potassium nitrate* (§ 56). Manganese gives rise to a deep green *fusion* of **alkali manganate,** which forms a *red* solution in water acidulated with *acetic acid.*

For the detection of **iron** a portion of the precipitate is dissolved in hydrochloric acid and the solution is tested with *ammonium sulphocyanate* (§ 56). If **ferric ion** is present a *blood red* solution results. For the reactions of the ions of *iron* consult §§ 67–70.

Sometimes **chromic hydroxide** is mixed with the residue left after treatment with alkali hydroxide (§ 57). To determine if this is the case the precipitate is dissolved in dilute hydrochloric acid. The solution is then neutralized with *sodium carbonate* (§ 56) and *sodium acetate* is *added. Basic chromic-* and *ferric acetate* precipitate on boiling this solution, and are separated as given in the text. The filtrate is tested with ammonium sulphide for **Mn··**. A pink precipitate of *manganous sulphide* forms if manganese is present. Consult § 56 for the method to be used if the precipitate contains $Mn(OH)_2$, $Co(OH)_2$ or $Ni(OH)_2$.

Aluminum and *chromium hydroxides* go into solution as aluminite and chromite respectively (§ 55). A chromite solution has a green color. On *boiling* **chromic hydroxide** precipitates, especially after *diluting* with water. For the reactions of *chromic ion* consult § 64.

To detect **aluminum ion** *ammonium chloride* is added to the solution *freed from chromium,* best in the solid form. **Aluminum hydroxide** separates on warming this solution. For the reactions of *aluminum ion* consult § 63.

Occasionally **ferric hydroxide** is mixed with the *chromic hydroxide precipitated on boiling* the solution. In this case the mixture (§ 57) is fused in a crucible with *sodium carbonate* and *potassium nitrate.* By this means **chromic hydroxide** is oxidized to *chromate* (§ 64), which dissolves in water to a *yellow* solution. Lead acetate precipitates yellow *lead chromate* from this solution after acidification with acetic acid. **Ferric hydroxide** remains undissolved by the above treatment and is identified by any of the methods mentioned above.

B. Detection of the Substances under *B* (§ 58).

A portion of the (washed) precipitate obtained with ammonia is used in the preliminary test for *phosphates.* It is dissolved in *nitric acid* and to the solution is added several times its volume of ammonium molybdate solution. (The solution must have an acid reaction.) **Phosphates** yield a *yellow* precipitate of *ammonium phospho-molybdate.* To detect *oxalates* and *fluorides* another portion of the original precipitate is heated with a solution of *sodium carbonate* for a few minutes. The resulting precipitate is filtered, and the filtrate neutralized with *acetic acid. Calcium chloride* is now added. If the anions C_2O_4'' or F' are present, a *white precipitate* forms, *insoluble* in *acetic acid, soluble* in *hydrochloric* or *nitric acid.*

(*a*) *Decomposition of phosphates* (§ 59). The precipitate is dissolved in concentrated *nitric acid.* This solution is then heated with *tin foil,* whereby phosphate ion is precipitated as *stannic phosphate.*

The cathions which were precipitated as phosphates are now found in the solution filtered from the *stannic phosphate* and *stannic acid.* The *residue* is extracted with dilute *hydrochloric acid* to free it from soluble salts, while the *filtrate* is *evaporated to dryness* to separate any *dissolved stannic acid.* The evaporation residue is then extracted with dilute hydrochloric acid and this solution is tested for cathions of groups III, IV and V and also for $Mg^{\cdot\cdot}$.

(*b*) *Decomposition of oxalates and fluorides* (§ 60). The precipitate is dissolved in *nitric acid.* This solution is then heated with an excess of *sodium carbonate solution.* The cathions which were previously contained in the *oxalates* or *fluorides* are precipitated by this means (as basic) *carbonates.* The precipitate is filtered, washed with water and dissolved in hydrochloric acid. This solution is then tested for cathions of groups III, IV and V and for $Mg^{\cdot\cdot}$.

(*c*) *Decomposition of a mixture of phosphates, oxalates and fluorides* (§ 61). The operations described under (*a*) and (*b*) are applied successively.

C. In case the precipitate is partly composed of *silicic acid* it is treated according to §§ 189–192.

D. If *sulphates* are present the precipitate obtained with ammonia also does not dissolve completely in acids. A *residue insoluble in acids* is treated according to § 183. These sulphates may also separate on oxidizing the filtrate from the second group before adding ammonia. In case this is observed the precipitate is filtered and treated in the above-mentioned way (§ 183).

GROUP IV. Reagent: **Ammonium sulphide** (§ 72).

A *portion* of the *filtrate* from the third group is first tested with *ammonium sulphide,* or by saturating the ammoniacal solution with hydrogen sulphide (§ 73). If a precipitate forms the fourth group ions are precipitated from the *entire filtrate.* On filtering, the precipitate often runs through the filter. This can be avoided, however, by warming the solution with ammonium acetate, which coagulates the precipitate. It may also be avoided by passing hydrogen sulphide directly into the warm slightly ammoniacal filtrate from the third group, instead of using a solution of ammonium sulphide. The following sulphides may be present in the precipitate: CoS, NiS, MnS and ZnS.

The precipitate is first washed with a solution of hydrogen sulphide or ammonium acetate saturated with hydrogen sulphide. Then very dilute hydrochloric acid is poured over it on the filter (§ **75**).

Zinc and **manganous sulphides** go into solution. An excess of *alkali hydroxide* is now added, H_2S being first expelled. **Manganous hydroxide** is precipitated and becomes brown through oxidation by oxygen of the air. For the reactions of *manganous ion* consult §§ **78–80.**

To detect **zinc ion** *hydrogen sulphide* is passed into the *alkaline* solution. A *white* precipitate of **ZnS** forms, becoming more distinct on *heating* the solution. Frequently the precipitate appears dirty white, owing to a content of FeS, CoS or NiS. *Ferrous sulphide* may be removed by adding acetic acid in small portions. CoS and NiS are best separated by extracting the precipitate again with dilute hydrochloric acid. For the reactions of *zinc ion* consult § 81.

Cobalt and **nickel sulphides** remain undissolved (compare § **76**, footnote). The residue is dissolved in concentrated nitric acid or in a mixture of hydrochloric and nitric acids. Excess of acid is driven off by evaporation. Then the solution is neutralized with *alkali carbonate* and acidified with *acetic acid. Concentrated potassium nitrite* solution is next added, precipitating **cobalt ion** as **potassium cobaltinitrite.** For the reactions of *cobalt ion* consult § 83.

To detect *nickel ion* the filtrate (freed from cobaltous ion) is gently warmed with alkali hydroxide. **Ni**·· gives a green precipitate of $\mathbf{Ni(OH)_2}$ (compare § 75). For the detection of **Ni**·· in the presence of **Co**·· by *potassium cyanide* and *bromine* and for the reactions of Ni·· consult § 84. [The potassium persulphate method may also be used to distinguish both Ni·· and Co·· together (consult § 75)].

GROUP V. Reagent: Ammonium carbonate (§§ 85–86).

Before making the tests *ammonium sulphide* is removed by heating the filtrate from the fourth group with acetic acid. Any sulphur that separates during this operation is filtered. The filtrate is then made alkaline with ammonia and is heated to boiling. *Ammonium carbonate* solution is next added drop by drop until the quantity of precipitate no longer visibly increases. The precipitate is filtered and the *filtrate* is tested by heating it, to see if any further precipitate is formed (§ **86**). In case any more solid separates the liquid is again poured over the filter.

The *carbonates* ($BaCO_3$, $SrCO_3$ and $CaCO_3$) are separated according to § 87. The precipitate is first washed and is dissolved in *acetic acid.* A portion of this solution is then tested with *dichromate ion.* If the solution contains **Ba**·· a *yellow precipitate* ($\mathbf{BaCrO_4}$) forms. In this case **Ba**·· is precipitated from the *rest* of the *solution.* For the detection of Sr·· and Ca·· *ammonium carbonate* is again added to the yellow *filtrate,* which is previously made

ammoniacal. This precipitates **strontium-** and **calcium carbonates.** After filtering and washing this precipitate is dissolved in hydrochloric acid.

A portion of the solution, *which is practically free from barium ion,* or of the solution obtained by dissolving the *original carbonate precipitate* (in case the test for $Ba^{··}$ with dichromate ion has given a *negative* result), is tested for **$Sr^{··}$** with a *solution* of *gypsum.* Since **$SrSO_4$** precipitates quite slowly, the solution must be *left several hours* before it is certain that all $Sr^{··}$ has been removed.

In case no **$Sr^{··}$** is found, the remainder of the hydrochloric acid solution is tested directly for $Ca^{··}$ by addition of *ammonium oxalate* and enough *ammonia* to give an *alkaline reaction.* $Ca^{··}$ yields a white crystalline precipitate of **calcium oxalate** (§ 94). If, however, **$Sr^{··}$** was shown to be *present, alkali sulphate* is added to the hydrochloric acid solution. The precipitate is *filtered after* it has been shaken for 20–30 *minutes* with the solution. Then **$Ca^{··}$** is detected in the *filtrate* as above given.

For the reactions of the ions of this group consult §§ **94, 97** and **98.**

GROUP VI.

The filtrate from the fifth group may contain the ions **$Mg^{··}$**, **$K^{·}$** and **$Na^{·}$** (§ 99). Before *carrying out* the *test* for **$Mg^{··}$** a portion of the filtrate is again tested with *dilute sulphuric acid* for **$Ba^{··}$** and **$Sr^{··}$**, and another portion with *ammonium oxalate* for **$Ca^{··}$**. If a precipitate forms the entire filtrate is treated with these reagents, successively when necessary. (In precipitating calcium ion as CaC_2O_4 the liquid must be made ammoniacal, since calcium oxalate is soluble in sulphuric acid.)

A portion of the filtrate is then taken for the detection of **$Mg^{··}$**. If necessary it is evaporated to concentrate it. Then *disodium hydrogen phosphate* is added, together with enough *ammonia* to give the solution a slight *alkaline* reaction. To avoid supersaturation the test-tube is shaken vigorously. **$Mg^{··}$** gives a crystalline precipitate of the composition **$MgNH_4PO_4$** (§ 102). The method to be used for the detection of **$K^{·}$** and **$Na^{·}$** depends upon the presence or absence of **Mg**-ion.

1. *The filtrate from the fifth group contains no* **$Mg^{··}$**. In this case the solution is *evaporated* to dryness. *Ammonium salts* are ex-

pelled by gently heating the evaporating dish over an asbestos plate, leaving a small air cushion between the vessel and the asbestos (§ 100). The residue is dissolved in a few drops of water. A portion of this neutral solution is tested:

For **K˙** by addition of *sodium hydrogen tartrate solution.* To remove supersaturation a small quantity of a dry mixture containing 1 per cent cream of tartar and 99 per cent sodium nitrate is introduced into the liquid. A white crystalline precipitate of **potassium hydrogen tartrate** forms. For the reactions of *potassium ion* consult § 103.

For **Na˙** with a solution of *dipotassium dihydrogen pyroantimonate,* saturated when hot and *then cooled.* **Sodium ion** yields with this solution a crystalline precipitate (§ 105), which may be easily separated by decantation (§ 100) from antimonic acid that is sometimes precipitated with it.

2. *The filtrate from the fifth group contains* **Mg˙˙** (§ 101). The solution is first treated as above given. The residue left after expelling ammonium salts is taken up with water and a few drops of hydrochloric acid. This solution is then heated with baryta, until the *filtrate no longer forms a precipitate on further addition* of baryta. From the filtrate **Ba˙˙** is precipitated by *ammonium carbonate* or *sulphuric acid.* Then the solution separated from this precipitate is tested for **K˙** and **Na˙** in the same manner as the solution containing no **Mg˙˙**, after expelling ammonium salts.

A separation of carbon on driving off ammonium salts, or a residue not entirely soluble in water, indicates that organic compounds are present. For this case consult § 193.

A special portion of the original substance or mixture is tested for **ammonium compounds** by heating it with alkali hydroxide, according to § 108.

TABLE II. TESTS FOR ANIONS.

Consult §§ 110 and 181 for the preparation of solutions to be used in making tests for anions. The following tables relate to *neutral* solutions.

	Reactions with			
	barium chloride	lead acetate	silver nitrate	Special reactions and references to the text.
Group I. The barium salts are practically insoluble in dilute hydrochloric and nitric acids (§ 111).				
Sulphate ion,[1] SO_4'', (§ 112)	white precipitate	white precipitate, soluble in ammonium tartrate (§ 14).	—	Hepar reaction, (§ 171). Lead sulphate is appreciably soluble in acids.
Fluosilicate ion, SiF_6''. (§ 113).	white precipitate	white precipitate	—	SiF_6'' forms a characteristic precipitate with K-ion, (§ 103).
Group II. The barium and lead salts are soluble in dilute nitric acid, insoluble in acetic acid (§ 114).				
Oxalate ion, C_2O_4'', (§ 116).	white precipitate	white precipitate	white precipitate	For identification in the presence of the ther anions see § 125.
Fluoride ion, F', (§ 117)	white precipitate	white precipitate	—	Separation of silicic acid on a ... ened glass rod (§ 117).
Chromate ion, CrO_4'', and dichromate ion, Cr_2O_7'', (§ 119)	yellow precipitate	yellow precipitate	brown red precipitate	With hydrogen peroxide the ...id solution forms a ...ble compound, ...ble in ...er, ...sily decomposed.
Sulphite ion, SO_3'', (§ 121)	...hite precipitate. $BaSO_4$ separates from the solution in HCl.	white precipitate	white precipitate. Silver separates on warming.	Sulphides ...ate SO_2 with dilute acids. Distinction from ...sulphates (§ 124). Hepar reaction (§ 171).
Thiosulphate ion, S_2O_3'' (§ 123).	...hite precipitate. Forms ... ...turated solutions readily.	...hite precipi...te, soluble in an excess of S_2O_3''.	white ...pi...te, soluble in an ...cess of S_2O_3''. Turns black and separates Ag_2S.	Thiosulphates evolve SO_2 with dilute acids and simultaneously separate sulphur. Hepar reaction (§ 171).

[1] A solution into which hydrogen sulphide has ben ...sed ...uld never be tested for SO_4'', for some sulphuric acid will always be present in consequence of the action of ...ic oxygen upon H_2S.

TABLE II. TESTS FOR ANIONS. (*Continued.*)

	Reactions with			
	barium chloride	lead acetate	silver nitrate	Special reactions and references to the text.
Group III. The barium, lead and silver salts are soluble in acetic acid (§ 126)				
Phosphate ion, PO_4''', (§ 127)	white precipitate	white precipitate	yellow [illegible]ite	Precipitation as $MgNH_4PO_4$ and $(NH_4)_3PO_4 \cdot 12MoO_3$ is also characteristic.
Arsenite ion, AsO_3''', (§ 132)	white precipitate	white precipitate	yellowish [illegible]ite pre[illegible]. The [illegible]tion in ammonia separates silver when warmed.	[illegible]ified solutions of arsenites yield a pre[illegible]ite of arsenic trisulphide with H_2S, immediately. Distinction from arsenates, (§ 134)
Arsenate ion, AsO_4''', (§ 133)	white precipitate	white precipitate	red brown precipitate	Distinction from AsO_3''', § 134. Distinction from PO_4''': precipitation by H_2S. The [illegible] from the H_2S-precipitate must always be tested for PO_4''' with *ammonium molybdate.*
Borate ion, BO_2', (§ 135)	white precipitate(?).	white precipitate	white precipitate(?).	With an excess of alcohol and a [illegible] H_2SO_4 borates yield a [illegible] which burns with a green flame. Behavior towards [illegible] paper.
Silicate ion, (§ 136).	—	—	—	In the phosphor salt bead silicates give a (silica) skeleton (§ 173).
Carbonate ion, CO_3'', (§ 137)	—	—	—	Dilute acids decompose carbonates with [illegible] of CO_2, [illegible] forms a [illegible] cloud with baryta solution [illegible] § 176).
Tartrate ion, $C_4H_4O_6''$, (§ 138).	white precipitate, readily forms supersaturated solutions.	white precipitate, soluble in ammonium tartrate.	white precipitate, separates a silver mirror on warming the ammoniacal solution.	Distinction from oxalates, § 139.

TABLE II. TESTS FOR ANIONS. (*Continued.*)

Reactions with: barium chloride	lead acetate	silver nitrate	Special reactions and references to the text.
V. The barium salts are easily soluble, the silver salts practically insoluble, in nitric acid (§ 140).			
—	white precipitate, soluble in hot water.	white [illegible] precipitate, [illegible]soluble in ammonia, CN′ and S_2O_3''.	[illegible] § 141 and § 145 for the distinction and [illegible] from Br′ and I′ respectively.
—	white precipitate	y[illegible] precipitate, [illegible] in ammonia, CN′ and S_2O_3''.	Chlorine water oxidizes Br′ to bromine, which dissolves in [illegible] and carbon [illegible] to a yellow brown solution. The color vanishes on adding an excess of [illegible].
—	yellow precipitate (§ 14).	bright yellow precipitate, insoluble in ammonia.	Chlorine water [illegible] I′ to iodine, which dissolves in chloroform or carbon disulphide to a *red violet* solution. [The color vanishes on adding an excess of chlorine water.]
—	white precipitate	white [illegible], soluble in ammonia, CN′ and S_2O_3''.	Sulphocyanate and Prussian blue tests (§ 146). Distinction from ferro- and ferricyanides (§ 150).
—	white precipitate	white precipitate, scarcely soluble in ammonia.	Ferrocyanide ion yields a blue precipitate with ferric ion (Berlin or Prussian blue).
—	—	orange precipitate, soluble in ammonia, CN′ and S_2O_3''.	Ferricyanide ion yields a blue precipitate with ferrous ion (Turnbull's blue). Distinction from ferrocyanides, § 149.
—	white [illegible] precipitate, [illegible] brown when heated.	white precipitate ([illegible], [illegible] § 151, (*3*).	Hypochlorite ion [illegible] iodide ion to iodine, even in a slightly alkaline solution.
—	—	[illegible], soluble in [illegible], CN′ and S_2O_3''.	Consult § 69 for behavior towards ferric ion. S_2O_3'' and CNS′ are sometimes mistaken for one another, [illegible] S_2O_3'' also gives a red coloration with $Fe^{\cdots}$. The color [illegible]. In case of doubt the behavior of the [illegible] towards

TABLE II. TESTS FOR ANIONS. (*Continued.*)

	Reactions with			
	barium chloride	lead acetate	silver nitrate	Special reactions and references to the text.
Group V. The barium, lead and silver salts are readily soluble in water.				
Nitrate ion, NO_3', (§ 155)	—	—	—	Reaction with forrous sulphate and [illegible]ic acid; iodide-, [illegible]de- and dichromate ions interfere and must be destroy[illegible]d (§ 156). Separation from nitrite ion, § 157. Distinction from [illegible]te ion, § 159.
Nitrite ion, NO_2', (§ 156).	—	yellow coloration	[illegible]ite precipitate, soluble on addition of water.	Oxidation of iodide ion in *weak* acetic acid [illegible]tion [illegible]re § 144).
Acetate ion, $C_2H_3O_2'$, (§ 158).	—	—	white leafy [illegible]rystalline p[illegible]cipitat[illegible] [illegible]le on addition of water.	Behavior [illegible]ds ferric ion; iodide ion interferes and must be destroyed. Cacodyl test. Formation of [illegible]thyl acetate with alcoh[illegible] and sulphuric acid.
Chlorate ion, ClO_3', (§ 159)	—	—	—	Formation of ClO_2 with concentrated sulphuric acid (violent reaction; caution!). After reduction with [illegible]i- or ferrous ion the soluti[illegible]n gives the reacti[illegible]n for Cl' with Ag-ion. Distinction from [illegible]'; if the solution originally contained Cl', this [illegible]ust be [illegible]ly removed by pre[illegible]tion with [illegible].

TABLE III. PRELIMINARY TESTS AND METHODS OF BRINGING SOLIDS INTO SOLUTION.

Preliminary Tests (§ 168).

I. Bulb-tube tests (§ 170). The most striking phenomena are:

A. *Loss of water* from salts containing water of crystallization and *fusion* (easily fusible salts).

B. *Formation of a sublimate.*

A **white sublimate** may consist of: *ammonium salts, arsenic-* and *antimony trioxide, mercury salts* (the latter become red when heated with sodium carbonate).

A **yellow sublimate** may contain: *arsenic trisulphide, mercuric iodide, sulphur.*

A **gray** to **dark sublimate** may be: *mercury, arsenic* or *iodine.*

C. A *change of color* is undergone by salts of heavy metals, many oxides and organic compounds.

D. *Evolution of volatile decomposition products:*

Oxygen	from	peroxides, HgO, nitrates, chlorates;
Carbon dioxide	"	carbonates and oxalates;
Carbon monoxide	"	oxalates;
Sulphur trioxide	"	sulphates;
Sulphur dioxide	"	sulphides, with access to air;
Nitrogen peroxide	"	nitrates;
Bromine	"	bromides in the presence of oxidizing agents;
Iodine	"	iodides in the presence of oxidizing agents;
Cyanogen	"	cyanides;
Arsenic (odor of garlic)	"	arsenic compounds;
Ammonia	"	ammonium salts;
Empyreumatic vapors	"	organic compounds.

II. Behavior on charcoal (§ 171).

A. *Metallic beads and sublimates.*

(*a*) *Metallic beads without sublimate* may be:

gold: yellow, ductile; *silver:* white, ductile;
copper: red masses; *tin:* white, ductile, fusible;
platinum, cobalt, nickel, iron, form gray magnetic masses.

(*b*) *Metallic beads with a sublimate* may be:
antimony: white, brittle, white sublimate;
bismuth: brittle, brown yellow sublimate;
lead: white, ductile, yellow sublimate.

(*c*) *Sublimates without metallic beads* may be:
arsenic trioxide, white (odor of garlic);
cadmium oxide, brown;
zinc oxide, white, yellow when hot;
mercury compounds.

B. *Fused masses of green or yellow color* may consist of manganates, chromates or sulphides; *green infusible masses* (chromic oxide) indicate chromic salts. To detect sulphur compounds the fused mass on the charcoal is triturated, moistened with water and its behavior towards metallic silver is tested (hepar test, § 171). Iodides interfere with this test.

C. *White unfused masses.* When moistened with cobalt nitrate solution and heated to redness in the *oxidizing* flame:
oxides of the alkali earth metals yield gray masses;
magnesium oxide flesh colored masses;
aluminum oxide blue masses.

Arsenites, borates, phosphates and silicates, when heated to redness and moistened with cobalt nitrate solution, yield blue masses, which are partly fusible, however.

III. For the Hempel method of reduction with sodium or magnesium consult § 172 and No. *VII,* this table.

IV. Phosphor salt and borax bead tests (§ 65). The colors correspond to compounds of the following elements (compare § 173).

Color	Phosphor Salt Bead		Borax Bead	
	Oxidizing Flame	Reducing Flame	Oxidizing Flame	Reducing Flame
Colorless	Most colorless salts. SiO_2 skeleton	SiO_2, Cu, Mn.	SiO_2	SiO_2, Mn, Cu.
Gray	——	Ag, Pb, Bi, Sb, Cd, Zn, Ni.	——	Ag, Pb, Bi, Sb, Cd, Zn, Ni.
Red	Fe, Ni.	Fe, Cu.	Fe, Ni.	Cu.
Yellow	Fe, Ag, Ni.	Fe.	Fe, Ag, Ni.	——
Green	Cr, Cu.	——	Cr, Cu.	Cr, Fe.
Blue	Co, Cu.	Co.	Co, Cu.	Co.
Violet	Mn.	——	Ni containing Co and Mn.	——

V. Flame reactions (consult §§ 106 and 174).

Yellow	*light* emanates from volatile	*sodium* compounds;
Red yellow	" " " "	*calcium* compounds;
Red	" " " "	*strontium* and *lithium* compounds;
Blue violet	" " " "	*potassium* compounds;
Green	" " " "	*barium* compounds, *boric acid;*
Pale blue	" " " "	*lead, arsenic* and *antimony* compounds.

VI. Preliminary tests for acids (§ 175).

A. *Dilute sulphuric acid* (§ 176) is poured over the coarsely powdered mixture, evolving:

Carbon dioxide	from	carbonates;
Hydrogen cyanide	"	cyanides;
Sulphur dioxide	"	sulphites and thiosulphates;
Brown vapors	"	nitrites;
Hydrogen sulphide	"	sulphides;
Acetic acid	"	acetates;
Hypochlorous acid	"	hypochlorites;
Oxygen	"	peroxides of the alkali- and alkali earth metals.

B. The substance yields with *concentrated sulphuric acid* (§ 177), either at room temperature or when warmed:

Hydrogen chloride	from	chlorides;
Hydrogen fluoride	"	fluorides;
Sulphur dioxide	"	sulphites or thiosulphates;
Acetic acid	"	acetates;
Chlorine	"	chlorides in the presence of an oxidizing agent;
Chlorine dioxide	"	chlorates;
Bromine (with HBr and SO_2)	"	bromides;
Iodine (with HI, SO_2, and H_2S)	"	iodides;
Brown vapors	"	nitrites or nitrates;
Oxygen	"	chromates, permanganates, peroxides;

Carbon dioxide from corbonates and oxalates;
Carbon monoxide " oxalates, tartrates, cyanides.

C. A further test is made with a mixture of *alcohol* and concentrated *sulphuric acid* for **acetates** (§ 158) and for **borates** (§ 135).

VII. Behavior of the mixture towards solvents (consult § 178).

Methods of Bringing Solids into Solution.

Concerning the preparation of a solution in water or dilute acids consult § 179, and for the preparation of a solution to be tested for *anions* see § 181.

With substances which are *insoluble in water and acids* it must first be determined whether they are halide salts, sulphates, cyanides, fluorides, sulphides, oxides or silicates. The various methods of bringing these into solution are given in §§ 182–192.

The *Hempel* method of reduction with *magnesium* or *sodium* is of great value in determining the nature of these compounds. The method of identifying *silicates* and *sulphates* is briefly stated in § 172. To test for *halide salts* the product left after heating the substance with magnesium or sodium is dissolved in dilute nitric acid and this solution is then tested with silver nitrate. Oxides and sulphides are often readily identified by their color, and sometimes by their behavior in the phosphor salt bead. When no sulphates are present, oxides may be distinguished from sulphides according to § 172.

INDEX.

The numbers refer to pages.

B.

C.

N.

O.

P.

Q.

R.

Lightning Source UK Ltd.
Milton Keynes UK
UKHW021209180219
337529UK00010B/456/P

9 781330 782095